The Confusion Caused By Being Your Own Twin

Also by Doug Baldwin:
Bugs, Blindness, and the Pursuit of Happiness, 2016
Consciousness: A New Slant on an Old Conundrum, 2017

The Confusion Caused By Being Your Own Twin

Consequences and Responsibilities that Arise from Dual Cognition

DOUG BALDWIN

THE CONFUSION CAUSED BY BEING YOUR OWN TWIN: CONSEQUENCES AND RESPONSIBILITIES THAT ARISE FROM DUAL COGNITION

ISBN-13: 978-1-5193-9729-4
ISBN-10: 1519397291

Cover and book design by Sarah E. Holroyd (http://sleepingcatbooks.com)
Cover illustrations by Terry LeBarr, Swanton, Ohio (gonefishn@yahoo.com)

Dedication

To aspiring egos
To souls on the path
To my family, those who have passed and those together still
To my friends in Mr. Roger's Neighborhood
To the great mystic poet Rumi who planted the seed for this book over 700
years ago:

Dedicated to those
whose work it is
to wake the dead—
Get up.
This is a work day.
~ Rumi

Janus, a Self-portrait by Sarah Perez (Saginaw, Michigan)

Acknowledgments

Karen Horwath: the PianoPoet, my editor and friend. Karen spent many hours reading and rereading this book. I am deeply grateful for Karen's insights, skills, and friendship. I often took Karen's editing and used her suggestions to rethink and rewrite sections of the book. It was at this point—during my post-editing rewrites—that errors may have crept back into the document. If there are errors and misjudgments in the book, they come from my mind and my pen alone.

Terry LeBarr: my book-cover illustrator and friend. Terry captures the essence of my books in a single image. I am very grateful for her skills and for the great cover designs on my books.

My children: Noah Baldwin (Mandy and Jared), Tyler Baldwin (Agatha and Thea), and Anna Baldwin (Brad). When your adult children and their spouses and children are among the smartest and kindest people in your life, then you can count yourself blessed—and I am. I realize, after having written three books on consciousness, that few people will ever read something as esoteric and so outside the mainstream as this book. However, as time passed I came to see how eloquent and talented my own children had become as adults. I realized that I was simply passing a body of knowledge to my kids and to my family. This passing of knowledge and emotions from father to children made me feel not only good but quite satisfied. The knowledge is in their hands and it will impact them as they go forward.

My extended family: my sisters and their families, and my mother who is still sharp and strong in her nineties. To those who have passed and those together still—their support and love have fueled my days. Family first is our motto.

My friends in Mr. Roger's Neighborhood: these are the people who travel through life with you and are always there for you—and you for them. These are the people who support your passions through encouragement and honesty. I am thinking especially of Duke Yost and LoLita Pfeiffer who were the first of my friends to encourage me to publish.

Daniel Kish: Readers of my first book, *Bugs, Blindness, and the Pursuit of Happiness* (2016), will remember my friend and colleague Daniel Kish. Daniel's mission, his steady passion and discipline, helped me turn my own mission into action—the content of my books originated from long hours of fascinating dialogue with Daniel.

My fellow explorers in the Esoteric Study Group at Saginaw Valley State University: Wayne O'Brien founded the Esoteric Study Group and has guided the dialogues with kindness and meditative calmness. Wayne read parts of this book early on and steered the energy in beneficial directions. Wayne has one mind in the academic world and another mind in the esoteric world. His aura and quiet guidance inspired me as it has inspired many others in our community. When I once asked Wayne about his belief system, he responded that belief was not reliable. He told me that he was fascinated by correspondences—as above so below. Wayne, myself, and our fellow esoteric friends are seekers who have yet to declare arrival—seeking is forever. I am blessed to be friends with Wayne and with the remarkable individuals in the esoteric study group.

My fellow philosophers in the Saginaw Philosophy Club have been a mainstay in my life every Saturday morning for many years: The Philosophy Club is a magnet for many of the brightest minds in our community. I go to listen, to learn, and to share. It is amazing how good energy draws good energy—it is no accident that a collective of innovative minds would act like a magnet and draw ever more intelligence and wisdom into a space—a shared commons—that feels more and more sacred as the years flow by.

My friends on a spiritual path in the Buddhist Study Group at Saginaw Valley State University: this is where my knowledge of Buddhist thought began to blossom, thanks to the guidance of my friend Professor Eric De Vos. The Buddhist Study Group is another gathering filled with wonderful, brilliant, and gentle souls. Buddhism has a core that is different from other Wisdom Traditions. The Buddhists have been obsessed with comprehending the human mind for over five thousand years. There is a vast amount of intelligence and wisdom at the heart of Buddhism that has attracted and then deeply influenced minds like mine and (through one path or another) yours.

Coleman Barks: Translator and poet, Dr. Coleman Barks has gifted the modern world with an English translation of the Sufi poet Rumi. The powerful poetry of Rumi flavors this entire book. An initial quote by Rumi in the dedication was the stepping stone that began my journey as a writer and your journey as a critic and fellow spiritual traveler. I will remind you often that this is a work day. It is your responsibility to wake the dead.

Contents

Introduction

You are a magician
caught in your own trickery.
~ "Up to the Neck," *A Year with Rumi,* 2006.

"Well, well, look who is back. It's good to see you again, Dutch. I thought you were done hanging out at the Third-Eye-Watching Vegan Café and Center-for-Perceptual-Strangeness."

"Hi Surge. I didn't intend to come back, but my allocentric mind insisted."

"That's good."

"You think? I'm not so sure that serious egocentric personalities are happy with dialogue as a suitable beginning for a serious book about the evolution of consciousness."

"Slow down, Dutch. The reader will adjust. It's okay to begin with whatever style works for you. Anyway, you might tell the reader what the book is about, that would help. Why write another book about consciousness? The market is flooded."

"The Rumi quote at the bottom of the Dedication Page contains the heartbeat of this entire book, so I already hinted to the reader what this is about. I planted a seed. I think human beings need more Rumi in their lives—indeed, I think egocentric minds lack *Essential Rumi.* That is a dangerous situation, this lack of *Essential Rumi,* given the current state of our ego-based world. In the ego, there is no poetry, no tolerance, no music, no mystery, no compassion, and no need for community or spirituality. The ego has no heart, no soul. It is not the job of the ego to be nice—the ego has another set of vital and necessary roles to fulfill."[1]

1

"That's common knowledge, Dutch. Let me get this straight. You intend to tell the world that the ego can be bad news, sometimes—that the ego and the soul are differently created. That should sell."

> "Very funny, Surge. Thank you for the sarcasm. Most people know about their ego and they have heard—somewhat incorrectly—that human beings have a left brain and a right brain, but what I am offering is something new. I suggest that we have two mutually exclusive minds. That is what Rumi knew over 700 years ago. For example, in his poem 'Two Kinds of Intelligence,' Rumi asserts that not only are there two kinds of intelligence, but they flow in opposite directions. There is intelligence that flows from outside inward and a second intelligence that flows from inside out. Anyway, I will explain more about our mutually exclusive minds a few pages from now."

"That's good Dutch, the reader needs a map. Of course, to be fair, it's not the role of the egocentric mind to compose poetry. That's the job of the allocentric mind. I am sure you will explain more about the ego and the self as the book unfolds. But why Rumi, Dutch? There are many great mystics and lots of good poets."

> "We have forgotten what Rumi and his fellow mystics knew many centuries ago: We are a confused species because we have two mutually exclusive minds. In poem after poem, Rumi shows the reader their dual-essence."

"Yes, that is true. However, at the Third-Eye-Watching Vegan Café we have *not* forgotten about humanity's dual-conundrum. We serve only the finest binary delicacies—fresh from nature's cosmic ovens. We blend the two primal ingredients into infinitely-varied original dishes. We specialize in complex food-for-thought, as you well know from your many visits."

> "What's for lunch, Surge?"

"Today's lunch is unusual. We have Hard-Boiled Mystery soaked in Sufi Vinegar for 700 years. Hard-Boiled Mystery is a bit hard to chew if you pontificate for a living."

> "Sounds acerbic and intimidating, Surge."

"Chew slowly. There is no need to hurry. However, to eat only bland, uninteresting food—devoid of mystery and intrigue—*that is not acceptable*. Human beings need rites of passage, a way to move into the higher realms where poetry is more real than donuts, so to speak."

"You see, Surge, if you keep talking about higher realms, professors will cluck their learned tongues and then go to lunch with their like-minded colleagues. You can't reach the village folk through language, especially language that is heavy with fairy dew."

"Actually, that reminds me, we have Fairy Dew Soup today. You gently sip it with a special spoon—full of holes."

"That makes no sense, Surge."

"Yes, of course. However, *sense* is overrated, Dutch. Belief in sensual reality becomes the collective opinion of fools at a certain level of consciousness. There comes a later time in your spiritual development when physical reality starts to unravel: clocks melt, space warps, and solid spoons are found to be porous. I know that's confusing for you, Dutch. But the reason it is confusing is because you are still evolving—your level of consciousness is relatively low, undeveloped. And that's okay, Dutch; everyone is cognitively evolving and everyone is somewhere along the spiritual path."

"I have no idea what you're talking about, Surge."

"Of course you don't, not yet. But be patient, Dutch. As you travel on your journey, and especially as you take the reader with you, experiences will transform you—and the reader, too. Be patient and trust your journey."

"You think I should try the Fairy Dew Soup?"

"Yes, I do. I think you should stop experiencing the world as solid and dependable, Dutch. The real world is far stranger and more fluid than you can imagine. Rumi tells us to stop hauling water around in small jars. Go down to the river where the water flows free. Poetry allows for strange images, like spoons with holes, but the egocentric mind finds such images to be nonsense, and asks 'What's the point? Get to the point. Analyze this in detail so that I can understand.' However, there is *no understanding* in the poetic world—it is not the job of the poetic mind to figure things out, only *to experience* life as it happens. If the poetic spoon has holes, and the metaphorical soup is dripping down your chin, so be it."

"Are you going to use the King's English any time soon? The reader is thinking of going for a donut at Tim Hortons and then playing some serious solitaire on the iPhone."

"Okay. Here it is, some King's English: The word silly comes from the German word *Selig*, which means *Holy*. To dwell in the realm-of-silliness is to be holy. A *Holy Spoon* is a silliness that can only arise within the allocentric,

3

poetic mind. However, the world of the egocentric mind is serious, rational, solid, and dependable. Contrary to the egocentric mind, the allocentric mind is not rational, and not solid. You have two minds—remember this. Furthermore, all humans are on a spiritual journey, but each person is at a different stage of the journey. Less evolved human beings are egotistic, intolerant, and spiritually barren. However, as human beings evolve they become more loving, peaceful, wise, joyful, and awake. There are many levels of reality, Dutch—not just one. Those below you on the spiritual pathway seem like fools at times, while those souls above you—with greater spiritual sophistication—seem like alien beings sipping from poetic spoons filled with holes."

"Do you think this book will have any value, Surge? If you keep talking about Holy Spoons and Fairy Dew Soup, the reader will vanish into Facebook-land, never to read another sentence."

"The book has great value, Dutch, but only for human beings who are ready for Fairy Dew Soup and Holy Utensils—for the poetic life. The rest of the common folk are stuck inside a world of routines and sleepwalking—they already put the book back on the shelf."

☙————☙

If you didn't put the book back on the shelf, thank you. You are still holding the book, and that is no accident. My poetic mind insists that you were led here, that you were invited to this expedition because our paths are meant to cross at this time and in this place.

In other words, you are here because you are ready to be here. You have been on a journey, and that journey has brought you face-to-face with this book and with the poet Rumi. And both Rumi and this book are about embracing duality, discovering your twin minds: sister and brother, rational and poetic, ego and self. In addition, this book also requests that you *respond* to the confusion caused by being your own twin. You have contributions to make—figure yourself out and find your voice. It's not enough to have knowledge or to have a backlog of experiences that brought you wisdom; you need to take action. Here is the 14th Dalia Lama saying the same thing in the language of Buddhism:

Karma means action.
So things change through action,
not by prayer
. . . not by wish.

This same message has come down to us over many centuries, passed on through the generations by a long heritage of saints and mystics. Let's look again at the Rumi quote that is in the dedication:

Dedicated to those
whose work it is
to wake the dead—
Get up,
This is a work day.
~ Rumi

Who are the people called to do the work of raising the dead? I know who you are. Rumi knows. The Dalia Lama knows. Maybe you also know that the so-called "dead" are those people who are sleepwalking through life. There are levels of consciousness and all those below you appear to be relatively unaware, less awake, and less spiritual—they seem spiritually dead compared to the level of consciousness that you know. Likewise, those above you on the spiritual path see you as relatively dead; it is their job to wake you up to a higher spiritual realm.

You have a talent that humanity needs, but your talent will be wasted if you remain stuck at a low level of consciousness. You are unique in all of eternity and across the whole of infinity. No one else has ever been born who has your abilities and potential. *Only you* can figure out how to be most helpful as you follow your path. So, this book is dedicated to you—because of your unique gifts. Your job is to wake up those who have a lower level of consciousness, so they might perceive from a higher vantage point, a higher realm of perception. So, *get up, wake up*, this is a work day. It is time to study your own cognitive evolution, to understand both your ego and your twin (the self); and it is time to help others with their spiritual journey. We are a community, a global community; we need and depend on each other. Those who have arrived at a certain level of consciousness, those who can perceive levels of consciousness in themselves and in their cultures—people like you—have to step up now. The planet is suffering and all of humanity needs your help.

Here's a reader map for you to follow in case you get lost in the abstractions:

I will return to the dialogues between Dutch and Surge at the beginning of each chapter. Using dialogue is a Socratic method for exploring difficult ideas. Dialogue is also used in esoteric communication to convey mysterious and controversial perspectives. I used this approach in my previous book *Consciousness: A New Slant on an Old Conundrum* (2017), and continue it here because I deeply enjoy the obtuse discussions between Dutch (my alter ego) and Surge, a spirit-guide who fronts as a waiter at the Third-Eye-Watching Vegan Café. Beyond my enjoyment, of course, there is a method to the madness.

I use dialogue as a literary device to demonstrate the duality inherent in our cognitive design. Our allocentric (poetic, spiritual) mind loves storytelling; it loves dialogue. Contrary to this, our egocentric mind wants rational arguments and explanations (like we are doing now). The dialogue between Dutch and Surge allows the egocentric mind to pause, reflect, and relax—the rational mind needs to chill out on a regular schedule so that the allocentric mind can express itself. I will elaborate about our twin minds—the allocentric (the self) and the egocentric (the ego)—in the pages ahead.

The ego is quite impatient with dialogue and silliness, and it bristles at having to "beat around the bush." However, the truth is in the duality: both our minds are wonderful and neither can be ignored. *We have two mutually exclusive minds locked inside the same body—two kinds of consciousness exist inside of us, but they are blind to each other's existence. This paradox defines the human condition.*

I invite you to ponder the Rumi quotes that begin many section headings. There is poetry strewn about, a voice arising from the non-egoic soul. That voice is trying to wake you up, to demonstrate your inherent duality. Through his words, Rumi is showing us that the ego cannot perceive the self (soul), and the self cannot perceive the ego. Among other delightful revelations, Rumi helps us experience this paradox.

The book is organized into eight chapters. **Chapter One, called The Dual Meaning of Life**, is a review of the significant conclusions reached in my previous book *Consciousness: A New Slant on an Old Conundrum* (2017).

In that earlier work, I offered a plausible explanation for the origin and evolution of human consciousness, especially our dual cognition. In that earlier book, I had intended to write about the confusion that arises from dual-cognition, and about the responsibilities that arise when we become aware that human consciousness evolves. However, it took three hundred pages to explain my theory of how the mind came about and how our inherent duality evolved. It became clear during the writing process that I had to write this second book to discuss the evolution of consciousness and the consequences (and confusion) of our duality. If you have read *Consciousness: A New Slant on an Old Conundrum*, you can skip ahead to Chapter Two—unless you want a quick review.

Chapter Two explores developmental levels of consciousness. Consciousness is a process that evolves over a lifetime. It is not a static entity. I use psychologist Susanne Cook-Grueter's developmental scale to compare the worldviews of individuals and cultures. Try to find yourself on the Cook-Greuter scale; it is enlightening to realize you are not the pinnacle of nature's great work. It was humbling for me to see myself from a broader perspective. Most people have no idea that there are levels of consciousness. I had great fun with this chapter—I hope you enjoy my irreverent look at our differences.

Our entire world has an average level of consciousness. Nation states also have levels of consciousness that arise from averaging the sophistication of the individuals who reside in these countries. In the United States, in the 2016 elections, Americans elected political leaders who spouted racist, sexist, anti-scientific, anti-intellectual rhetoric. It was a shameful time for the nation (for human evolution), an embarrassment on a global scale (it still is, as I write this). The level of consciousness of the populous that elected the 45th President of the United States—on average—was remarkably low. The 2016 election demonstrated a dangerously unevolved culture.[2] This is my personal perspective, of course, although I share this opinion with many others on the spiritual path.

In Chapter Three, I look at poetry and ask what creative minds have done with their inherent duality. Poets use their allocentric minds so splendidly they stand in stark contrast to the great thinkers in philosophy and psychology. I make the point in Chapter Three that poets—actually, all true artists—sense their duality keenly. There is evidence for two minds throughout our literature. What our poets and writers describe, understandably, are two frames of reference, two ways to pay attention, two ways to know reality. The poets discovered angst and awe as they came to see that human beings are actually dual creatures.

In Chapter Four, I examine duality through the perspective of religion. There is a Wisdom Tradition that cuts across all religions, Eastern and Western. This perennial philosophy, as it is called, offers the very same explanation of human duality regardless of the geographical region or the time period. Human beings have discovered, over and over again, that they have a fundamental mysterious duality. One of their minds is hidden and sacred. According to the Wisdom Traditions, this soul-mind must be recovered, nurtured, and allowed to evolve. If we are to bring peace and love into the hearts of individuals, we will have to strengthen and appreciate our spiritual mind (the allocentric mind).

Chapter Five looks at Western philosophy and duality. In the abstract battleground of Western philosophy we find one of the minds, the egocentric, struggling to understand itself but failing to recognize that its twin allocentric mind is worthy of study. The point again is to explore how the philosophers introspected their way to an understanding of individual and cultural duality. They discovered what religion had concurrently discovered: that there are two frames of cognitive reference, and these cause conflict and confusion when we attempt to understand how we think and act—both individually and culturally.

In Chapter Six, I discuss the psychologists, especially Carl Jung, William James, and Ken Wilber. Psychology is the profession wherein the modern-day study of consciousness has played out. Here we find that the great thinkers in psychology discovered what religious leaders and philosophers also found: we have an inherent duality, two ways to behave, two contrary minds that befuddle our attempts to understand who we are and how we should act. I examine perspectives that come from psychology and show how two frames of reference—egocentric and allocentric—are at the heart of the confusion.

Chapter Seven explores science and duality. The scientific method has stood the test of time, but scientism and materialism have been justifiably under attack for a hundred years. The conflict between our two minds is at the heart of the conflict between science and religion, reason and intuition, reductionism and holistic thinking. We are in the age of robotics, gene manipulation, and virtual reality. If our scientists believe that the egocentric mind is all there is, then they will build their robots and genetically modified cyborgs in their own egoic image—that is a recipe for generating masculine, autistic, dull, and half-assed robo-beings; the future looks bleak given this scenario. But the future doesn't have to evolve this way, especially if we can convince scientists that they have dual cognition.

In Chapter Eight, on culture and duality, I consider masculinity and femininity. There is an elephant in every room. That beast is the male of our species—almost the entirety of human suffering occurs at the hands of one sex (one end of the gender spectrum). Duality has played itself out in the battle of the sexes, and now a most extraordinary shift to androgyny seems to be slowly creeping forward. The allocentric mind and the egocentric mind are differently developed in the sexes—or so I contend (with trepidation and caution). Cultures need to rebalance, to become more allocentric—more spiritual, more poetic, and more compassionate.

Chapter Eight also examines education and duality. I worked in special education for 33 years and saw firsthand the conflict between a patriarchal, hierarchical academic world, dominated by the egocentric mind, versus the innovative, creative, intuitive challenge that was constantly welling up from the allocentric mind. Teachers are caught in a cross-fire because education, at an administrative level, has become the champion of the egocentric frame of reference. It is no wonder that our educational culture is such a mess, given the obsession with the development of the egocentric mind.

In Chapter Eight I conclude with a look at Music and Duality. Music is somehow linked with developmental levels of consciousness, with the vibrational universe that gave rise to our duality, and with spiritual concepts like the chakras (energy centers) of the human body. I could only touch on this complex and important subject, but I also could not leave it out of the discussion.

In each of the chapters, I looked at a few examples that supported my supposition that duality is found in every discipline. I was not able to look at each discipline in depth—I am not an expert in music theory or gender studies, for example. Experts in the professions will undoubtedly find reasons to be critical, especially as I speculated and reached to draw parallels. I accept all responsibility for errors of judgement as I attempt to carry my logic to the edges of the universe. I welcome comments and dialogue.

One

The Dual Meaning of Life

Jars of spring water are not enough anymore.
Take us down to the river.
~ "Jars of Spring Water," *A Year with Rumi*, 2006

Something has brought you to this book. Perhaps you are struggling to understand your own mind—which is no doubt quite sophisticated—and yet you are often confused by this unique and wonderful gift. Maybe you are fascinated by a profession like neuroscience or neuropsychology. Perhaps you want to contribute to the dialogue about the evolution of consciousness. However you got to this page, I am pretty sure that the meaning of life is a topic you find worthy of dialogue.

One thing is clear: you know your own mind better than anyone else— no one is ever going to solve the riddle that is uniquely you. Indeed, *only you* have the evidence needed to solve your own mind. On the other hand, there are common features and patterns of development that all minds share. We can and we must learn from each other.

Besides being the world's expert on your own cognition, you probably also have a wealth of data about the mind. The media is constantly feeding us information about how we are cognitively designed and why we behave as we do. We can even buy brain development kits that purport to enhance the mind.

Unfortunately, there is a basic problem with our rushing ahead to enhance or alter our cognition: much of our current knowledge is flawed. We are awash in what I call *scientific memes*, common language that is too vague and, in some cases, simply wrong. We can't move forward on this path of self-understanding unless we pause to study the map. We need to look at all the false trails that lead us toward cul de sacs.

Therefore, it is necessary—as we begin this journey—to have a quick look at some false (or dubious) assumptions commonplace in our Western culture. We have been unable to comprehend consciousness in its entirety because we hold many misleading ideas. These questionable assumptions are accepted in our culture as reality—mostly because they are often repeated as fact. These common scientific memes are false (or incomplete and misleading) and they have thrown us off the trail countless times:

- A commonly accepted meme is that we have one mind and one kind of consciousness. In *Consciousness: A New Slant on an Old Conundrum*, I provided logical, anatomical, and physiological evidence that we have two minds and two kinds of consciousness.

- Another common assumption is that consciousness is a thing, a noun. However, *consciousness is an evolutionary process.* Just as we age and change developmentally, so also does our consciousness evolve and change as we attain knowledge and wisdom. *Knowledge* and *wisdom are not the same thing*—they are the memory banks for two different minds. The egocentric mind gathers and remembers knowledge, while the allocentric mind evolves wisdom through life experiences.

- A very common meme is that we have five senses; Aristotle said so 2500 years ago. Ask anyone on the street how many senses human beings have, and without thinking—just repeating the meme—they will say *five.* However, that's the wrong answer. There is a hidden sense called proprioception. If we leave proprioception out of the discussion, we will fail to understand the derivation and evolution of consciousness—a rather serious oversight.

- Another popular meme is that we have a right-brain and a left-brain. According to current usage, the two anatomically-obvious brain lobes became a left mind and a right mind—a rational mind and an intuitive mind. Because of this fallacy—a partial, incomplete concept—we have failed to look for other reasons why we might have an inherent cognitive duality.

- Another meme is that human beings have separate sensory systems. For example, we believe that we have a vision system, a hearing system, an olfactory system, and so on. From time to time, I will also speak as if this was true because it is a convenient way to communicate and to analyze research. However, there is *never a time* when the so-called vision system

11

functions in isolation, as if disembodied. *None of the senses can ever exist independently of the whole body.* Every action, every behavior, is embodied. There is a balance, a synchronicity, and coherence to the entire human form. Failure to understand this embodiment has allowed us to isolate and reduce behaviors as if they had no connection to our body as a whole or to the specific domain wherein we are designed to function.

- A deeply held meme is that we have a conscious mind and an unconscious mind. I believe this perspective is misleading and has thrown us off course. Instead, what we have is an egocentric mind (an egocentric frame of reference) and an allocentric mind (an allocentric frame of reference). These two minds have a mutually exclusive perspective. The so-called *conscious mind* has a frame of reference that is egocentric—wherever we go, we are the center of all activity. I suggest that the so-called *unconscious mind* could plausibly be the dream-state or memory system of the allocentric mind—I will explore this when I discuss psychologist Carl Jung later in the book.

- Another common meme is that there is an objective world *and* a subjective world. This seems intuitive—we apparently have a mind that resides inside our bodies, and that mind and body move through an actual world. However, this is a purely egocentric perspective. From an allocentric perspective, the mind is not confined to the body. Furthermore, the duality needed for the egocentric mind to operate is unnecessary for allocentric processing of gestalts. This will become clearer as we move through the dialogue.

- A recurring meme (assumption) is that the mind is in the brain. This line of reasoning leaves out the obvious evidence that the brain is connected to a nervous system. If we dissect the brain from its nerves, we end up with a control system that has nothing to control—or no way to control the body that it is incased within. The mind is not just in the brain; the mind is located where the nervous system resides, in the body as a whole. This is a necessary starting perspective for understanding that the mind reaches well beyond the brain. With an understanding of proprioception, we soon realize that "the mind" extends into the environment. Proprioception maps spaces, just as it maps the body. I will explain this is more detail later in this chapter.

- We are all one. This is a New Age meme that many people find valuable, including me. It is valid on different levels of understanding, but it masks a fundamental truth about our inherent duality. We are actually binary creatures crafted using a quantum duality that is inherent in the physical universe. Confusion arises because the egocentric mind requires duality to function, while the allocentric mind does not require duality to function. For the allocentric mind, the meme *we are all one* is a fundamental truth, but for the egocentric mind the meme seems false.

- A commonly accepted scientific meme is that human cognition reaches a plateau (stops evolving) when adulthood is reached. Pioneering developmental psychologist Jean Piaget documented an initial *adult level* stage of cognitive ability he called the Formal Operational Stage. This stage is the age of abstract thought and it first appears during the teenage years. After rapid development in the prefrontal lobes during adolescence, it was assumed that an adult stage of cognitive competence had been reached. However, modern developmental psychologists postulate that there are levels of cognitive maturation that occur over a lifespan. These advanced levels of mental development evolve through experience and knowledge acquisition. The result is that human cognition evolves through stages over a lifetime—there are levels of consciousness.

Belief in anyone of these scientific memes is sufficient to derail our attempts to comprehend the evolution of consciousness. However, the very first supposition—assuming that we have only one kind of consciousness—is a fatal flaw right off the bat. This book is based on the assumption that human beings are cognitively dual creatures. Therefore, it is important to clearly show the logic that enabled this perspective. There is a four-part logic that supports this assumption of dual cognition:

- Two mutually exclusive neural-control systems (minds) are necessary for navigation to evolve. I call these control systems the *egocentric mind* and the *allocentric mind* because "egocentric" and "allocentric" are navigational terms used in my field of expertise (orientation and mobility) to denote spatial frames of reference. In other words, to accurately and efficiently move straight-ahead through space, two neural systems *had to* evolve. Therefore, the *evolution of navigation* in animals slowly gave rise to two different kinds of minds.
- Because of the dual requirements of navigation, human beings eventually evolved two sophisticated kinds of consciousness. The two kinds of

consciousness that resulted from this inherent, mutually exclusive sensorimotor design are biological/quantum twins looking in a mirror at each other—but they don't see a mirror image; they see a being that is opposite in every conceivable way.

- The collection of inner senses called *proprioception* enabled both kinds of consciousness. In my earlier book *Consciousness: A New Slant on an Old Conundrum*, I made the statement that "eyes are the portals for vision, ears are the portals for hearing, and proprioception is the portal for consciousness—both kinds of consciousness, allocentric and egocentric." In other words, vision is a sense, hearing is a sense, and consciousness is a sense.

- The primal origin of duality came from the quantum world—we are quantum creatures because the universe is a quantum duality. We can see our quantum heritage in the design of the senses. For example, the design of the eyes—nature's bioengineering of vision—appears to follow Heisenberg's Uncertainty Principle (a mathematical construct fundamental to quantum physics).

In this chapter, I will look at the four-part logic summarized above and provide an overview. Readers who want more detail should refer to *Consciousness: A New Slant on an Old Conundrum*. Let's turn now to the four conclusions that must be understood before we turn to levels of consciousness in the next chapter.

The Evolution of Navigation in Animals Gave Rise to Two Different Minds

There is a moon inside every human being.
Learn to be companions with it.
~ "Listening," *A Year with Rumi*, 2006.

I worked for over 30 years as an orientation and mobility specialist in a public school for students in special education. My job was to teach travel skills to children with navigational disabilities. All during my teaching career, I was fascinated with the brain and nervous system. Consequently, I tried to understand how each of my student's impairments affected their ability to navigate.

After I retired, I had time to reflect and to write—a luxury I never had when I was working. Thus, I began the journey to deeply understand the neuroscience of navigation. I had time to do library research and to follow logical speculations. I had no idea where my journey would take me, but I loved to write and to think, so I started getting up early most mornings to organize my thoughts. It wasn't long before I discovered a foundational biological understanding. There is a single awareness in neuroscience that became the underpinning for all my subsequent thoughts: *The brain and the nervous system evolved to enable purposeful movements.* Therefore, the brain and nervous system didn't evolve so we could communicate, or think, or socialize, or emote; the evolution of the brain and nervous system began because organisms needed a way to move with a purpose.

After some reflection, I realized that it wasn't just purposeful movement that caused the evolution of brains and nervous systems; it was primarily the ability (the evolution) of navigation that held the key to comprehending the mind.

The next logical leap of understanding was that navigation requires two mutually exclusive neural systems. This is quite clear when we look at the muscular (neuro-motor) system, but it is also clear when we look at the sensory system—the two work together, of course, they are embodied. In other words, motor systems never function without sensory systems, and vice versa—that's why we use the term *sensorimotor*.

Motor Duality

To navigate it is necessary to stabilize a set of muscles at the same time that a complementary set of muscles are moved. For example, the right side of the body is moved while the left side of the body is stabilized during the act of moving forward. Then the left side is moved while the right side is held stable. This alternation requires a very smooth and very fast neurology. In other words, what we call "mind" is a control system that manages the *activation and simultaneous inhibition* of corresponding sets of muscles. This is a mutually exclusive design—each muscle set is doing the exact opposite of its twin on the other side of the body. The neurological basis for this design is *inhibition versus excitation* (on and off); one system is activated but only when a twin system is inhibited. We find this design throughout the animal kingdom.

Sensory Duality

To navigate it is also necessary to differentiate figures that stand out from a background. This is best understood when we look at the vision system, although all the senses are doing the same thing at the same time. The vision system generates a background scene. In this scene are forms, such as objects and people. What this implies—and what we see anatomically and physiologically—is that there are two vision systems, one which creates backgrounds and a second which creates the figures that populate backgrounds. These are actually mutually exclusive twin systems. It is not possible to closely examine a form (feature, figure) at the very same time we take in the whole scene. There is a rapid alternation that requires a mind—a neurological control system—to pay attention either to a form or to the background that holds the form.

We would be dizzy and confused if there was no background steady-state. Indeed, we would be unable to move if the background wasn't relatively stable. Figures could not exist without a background. Could we paint a picture without a canvas? No, we could not. Could we read words on a page without the page? No. Could we wallpaper a wall without the wall? No. Therefore, the first sensory pre-requisite for efficient navigation is the existence of an invariant background scene to walk around upon and within. Forms—features and patterns—could not exist without a dependable canvas. On the other hand, a canvas with no features would be a blank space containing no information.

Sensorimotor Duality

If we combine the motor oscillation and the sensory oscillation, we can see that their common bond is the *inhibition versus excitation system*. The sensory system that perceives form (the egocentric mind) corresponds to the motor system that stabilizes. The sensory system that perceives flow/motion (the allocentric mind) corresponds to the motor system that activates muscles. In other words, when the egocentric mind is paying attention to details in the environment, there is relative stabilization of specific muscles and an inhibition of the allocentric mind. Contrary to this, when the body is moving through a scene, the allocentric mind is active and the egocentric mind is relatively inhibited.

Anatomically and physiologically, the evolving brain and nervous system had no choice but to develop two processing systems. One system creates a background using allocentric (parallel) processing. The other system, using egocentric (serial) processing, creates the patterns that arise out of the background.

Leaps of Logic

I was quite pleased with my small contribution to the discussion about the origin of the brain and nervous system. Indeed, it was quite plausible that we had two overall nervous systems that were both synchronous and opposite—it is a valid hypothesis to suggest that navigation was a prime catalyst in our evolutionary unfolding.

Eventually, I wondered how the evolution of two mutually exclusive and codependent nervous systems would change over millions of years. What if our current psychological duality came from these two contrary nervous systems? That would explain much of the strange behavior of human beings—it would give our contrary "human natures" an anatomical and physiological underpinning. If we had two balanced neural systems that rapidly alternated, then perhaps eventually they became two sophisticated minds. This seemed like a better explanation than the earlier right-brain versus left-brain speculations about the origin of our behavioral duality.

In the early mornings, the idea that navigation was the origin of two minds seemed like a brilliant supposition, but by evening I often felt like I was writing science fiction. I eventually decided to call my insights *cognitive philosophy* because that seemed to allow for theory, and yet also allow room for speculation.

As I was doing library research about consciousness, I came across the work of several academic pioneers who had defined a field of study called dual-process theory[3]. This discipline concluded boldly that human beings had two minds. Consequently, I felt much better about my own hypothesis that postulated two kinds of consciousness—I wasn't proclaiming anything new that the academic world had never considered. One of the leading researchers in dual-process theory, Emeritus Professor of Psychology Jonathan St. B. T.

Evans from the University of Plymouth in the UK, had actually published a book about dual-process theory called *Thinking Twice: Two Minds in One Brain* (2010). Unfortunately, I didn't know about this book when I wrote *Consciousness: A New Slant on an Old Conundrum*. I did use Professor Evan's insights from an earlier book *In Two Minds: Dual Processes and Beyond*, which he co-edited with European philosopher and writer Keith Frankish in 2009. Dr. Frankish, Adjunct Professor with the Brain and Mind Programme at the University of Crete, had also published a book called *Mind and Supermind* (2007) that made the case for a dual biological substrate in human beings. Frankish and Evans also wrote an online article which summarized the history of the idea that we had two minds. The article is called "The Duality of Mind: An Historical Perspective." Therefore, there is ample evidence to support the hypothesis of dual-cognition. I didn't need to prove this because others had already done the supporting research.

The next logical hurdle for me was the notion that two kinds of consciousness came from two mutually exclusive nervous systems—essentially, dual consciousness evolved from two oppositional minds. Defining consciousness as different from the mind posed a challenge until I realized that proprioception might be the sensorimotor system that turned the two minds into two kinds of consciousness. The more I ran the logic over and over again, the more convinced I became that proprioception enables consciousness. You can decide for yourself—I offer it here as part of an overall theory.

Later, I hit a cognitive wall when I realized that primitive organisms—with no bilaterality—motored about their domains perfectly well without neurons, muscles, or eyeballs. It was then that I found myself knee deep in quantum theory—as a possible explanation for how these primitive creatures move with a purpose. After many mornings of musing and doing academic research, I slowly came to see the connection between quantum philosophy and cognitive philosophy. That is where you find me as I write this new book. Wading into the complex sea of quantum theory moves the discussion deeper toward speculation, but the logic, for me, holds together, and so I share my insights here.

I invite you to ponder with me not only the insight that we evolved two kinds of consciousness, but also to consider the consequences of dual cognition. Our understanding of consciousness is upended by dual-process theory—we are faced with a set of serious responsibilities as we ponder the consequences of dual cognition.

Human Beings Evolved Two Mutually Exclusive Kinds of Consciousness

Good and bad are mixed.
If you don't have both,
you do not belong with us.
~ "One Transparent Sky," *A Year with Rumi,* 2006.

As I discussed above, two mutually exclusive neural systems evolved to enable straight-ahead navigation. This suggests that dual-process theory is not only plausible but inevitable. Keep this duality in mind as we explore ways to define the term *consciousness.*

There are actually four ways human beings have defined consciousness. If we don't understand this, we won't realize why there is so much confusion and debate about the concept. Those readers familiar with the philosophy and many books of Ken Wilber will recognize his integral theory of knowledge. I will discuss the ideas of the remarkable Ken Wilber later.

Let's look at the four ways to define consciousness using Ken Wilber's Integral Theory.

From a *personal, subjective perspective,* consciousness is an individual and isolated phenomenon. It is the sum total of all the mental activity that is occurring for a specific individual. This is a private universe that can never be entered or understood by another human being. Subjective beauty is different for each person. What we like, what we love, what we crave, what we hate or fear, all are housed within this subjective cosmos. This subjective definition of consciousness is entirely about personal *experiences,* and is a purely allocentric way to look at consciousness. Novelists and poets often explore from within this perspective, as do mystics and esoteric researchers. This is also the domain of many social scientists and cognitive philosophers.

A second way to define the term consciousness is to consider *collective (cultural) experience.* Instead of being about *me,* consciousness is about *us.* Groups of human beings define morality and justice to suit the worldview of the members. What is good and what is evil is collectively agreed upon. This way to define consciousness is also allocentric; it is about the quality of relationships. There are levels of consciousness, degrees of spiritual sophistication that can arise from within this domain. Social scientists, historians of the mind, philosophers, and psychologists often do research from this perspective.

A third way to define consciousness is from a *materialistic, scientific perspective*. The starting point here is the physical world, the physical body, and the laws of nature. Consciousness is not about *me* or *us*. It is about *it*, an impersonal world devoid of emotion and human relationship. The search here is not for what is beautiful, good, or desired, but for what is logically true. This third way to define consciousness is also about individuals (not groups) and scientific research seeks to find the neuro-correlates to subjective experiences. This perspective is egocentric. Neuroscientists and neuropsychologists do research from within this perspective. Within this domain, we have debates about quantum effects and biochemical interactions.

The fourth way to consider consciousness is through *systems thinking*, theories and philosophies that create and define societies. Here, the concern is not about me, or we, or it, but is rather about *its*. For example, when I proclaim that consciousness can best be understood using dual-process theory, I am using this fourth way to dialogue. I am offering *a system* for understanding the evolution of consciousness. Other systems thinkers have also created coherent (logical) systems that they feel help—in some way—to define the complexity that we call consciousness.

Notice that subjective definitions of consciousness deal *with experience*—individual and collective—and are about relationships and values, about what is beautiful and what is good. However, scientific and systems-thinking are about *understanding*, about gathering and examining knowledge.

Heated arguments can occur when people debate using these four different definitions of consciousness. I have added an additional wrinkle to the complexity of this debate by proposing that human beings (all animals, actually) evolved both allocentric and egocentric minds—which are mutually exclusive. Indeed, I would suggest that the *me* and *we* division is the work of the allocentric mind, and the *it* and *its* perspective is the work of the egocentric mind.

I am advocating my version of dual-process theory from within the *it and its* perspective. I am talking about neuroanatomy and neurophysiology from within a particular systems-theory. Later in the book, when I speak of the duality inherent in our literature—especially poetry—and within religion, psychology, and philosophy, I will be changing perspective to talk more about the subjective realms.

⟳————————⟲

For the purposes of this discussion, I will define *the mind* as a dual neurological control system that uses an *inhibition versus excitation* system for maintaining bodily integration (balance, harmony, and synchronicity). I speak about two minds because the neurological system *that inhibits* and the neurological system *that activates* are mutually exclusive—when one is on, the other is off. One mind, the allocentric, deals with flow (movement management), and the other mind, the egocentric, deals with no-flow (stabilization management). But if the mind is a set of neurological control systems, what is the physiological origin of consciousness?

This question is similar to asking: "If the brain and nervous system evolved to be a neurological control system, then *what is vision?*" Or *what is hearing?* Vision and hearing emerged from sensory portals, eyes and ears. The question then becomes "What sensory portal enables consciousness?"

The answer I found is that the portal for consciousness is proprioception, which is an internal sense. Proprioception has both an egocentric component and an allocentric component. I will discuss this in detail below.

Our two kinds of consciousness gifted us with two ways to pay attention and two ways to remember. These two cognitive systems essentially resulted in two language structures. Therefore, we can find ample circumstantial evidence for our two kinds of consciousness when we look at language. There are two closely related words for every concept, and often we find starkly contrasting, oppositional-words that are mutually exclusive.

If we hold up a list of all the attributes of the egocentric mind, and then extrapolate the opposite, we arrive at a definition for the allocentric mind. For example, if the egocentric mind is verbal, the allocentric mind is nonverbal. Whereas egocentric processing is serial, and its purpose is *to do*, the allocentric mind, in stark contrast, processes in parallel, and its purpose is *to be*. There is a long list of contrasts in our language that provide evidence for an egocentric mind (the ego) and an allocentric mind (the self). Here are a few—after you get the idea, you will be able to add to the list:

21

- The ego *arises from an egocentric frame of reference* (everywhere you go, the world revolves around you); the self comes from an *allocentric frame of reference* (you are part of a total gestalt—there is no egocentric you).
- The ego *believes*; the self *knows*—the ego uses physical evidence to draw conclusions about reality; the self has faith and understands the world intuitively (experientially).
- The ego has *personality*; the self has *essence*.
- The ego has *discussions* (pontificating from an egocentric frame of reference); the self has *dialogues* (as an equal member of the community).
- The ego is *sympathetic* ("I" am sorry); the self is *empathetic* (there is a whole-body sharing of emotion—there is no "I" to take credit for being sorry).
- The ego uses *deductive* reasoning; the self uses *inductive* reasoning.
- The ego uses *explicit* memory (episodic, knowing that, declarative); the self uses *implicit* memory (knowing how, habits, procedural).
- The ego remembers *knowledge* and this creates intelligence; the self remembers *experiences* and this creates wisdom.
- The ego values *efficiency*; the self values *kindness*.
- The ego *reduces* (takes puzzles apart to examine individual pieces) and differentiates; the self *assembles* and *integrates* (creates gestalts).
- The ego can be understood as *diversity within a unity*; the self can be understood as a *unity filled with diversity*.
- The ego draws energy from what quantum physicist David Bohm called the *explicate order*; the self draws energy from what Bohm called the *implicate order*.
- The ego is concerned with *analysis*; the self is concerned with *synthesis*.
- The ego is concerned with *either-or* decisions; the self is concerned with *both-and* decisions.
- The ego uses *declarative knowledge* (pattern analysis); the self uses *procedural awareness* (process or flow analysis).
- The ego is *secular*; the self is *spiritual*.
- The ego *reacts to patterns*; the self *responds to experiences*.
- The ego *reacts to variance*; the self *responds to invariants*.
- The ego *absorbs energy/information* (the ego is the *in-breath, inhalation*); the self *transmits/radiates energy/information*—the self is the *out-breath (exhalation)*.

- The ego employs *top-down neurological* feedback systems; the self employs *bottom-up neurological* feed-forward systems.
- The ego is part of the brain's *executive control center*; the self is part of the brain's *default mode network*.
- The ego internally *reflects* over a long-term time frame; the self immediately *predicts and projects* within a spatial domain.
- The ego uses focal (head-bound) *attention*; the self uses whole-body *awareness*.
- The ego is *digital*; the self is *analog*.
- The ego *trains*; the self *educates*.
- The ego uses *causal systems*; the self uses *acausal systems*.
- The ego generates *individuality*; the self generates *community*.
- The ego is related to *de-coherence*; the self is related to *superposition*.
- The ego measures *quantities*; the self experiences *qualities*.
- The ego is *local*; the self is *non-local*.
- The ego is *deterministic*; the self is *stochastic*.
- The ego is *exoteric* (searching for what is to be done); the self is *esoteric* (searching for how to be).
- The ego seeks *pleasure*; the self seeks *happiness*.
- The ego creates the entity called *time*; the self creates the entity called *space*.
- The ego conceives of *eternity*; the self conceives of *infinity*.

The last two entries on the above list, concerning time and space, and eternity and infinity, need further explanation. This division in terminology is an insight that came to me about half-way through writing *Consciousness: A New Slant on an Old Conundrum*. I was pondering that color doesn't exist in the physical universe—the mind generates (creates) colors using various wavelengths of light. I reasoned that a mind that could create colors was probably also creating other things that didn't actually exist in the physical universe. That's when it occurred to me that space and time were both fabrications of the mind. From the idea of unlimited space came the speculation we call infinity. From the idea of unlimited time came the speculation we call eternity.

After many eons of evolution, the egocentric processing system created the entity we call time. However, the egocentric mind had no way to comprehend space because the ego is hard-wired to process things in a series. Contrary to this, the allocentric mind, after eons of evolution, created our sense of space. However, the allocentric mind had no way to comprehend time because the self is hard-wired to process things all-at-once, in parallel. Notice the mutually exclusive contrast between the two minds: space is an allocentric creation, while time is an egocentric creation. In a sense, we can say that nature *had to* evolve one mind to create time (a time-mind) and had to evolve a second mind to create space (a space-mind). In other words, one kind of consciousness (egocentric) is charged with the job of freezing time into events, or objects, and the other mind (allocentric) is charged with generating the spaces that make up our domain—a place for objects and events to manifest.

One mind creates space, the other mind creates time. *Unfortunately, these two cognitive processing networks don't know they have a twin—that is the great paradox of our existence as human beings, the irony of it all.* The twins cannot perceive each other clearly enough to be sure that all the clues about duality are valid. Therefore, our two minds take the stage in a very fast alternation that gives a false sense of wholeness. Whenever they are alone on stage, each is sure they are the only reality, but this is an illusion. Space cannot know time, and time cannot know space, mostly because they can never manifest together; one appears only while the other is in hiding—being suppressed or inhibited. These two processing streams had to find a way to coexist, to trust that the invisible other was real. Of course, during evolution, they did find a way to overlap and oscillate. Otherwise, I wouldn't be talking to you.

In our current world, the egocentric mind is dominating the allocentric mind. Only a relatively few individuals are able to articulate their duality—most human beings on the planet still believe they are just egocentric beings. There is a good reason for this, of course. The egocentric creation of time was the beginning of a sense of the future and a deep appreciation for the past. The future was an especially thrilling invention of the egocentric mind. Peering into the future gave birth to the emotions called hope and dread. It also gave us a thirst for ever greater adventure, ever more pathways to follow. Egocentricity—because of its ability to review the past and speculate about the future—is the new kid on the block, the latest and greatest invention of nature. Consequently—at this juncture of evolution—the egocentric *time mind* is overpowering the allocentric *space mind*. When we sit in meditation, trying to *experience* the allocentric spatial mind—in the pure state that the saints and mystics describe—we should be sitting as if time is non-existent.

The new ego-kid on the block is wildly enthusiastic and idealistic—a young bull in the flower garden. The bull is ripping out all the flowers, obsessed with details and with the patterns inherent in nature. Mental bulls are suddenly everywhere, dissecting the flowers, pressing them into books, totally ignorant of the background scene, the lovely gestalt that makes a garden the wonder that it is. Meanwhile, poets, artists, and gardeners—the delicate souls—are abhorred by all the smashed flowers, by the clumsy, destructive behavior of the ego. There is a backlash in our time as our allocentric minds are hurrying to find ways to communicate the damage; to slow down the manic energy of the oblivious ego.

The allocentric mind is about experience, about the flowing-moment, the poetry, the music, the dance of life. It is *the* source of compassion and creativity; it will not yield to its brother's lack of empathy or to the ego's systems-oriented obsession with lists and accumulations.

Because there is no *ego* in allocentric consciousness, no focal power of attention, the feeling we have when using the allocentric mind is like being the ocean, rather than being a droplet separate from the ocean. Allocentric feeling is like living in an eternal present where there is no time—because for the allocentric mind *there really is no time.* Dwelling within allocentric consciousness results in a great sense of relief and calm because guilt and grief held over from the past, and anxiety about potential futures, simply do not exist. Meanwhile, the egocentric mind is busy doing one thing after another, going from event to event, task to task, with no sense of space. This is because egocentric consciousness actually *has no sense of space.* Space perception evolved separately from time perception. *Blindness to total reality is a given in the human perceptual system:*

> *He is like a pearl on the deep bottom*
> *of the ocean,*
> *wondering,*
> *inside his shell,*
> *Where is the ocean?*
> - "A BASKET OF FRESH BREAD," *A YEAR WITH RUMI*, 2006.

We know these two realities exist within us. There were rare times in our lives when we felt the joy of being the ocean, when we felt good for no reason. Childhood might have been the last time we had this sensation because sophisticated egocentricity evolves later in development, normally in parallel with chronological maturation. We also know, especially as adults, what dread feels like. We know about anxiety and existential angst. We know very well the stress of living in our modern age.

The mind that *knows that it knows* is egocentric. The allocentric mind does not know that it knows. The allocentric mind has no sense of time, no thought; *it just is.* The egocentric mind has language and the will to understand, but it is bewildered by the notion that a separate mind, an equal twin, can exist. The ego cannot locate and examine the allocentric mind and, therefore, must conclude that dual cognition cannot be proven.

I know this is complicated and new for most readers. A more detailed explanation is available in *Consciousness: A New Slant on an Old Conundrum.* It is also true that this complex new perspective still needs further thought and better explanations—it seems we are constantly rethinking and reframing dual-process theory. I welcome your comments and insights.

Proprioception Enabled Both Kinds of Consciousness

The central generation of movement and the generation of mindness are deeply related; they are in fact different parts of the same process. In my view, from its very evolutionary inception mindness is the internalization of movement.

~ I OF THE VORTEX, *FROM NEURONS TO SELF,* 2001.

The quote above by neuroscientist Rodolfo R. Llinás says "that mindness is the internalization of movement." As complex as this thought is, it is essential to understand. Dr. Llinás uses his own term *mindness*—avoiding terms like *consciousness* and *cognition*—because he sees that most terms that purport to describe mental states don't include an understanding of the brain and nervous system as agents of purposeful movement. Movement sequences (routines, behaviors, habits) became established over eons of evolution. In other words, *the mind evolved as a control system, to activate or inhibit movement routines.*

Notice that navigation (purposeful movement) requires precise control of muscles. Therefore, whatever sensory system monitors the activity of muscles (movement sequences) is a candidate for enabling the evolution of consciousness. We know that the portal called *the eye* enables vision, and the portal called *the ear* enables hearing. So, what is the portal—related to muscle memory—that enables consciousness? Rumi asked the same question, in his own poetic way, 700 years ago:

> *A horse is moving beneath the rider's thighs,*
> *yet still he asks,*
> *Where is my horse?*
> ~ "A Basket of Fresh Bread," *A Year with Rumi*, 2006.

Here is your horse: *proprioception*.

It is your *animation* that holds the key to understanding how consciousness works. You are not a statue; you are always moving (micro-movements are occurring even when you think you are still). We call repetitive movement-patterns *behaviors*, or *communication*, or *navigation*. We know what we are doing moment-by-moment because there is a neurological whole-body system that is designed to manage and monitor movement patterns. This system is called proprioception.

I use the perm *proprioception* very broadly so that it encompasses all terms used to describe internal processing of body states—other terms like interoception, kinesthesia, and vestibular processing, are subsumed (for me) under the heading *proprioception*. Furthermore, I am using the term proprioception so that it includes more complex fields of study like embodied cognition, embedded cognition, and enactivism. For an overview of these more complex ideas see the books of Professor Shaun Gallagher, especially *Enactivist Interventions; Rethinking the Mind* (2017).

As I wrote and reflected about consciousness, I slowly began to understand the importance of the inner senses—the so-called *hidden senses*. It eventually dawned on me that these not-so-obvious senses were the source of both the allocentric mind and the egocentric mind. Proprioception as the origin of consciousness is a new theory—as far as I know. I will summarize below the insights that evolved as I formulated this theory:

- Proprioception is muscle memory. It is an internal perceptual system responsible for remembering the motor patterns that enable purposeful movement and navigation. In the brain, neural networks called *brain maps* evolved to contain the instructions for repetitious movement patterns. For example, every morning you reach for your coffee or tea and successfully bring the brew to your mouth. When you were a child, this drinking behavior was not yet laid down in your brain, but as your brain remembered the motor pattern for drinking and played it back over and over again, you eventually got proficient. You could lift the cup, bring it to your mouth, tip the cup to release the beverage into your mouth, and then return the cup to the table—all smoothly and without accident. All behavior follows this kind of motor learning and recall.

- These blueprints for motor behavior are embodied—they can never occur without the synchronous firing (excitation or inhibition) of all the body's sensory and motor systems. Therefore, motor patterns are always recalled in exact harmony with sensory activity. The feeling of the cup in your hand, the taste of the beverage, and visual/auditory activity is synchronous with the muscle patterns needed to take a drink.

- The activity of the muscles that enable speech—the muscles in the lips, tongue, larynx, and mouth—evolved to enable language and communication. Most importantly for the discussion of consciousness, the sensory-motor memory of speech—how to talk—is laid down in brain maps. Evolutionary forces eventually enabled *inner speech*, which is based on the memory of the motor behaviors that enable actual speech. Thus was born our ability to talk silently to ourselves. *The inner voice is proprioceptive recall.*

- After many additional eons of evolution, the brain maps that enabled inner dialogue were consolidated into super-maps, which then created what we call self-awareness, self-consciousness, and awareness of being a separate entity with a mind. Therefore, *consciousness is the result of memory of memory.* A *brain-map fractal* eventually constructed ever more sophisticated layers of internal proprioception.

- Beyond self-consciousness, the brain-map fractal built super-maps of super-maps; the result was a witness, a mental watcher that could silently observe inner speech. There is no reason why this proprioceptive fractal system can't build ever more sophisticated layers of witnesses. In other words, what we call the "evolution of individual and collective consciousness," has a physiological basis. That there are varying degrees of internal sophistication appears to be undeniable. As I explore levels of consciousness in later chapters, realize that these cognitive divisions have an anatomical and physiological underpinning that can be traced to proprioception.

- Our internal voice ages more slowly than our body ages. That is why we still feel young even as our bodies decline. Even though our aging voices lose volume, the inner voice remains clear and of constant volume for much longer before cognitive decline sets in. This makes the creative act of writing an ideal enterprise for the elderly—for which, as a senior myself, I am very grateful.

- The potential for ever greater levels of mental sophistication is a birthright of human beings. We can evolve ever higher levels of consciousness. The trick is in knowing how to stimulate the brain map fractal to construct ever more sophisticated super-brain maps—witnesses of witnesses. The mind has the amazing capacity to change itself. Meditation can rewire the brain and change the biochemistry of brain maps. Indeed, meditation is the tool mystics have used over the centuries to self-evolve. This tool for self-evolution is available for any human being who wishes to create a more loving, peaceful, joyful, and balanced cognition.

- Not only can the proprioceptive system record the muscle patterns of its own body, but it has learned how to mimic the motor patterns of others. The infant and child watch, listen, and then copy behaviors (animations) that are observed. Developmentally, therefore, children can lay down brain maps—memories of motor sequences—that come from outside themselves. This is called mirroring. We can literally copy and then recall the behaviors (muscle patterns) of others as if they were our own memories. In other words, we can incorporate others into our physiological make-up. Our consciousness is in part crafted by the people we come into contact with, by our relatives and friends, and by the culture at large.

- Besides mapping body movements, proprioception also maps the environment. Tools, for example, are mapped as if they are extensions of the body—think of driving a car. The car, as a tool for transportation, becomes an extension of your body—that's why you don't run into things while driving (hopefully)—because the car is actually you, as far as the mind is concerned. Spatial areas, like rooms, are also mapped by proprioception. For example, rooms are mapped so that we become inseparably part of the enclosed space. The room becomes part of us just as tools became part of us. So, here is a paradoxical and bewildering thought:

If the allocentric mind *manufactures space* (if we project our spatial reality), then the information embedded within spaces is at least partially manufactured by the allocentric mind. In other words, perhaps space is a combination of self-generated information—a matrix built by the mind—plus channeled messages coming from a universal consciousness, available for the asking. You don't have to believe this, I am only asking that you speculate and muse as we travel together through the book.

- The ability to internally map external space (to "manufacture" space) led to the ability to imagine (project) the ego into different locations beyond the body. Buddhist meditation has a technique—to merge compassion with action—that requires the mind to project an image (the White Tara, for example) in space opposite the body, looking back. Using this technique, the ego can conceive of mirror images, including the mirror image of a twin mind staring back. We can also "put ourselves in the shoes of others" when we want to feel what it is like to be that person. It is as if allocentric proprioception has the ability to move the egoic twin through space—as a tool for imagination and prediction.

- If our two minds developed to enable navigation, then it follows that the 3-D image that seems to project from the eyes is actually an *animated brain map of the environment*. In other words, what we see ahead of us as we move is a projection of an internal map onto the external world.

- I often joke that I am not writing my books, that an alien is helping me, or my ancestors are helping, or local gods are crafting my thoughts, or book-nymphs are whispering wisdom into my essence. Many artists make similar observations. Saints, mystics and the supporters of the paranormal also claim to get help from external voices that are more or less realistic to them. I was quite surprised when I had a peculiar insight one evening while walking home from a friend's house. My insight was this: Once you have an inner voice, once the brain map fractal has successfully laid down a witness, you have a mechanism for receiving information from a conscious universe. If the universe is intelligent, if things like the Akashic field[4] are somehow real, then our inner recording system can be used to receive instructions and insights[5]. In other words, the universe now has a way to use our own language to talk with us or through us. I rushed home and wrote this section of the book—as if I had just received a text message from the universe.

- I was quite pleased with the insight above—that our inner voice could theoretically be used by outside intelligence to talk directly to us—or through us (for example, I use automatic writing to receive insights). Then I realized that vision also has a proprioceptive component. There are six skeletal muscles in each eye. This means that 12 muscles must precisely fire in harmony to enable visual tracking and fixation. Because of my education as an optometrist, I also know that the eyes must always oscillate in unison. Tiny synchronous eye movements, called micro-saccades, are necessary for perception. Experiments show that stopping this micronystagmus induces temporary blindness. The tiny synchronous eye movements are necessary for visual perception. If there is no eye-movement, then there is no vision. Putting these insights together, visual *proprioceptive memory of images*—overlaid with muscle memory—provides an avenue for *image channeling*. Visual proprioception is a memory of oculomotor muscle sequencing. Therefore, outside channeling of information also has an avenue for transmitting images as well as words directly into our brains (bodies). Memories stored in the holistic allocentric mind are called *dreaming*. Dreaming is not just an internal playback system, it is also a potential avenue for receiving images associated with words—the universe has evolved a way to converse with our inner essence. The significance of this insight for a hard-nosed scientist, or a skeptic, is that understanding *layers of proprioceptive memories* offers biological credence for channeling,[6] to understanding the brain as a receiver of subtle outside frequency codes. I know that I just stepped into the twilight zone (from a materialistic perspective), but I was brought to this conclusion through a logical progression. Both words and images—proprioceptive playback systems—can be used by ethereal book nymphs to inform writers. (I am grinning as I write this—for several reasons).
- The phrase *monkey mind* is often used to describe a low level of internal proprioceptive consciousness. Meditative techniques for calming the monkey mind are actually strategies for lowering the barriers that inhibit reception of insights from an intelligent universe (so say the booknymphs). The uncontrolled voices in the brain that rant and whine, rage, worry, and grieve are a kind of neural noise. This noise interferes with the mind's ability to pick up signals from the surround. Using meditation, the mind can calm this noise so that reception from outside the body

becomes clearer. Meditation is a way to tame the noise-obsessed mind so it can listen more deeply to signals that come from outside the egocentric mind.

- When the brain builds super-sophisticated layers of proprioceptive memories of memories, a point is reached at which this inner reality is very acute, more vivid and more realistic than everyday reality. I first discovered this when I was experiencing hemispheric synchronization at the Monroe Institute in Virginia. During meditation, I actually heard trumpets that seemed to come from some heavenly place (and I'm not that religious in a traditional sense—I wasn't primed for receiving transmission). When I reflected on where my vivid sensations might be coming from, it occurred to me that out-of-body experiences and near-death experiences might be proprioceptive hallucinations. I don't claim paranormal experiences are actually proprioceptive hallucinations, although they could be, but this proposition needs to be ruled out before we accept paranormal activities. However, given the above speculations, something beyond our capacity to comprehend might actually be connecting and communicating with our allocentric mind.

- There is an additional insight—a bit of speculative philosophy—that arises from the awareness that proprioception maps the environment and everything (and everyone) within an environment. Suppose proprioception could be linked to quantum entanglement. This would lead us to the proposition that experiences entangle us in a quantum network. We affect and are affected by all that we experience—forever. In other words, everyone we meet becomes a part of us—whatever we do from that moment forward generates actual causal and acausal effects. Logically, it would follow that all humanity is entangled in a quantum universe.

- When a human being dies, the body is totally still—proprioception is completely shut down. This lack of animation, of muscle tone, of purposeful movement, is unnerving for those who are left behind to experience the loss. The life force, the soul (the self) is missing in a body where proprioception has ceased to function. We witness the loss of embodied allocentric consciousness at death—this is what feels unnerving to us as we look at the stillness of a body without the presence of a self (soul).

- Something came before the sophisticated set of internal systems called proprioception. For example, evolution had crafted an internal dual-processing system for single cells several billion years ago. These early cells had two fundamental kinds of protein: receptor proteins that sense the environment; and effector proteins that react to conditions. The balance between these two proteins evolved into the balance we see today within our own biology.

- Proprioception works primarily by monitoring the large skeletal muscles that move the limbs. This makes sense for a bilateral creature that moves powerfully forward, straight ahead, with purpose. However, notice that there are few skeletal muscles in the head. This causes a strange phenomenon: we feel like we don't have a head. Think about it for a moment. We cannot see ourselves as we move about—we are blind to our own body most of the day. From a personal sensory standpoint, we are headless creatures. The neck feels like it is holding up a vision system, not a bowling ball with eyes. What might this suggest about consciousness? The mystic G. I. Gurdjieff would say that we don't remember ourselves. We go about our moments blind to our very essence. We have a talking-head-egocentricity, but no sense of a self (a whole body). Our allocentric mind is not only metaphorically hidden from view, it is also literally, physically hidden.

- Here is the bottom line: *just as the eyes are the portals for vision, and just as the ears are the portals for hearing, proprioception, as a global term for the combination of all the internal senses, is the portal for dual consciousness. In other words, consciousness is a sensorimotor system, a sense. We have a sense of vision, a sense of hearing, and a sense of consciousness.*

The theory that *consciousness is a sense* is a new paradigm and, like any new paradigm, it is so unusual that a first reaction is often resistance. Let me elaborate (reiterate, summarize) further in case your mind is recoiling from this new perspective.

The eyes and ears are external distance receptors. Information from the environment strikes the sensory apparatus inside eyes and ears, and the process of vision and audition results. Unlike the eyes and ears, proprioception begins as an internal sense, not external—proprioceptive sensations come from within the organism. Proprioception senses frequency information from the muscles of the entire body—therefore, the sense of (allocentric) consciousness is the result of this embodied neural processing.

Dynamic brain maps evolved to monitor the relative movement of muscle groups. Every behavior we have is actually a muscle sequence. We use these programmed movements to manipulate tools, to navigate, and to communicate. Our animation is species-specific; specific movement behaviors define us as humans. Consciousness is this sense of animation, of being alive.

Thus, proprioception is a *mapping system* that registers the incessant activity (flow) of every muscle in the body. However, not only does proprioception make maps of our body parts, not only does it keep track of our relative location in immediate space, but it also maps the environment in the very same way. Our minds can map that which lies beyond our body, incorporating these maps into what we perceive to be us. For example, a blind individual with a cane has an actual brain map that includes the cane as part of the arm. The incorporating of tools as extensions (actually as part of our body) is called *peri-personal space perception*. Here is another perspective: when a child is born out of a woman, that child is inherently "mapped" by the mother. In other words, the baby that was moments ago an integral part of the mother's body, is now outside the body of the mother, but the mother's body still maps that child as being part of her. Parents, fathers and mothers both, include their children in their body maps.

As we extend our allocentric mind into spatial regions beyond our bodies, we are also moving through states of consciousness—we are expanding what we mean by self or soul. Therefore, what we experience—as our soul—can be expanded with meditative practice. We are transformed as we *map and include* ever more distant spaces into our self-image. This journey alters our chemistry, our health, our power, and our wisdom. It is a perceptual journey that is also a spiritual journey. When we have pushed our proprioception to infinity and have included all of the universe within our mapping system, we arrive at the conclusion repeated by the mystics through the ages: *we are one.*

The Quantum Origin of our Two Minds

Pay attention (be aware):
there are those who want to change so badly
that they dissolve
into loving-kindness and freedom.
~ "A Dying Dog," A Year with Rumi, 2006.

Most authors who write about consciousness sooner or later get around to addressing the role of quantum phenomenon. Somehow the basis of the evolution of consciousness must come down to basic physical principles, perhaps to a fundamental fractal.

The importance of the quantum world is mostly an intuitive awareness. Hard facts are lacking although supporting research data is trickling in. Therefore, writers use their intuition and whatever logic they can muster to make the case that consciousness has a quantum basis. I followed shallow footsteps as I made my way through the speculations of other thinkers. However, if you muck around in a soup of quarks long enough, your mind becomes entangled in complexity and weirdness. That was my state when I went to lunch at the Third-Eye-Watching Vegan Café and Homeland for Quantum Queerness. I visit Surge whenever I am unsure or anxious:

"Okay, Surge, I need some plain talk about the quantum question."
"We've been down this road before, Dutch."

"Yes, I know, but I still have angst. Sometimes I am pretty sure of my conclusions, but then later I am not-so-pretty-sure. For example, I am pretty sure that the phenomenon we call *wave* simply means *movement*. And the phenomenon we call *particle* simply means *no-movement*. Mathematics is a smoke screen and physics is spinning on the head of a pin. I think *wave* means *to change* and *particle* means *no-detectable-change*. Stop and go—that's our perceptual universe."

"I remember you said that in your previous book—very clever bit of imagining, Dutch. But you look a little thin and pale today. I think you need some food for thought with sprinkles and whipped cream. Is that it?"

"Yeah, that's it. Talk to me, Surge. I am hungry for dialogue and musing. My mind got entangled in its own web, and now I am looking at my original face in a Funhouse mirror. You see, I know intuitively that we are quantum-powered creatures, made of the same stuff as all the rest of the stuff, but I don't know how to prove it. How do I convince the reader that they are not just skin and bones?"

"First of all, Dutch, any reader of a book like this doesn't need convincing—the reader has the same intuitive knowing as you. So forget about proving anything; just muse and speculate—do your cognitive-philosophy dance. By the way, Dutch, have you tried our award-winning Now You See It, Now You Don't Quantum Upside-Down Cake?"

"No. Should I?"

"Sounds to me like you should."

"Okay, but why? What does cake have to do with this?"

"Have you heard the meme that you can't have your cake and eat it too?"

"Yes, of course."

"Well, that's an old grandmother's meme. Grandma-wisdom claims that you can either admire your lovely cake on the table, or you can eat the cake and leave the plate empty, but you can't eat the cake and not eat the cake—you have to choose. This ancient meme seems like common sense, but it's a lie of sorts—a misdirection at best. Where is your cake, Dutch?"

"I ate it, so the cake is in my stomach, right? You are starting to lose me here, Surge. Is this more poetic license, like Holy Utensils?"

"All we serve are tasty metaphors and savory similes. Don't analyze your food, Dutch, just enjoy eating."

"All I am asking for is clarity, Surge. I don't need poetic mysteries. I need help formulating clearer thoughts."

"Then you should go to the library, Dutch. This is a Third-Eye-Watching establishment. Anyway, I can help. Go ahead and state your case. Give me five speculations about consciousness and quantum phenomenon. Talk some quanta with me, Dutch."

"I can't. That's why I came here in the first place. I came in for a meal, not another faux-meal."

"Just get in a flow, Dutch. Readers know this is a dialogue, not an essay. The reader knows you are thinking things through on the fly, learning as you go. You and the reader are traveling companions, exploring as you go from one cognitive restaurant to the next. Five speculations, Dutch; that's all I am asking. Stop mumbling and state your case."

"Okay, but I might ramble."

"Nothing new there. Ramble away, Dutch. What's behind door number one?"

"The retina of the eye."

"Go on."

"The philosopher Goethe said *the eye was sculpted by the light*. The eye is a creation of light, and light contains quantum information. Therefore, photon patterns designed the vision system. Quantum codes are dual, so the retina is a quantum processor that decodes a physical duality. Rod cells in the retina process movement; cone cells process no-movement."

"Was that so hard, Dutch? All you are doing is planting seeds. No proof is required. It's all just food for thought. What's behind door number two?"

"Embodiment. The vision system never acts alone—all the senses work in harmony. The human body is coherent, a single unit. Whatever vision is doing, the rest of the body is also doing. If vision decodes dual quantum information, so also are the rest of the senses cooperating in this process. The whole body is a quantum processor."

"Door three?"

"*Heisenberg's Uncertainty Principle*—a mathematical formula derived from quantum theory—is behind door three. *Heisenberg's Uncertainty Principle* states that we can either know where something is, or we can know at what velocity it is moving, but we cannot measure both at the same time. Think of this analogy: there is an ocean wave moving toward the shore. If you want to measure the velocity of that wave, it is easy to do. However, if you want to measure the exact location of a particle in the wave, it is not possible so long as the wave is flowing—the particle is forever not where you are about to look. You have to freeze the wave, and then you can look at a particle. But you cannot "freeze a wave" and "not-freeze a wave" at the same time. We are stuck with a paradox; we cannot measure the velocity of the wave at the same time we look at the location of a particle in the wave."

"Why is that relevant?"

"Because the eye has two kinds of photo-receptors—rods and cones—and these appear to follow quantum laws. Rods are wave translators and cones are particle translators. Rods measure relative movement, cones measure no-movement. Rods measure flow; cones measure stationary patterns. The retina is a quantum processor and so are all the other senses."

"How does this relate to your proposition that we evolved two minds?"

"Allocentric processing *can measure velocity*—it is very motion sensitive—but it cannot measure where something is located. On the other hand, egocentric processing *cannot detect velocity*, but it can calculate where something is, right now, from the viewpoint of an ego. Consequently, our two minds have their origin in the very structure of the quantum world, where "waves" and "particles" appear to exist, one or the other, but not both at once. Just as there are particles, or there are waves, we can analogously surmise that we can either pay attention using our egocentric mind—processing one-thing-at-a-time—or we can pay attention using our allocentric mind—processing everything-at-once."

"See, Dutch, this is making sense. No big deal. Just say what's on your mind. Is there a door four?"

"Behind door four is a gradual awareness—as research unfolds—that quantum weirdness at the scale of the very small is also true at a macro scale; as below, so above. Frequency strangeness—quantum tunnelling and non-locality, for example—occurs not only in the tiny world of quarks and photons, but also at the level of atoms and molecules. Human beings are made of strangeness and mystery. We are not simply stated, not easily defined and pigeon-holed as biology first surmised. Research has already revealed cracks in our biochemical reality. We are starting to look for quantum evidence, and we are beginning to catalog that evidence."

"I am proud of you, Dutch. You are finally losing your rigid mind, so to speak. The world of solid spoons and dependable reality is beginning to unravel for you. Perhaps upside-down cakes can be eaten and not eaten. Straight ahead mystery and poetry appear on the horizon. What is behind door five?"

"The evolution of technology and the analogies we draw from technology frame our dialogues. The newest revelations coming from science provide us with the metaphors for comprehending our cognition. There was a time when clay-based cultures defined human creation—we were made of the elements—of earth, air, water, and energy. Then came an era in which fluids and pressures crafted animal forms—spirits and vapors created human beings. Later came the light microscope, and the cellular world was exposed—we were blood and guts, and we were crafted from tiny cellular factories. It makes logical sense that we are now claiming our origin is to be found in the recently exposed world of quantum phenomena. We now understand ourselves as frequencies within frequencies. The irony and the joy is that all of the perspectives are true—but on different scales. Anyway, in our timeline, we need to think of ourselves as quantum creations. Of course, this thinking will blend with vir-

tual reality, with artificial intelligence, and with holographic speculations as technology evolves."

"Fair enough, Dutch. Now what? Are we done making the case for dual cognition?"

"That's all I want to say, Surge, except this book is about consequences and responsibilities that arise from the acceptance of our
duality. Therefore, I need to say a few words about the meaning of
life that arises from this line of logic. And I need to look at the two
journeys that exist side-by-side as we move through life—the journey of the hero and the journey of the saint."

"Go for it, Dutch. We have your back. To cap our dialogue, here's some Rumi
duality:"

Out Beyond

> Out beyond ideas of wrong-doing and right-doing,
> there is a field. I will meet you there.
> When the soul lies down in that grass,
> the world is too full to talk about.
> Ideas, language, even the phrase "each other" makes no sense.
> ~ "OUT BEYOND," *A YEAR WITH RUMI*, 2006.

The Meaning of Life: Both of Them

> There is a learning community where the names of God
> are talked about and memorized,
> and there is another residence
> where meaning lives.
> ~ "HOOFBEATS," *A YEAR WITH RUMI*, 2006.

Does human life have a purpose? Is there any meaning to what looks and
feels like any other bug's life: we are born, fly around innocently through
our youth, sleepwalk through our career with occasional shocks, and then
we get squashed like an ant on the sidewalk—the end. The time of our
death will be a surprise.

This bug's-eye viewpoint is depressing, so we have no choice except to
look for a purpose to our lives. Even pretending there is a purpose to the

whole biological freight train seems better than accepting this depressing insect view of existence.

I am not alone in this angst; most seekers feel there is a reason for their searching, a goal to be accomplished, a justification for getting up every day and doing it all again. Many philosophers and religious figures have weighed in on the meaning of life:

> Aristotle thought so [that life had a purpose]. He thought the telos [purpose] of human life is happiness, a point disputed by other philosophers throughout human history. St. Augustine, seven centuries later, thought the telos of life is to love God. To a twentieth-century existentialist like Martin Heidegger, man's telos is to live without denial of the true human condition, particularly death.
> ~ *Plato and a Platypus Walk into a Bar*, Daniel B. Klein, 2007.

Aristotle's idea of happiness is similar to that of the Buddhists. Aristotle thinks of happiness as a balance between extremes, equanimity. The Buddhists also speak of the Middle Way, a similar balance between extremes. Therefore, what both Western and Eastern viewpoints suggest, from the perspective of dual-process theory, is that happiness is a product of a balance between the allocentric mind and the egocentric mind. Another way to interpret the Middle Way and Aristotle's balance is to realize that *our minds are almost always out of balance* because our ego is too strong and it dominates the allocentric mind—this is certainly true of the era in which we are now living.

Therefore, if we wish to restore balance, it will require understanding and cultivation of the allocentric mind—we will need to bring poetry and love back into our lives. This means that happiness is accomplished—for modern human beings—by developing the allocentric mind. The ego *seeks pleasures* that come and go with mood and circumstance. However, *happiness is whole-body stability*, a sense of well-being and soul-health. Happiness endures while pleasures do not. For Aristotle then, the meaning of life is understood only when human beings cultivate the allocentric mind (to use my terminology).

From the beginning of self-awareness, human beings have wanted to understand their purpose. What's the point of tooth fairies and bowling alleys? Why read Aristotle or philosophy? Why follow the Buddhist Middle Way? What are you and I doing on this planet?

It has now become clear using dual-process theory and the conclusions reached in this book—and in my previous book on consciousness—that *our purpose is to navigate*. Navigation begets consciousness. Therefore, our highest purpose is *to move* toward ever higher levels of consciousness—this spiritual flow is what mystics call *love*.

We know that navigation requires two mechanisms, so there are two ways to navigate through life. We are programed dually:

- To *explore*, to attain and discover. To attach. To self-fulfill. To follow the mandate of the egocentric mind.
- To *experience*. To love. To participate. To follow the mandate of the allocentric mind.

Therefore, we have *dual navigational life purposes*: first, *to explore and figure out, because that is the evolutionary job of the ego*, and second, *the job of the self is to experience moments as they arise along our journey*: to soak up the awe, terror, joy, grief, angst, peace, and awareness of being alive moment-to-moment in communion with other sentient creatures.

When we find a life mission, we feel centered and at peace because we are fulfilling the mandate of the ego. On the other hand, when we use our allocentric mind to simply have experiences, we stop trying to frantically reach a goal—we stop clicking off milestones; we stop racing toward the next finish line, and the next. For the allocentric mind, life is a dance, life is music. Life is about participation, not problem solving and judging. Those among us who live in the moment say that enlightenment naturally arises from the cultivation of the allocentric mind; we eventually become awash in love for ourselves and for others.[7]

The egocentric mind exists to ask questions and to probe for answers. We are creatures who are driven to know as well as to experience. This dual quest is our birthright. Here is how author and philosopher Owen Flanagan expresses this birthright:

Speaking platonically, we humans show persistent signs of relishing the adventure of trying to track down what is good, *what is* true, *and what is* beautiful.

~ THE REALLY HARD PROBLEM, MEANING IN A MATERIAL WORLD, Owen Flanagan, 2007.

I agree with Flanagan, we relish the adventure—we constantly ask ourselves whether something is good, true, or beautiful. We make judgments every day based on how we answer these fundamental questions.

If the egocentric mind believes that a larger, more powerful ego—a god of all egos—holds the answer to our purpose, then the meaning of life will have to wait until that entity comes forth with the answer. However, if the god of the universe is allocentric, that god is probably quite pleased when we simply sing, dance, and are joyful—in harmony with others and with creation.

The meaning of life is not a noun; it is both *a striving* to know and *a flow* of love. There is joy and fulfilment when both our minds are given full expression. The trick is in the alternation, and that too is part of our heritage: we are meant to become experts in this balancing act.

Also, just for the record, human beings found that they were happier, felt more complete, and were more efficient if they helped each other. They could navigate to more places, more often, and faster if they were part of a community. Members of the community also experienced life in different ways, so it became possible to have shared experiences through the senses of others—vastly enriching a singular life. Here are a few lines from the poet and artist Oliver Loveday about shared pain:

Crazy Wanda was off her meds
Sitting in an AA meeting seven years ago
Blowing through a dozen personalities
Faster than a mockingbird could sing

For some reason the crazies have always sought me out
Stretched too far and suffering in their anguish
They speak the Truth in the middle of their psycho-babble
Like angels too tender for this world they have nothing to lose
- "The Merits of Suffering," 2017

When we witness our mind at work, what we observe are thoughts, sensations, and emotions that come and go like storm clouds. Sometimes we are witness to the storm-cloud emotions of those around us, while at other times we behave unconsciously like Crazy Wanda, blowing through our various personas. We affect each other, we need each other, and all of it is fascinating and bewildering.

The egocentric mind tries to figure out life, while the allocentric mind stays immersed and connected in the fabric of the moment. Through it all, there are emotional moments when we feel like angels too tender for this world.

The Journey of the Hero and the Journey of the Saint

Remember, he who lives today is responsible for the events of tomorrow.
- EINSTEIN AND THE POET: IN SEARCH OF THE COSMIC MAN, William Hermanns, 1983.

In a Great Courses lecture entitled *Philosophy, Religion, and the Meaning of Life*, Georgetown University Professor Francis Ambrosio explores the meaning of life from the perspective of Western culture. As a professor of philosophy looking back over 5,000 years of recorded human history, Professor Ambrosio speaks of a cultural genome. It is as if the genetics of humanity has also played out in cultural evolution, as if there is a cultural heredity that parallels biological heredity.

Not surprisingly, from the perspective of dual-process theory, Professor Ambrosio sees two fundamental archetypal paths that define Western philosophy and religion: the Greek world view, the journey of the hero, and the worldview of Abraham, the journey of the saint. According to Dr. Ambrosio, there are:

> . . . two parallel but distinct genealogical histories of cultural heredity. One line of descent originates in Greek culture with the figure of Socrates. The other line of descent traces from Abraham, "the father of faith."
> *- THE GREAT COURSES LECTURE SERIES ON PHILOSOPHY, RELIGION, AND THE MEANING OF LIFE, Francis Ambrosio.*

Both the hero and the saint have explored the mystery of mankind's purpose and have come to opposite conclusions:

Is meaning to be found in life, if it is to be found at all, by trying to be a hero or a saint? Or, perhaps more modestly, is it to be found by at least trying to learn from heroes or from saints? Should I try to fulfill my own potential as a person, to be all that I can be, to be a hero, in other words, or should I live for loving others and give my life as a gift to them? What do I owe myself? What do I owe others? How is it possible (is it possible?), to weigh and balance both sides of the scale?
~ *The Great Courses lecture series on Philosophy, Religion, and the Meaning of Life*, Francis Ambrosio.

In other words, is the meaning of life about fulfilling the quest of the egocentric mind, the hero's journey? Or is the meaning of life about our allocentric mind, our relationships and responsibilities to the whole community, to the cultural gestalt, the journey of the saint:

Are these two paths compatible, and to what extent? Or do they ultimately pose an implacable choice to be made to live either for the sake of the self or for the sake of the other person?
~ *The Great Courses lecture series on Philosophy, Religion, and the Meaning of Life*, Francis Ambrosio.

It seems we are gifted with two quests: to be heroes on a journey through life, and to be saints who spread joy and love as we travel the path. It feels to me that we can strive to be both a hero *and* a saint, as long we know which role we are playing moment-to-moment. It is a matter of mindfulness, intention, and paying attention to our various states of consciousness. In other words, life is both a *striving toward ego-fulfilment* and a *flowing-forward* through each moment filled with love. Life is not about actually attaining hero status or gaining recognition as a saint—life is a process.

Two Kinds of Emotion

When I am with you, we stay up all night.
When you are not here, I can't go to sleep.
Praise God for these two insomnias
and the difference between them.
~ "When I am with you," *A Year with Rumi*, 2006.

It comes as no surprise that there are two sets of emotions that correspond to our two minds. The egocentric mind needs an object-of-regard to function. The egocentric mind must pay attention *to something*. It exists as an ego that looks out from within itself into a separate environment filled with objects and life forms. The egocentric mind knows only a world where a *seer* looks out and *locks onto* the *seen*. Therefore, for the ego, emotions are about attachments and aversions. The ego assigns meaning to the "seen" and then attaches emotions. If the emotion is positive, the ego is attracted to the object-of-regard. If the meaning is perceived as negative, the ego is repelled by the object-of-regard. The egocentric mind is always concerned with the past and the future; attached emotions are remembered and they have an effect on later actions. The ego's bag of emotions include anger, greed, jealousy, envy, dependency, dread, fear, grief, envy, lust, and so on—a rather sad collection of emotions that have brought untold misery to humanity. There is something fundamentally and socially wrong about ego attachment.

In contrast to the egocentric mind, the allocentric mind *is aware* from a very high overhead perspective. As it pulls back ever higher, this mind becomes absorbed into the scene—it becomes the background and loses all contact with egocentricity. From this broad, all-inclusive overview, the allocentric mind can sense the vast interconnectivity of all humanity. The fabric of the earthly domain can be seen as a single entity; all living forms are part of an animated natural world. There is no seer and seen for the allocentric mind. It cannot separate itself out from the gestalt. Therefore, the allocentric mind is about relationships, the awareness of oneness, of shared existence; there is no attachment, just acceptance and gratitude. The emotions that arise from this sense of oneness, interconnectivity, and relationship are positive. These include love, peace, and joy.

Given the contrast between our two kinds of emotion, it is not surprising that the world's Wisdom Traditions, for example, the Christian Gnostics and the Muslim Sufis, have stood for the evolution and ascendancy of the allocentric mind. Unfortunately, egocentric, patriarchal religious practice has dominated our current era and overshadowed the loving heart of allocentric religious practice.

It is helpful to use language that highlights the contrasting modes of perception of our two minds. Just as we differentiate *self* from *ego*, we can make this distinction:

- The egocentric mind *reacts* rather than responds. Reactions have a calculated, judgement-heavy component—me against the world.
- The allocentric mind *responds* rather than reacts. Responses are loving, immediate, and without judgment or calculated oversight—us together in the world.

There is another fascinating aspect to the duality of our emotions: The allocentric mind has a physiological *response* in the moment, to an immediate experience, which is its biological mandate. Thus, if a powerful fight-or-flight situation arises, the allocentric mind responds without consulting the egocentric mind. Only *after* the body—the allocentric mind—has made an initial adaptation or response to a powerful experience is the egocentric mind informed. Therefore, egocentric emotions are what the egocentric mind makes of the already past response of the body. In other words, the allocentric mind responds and adapts before the egocentric mind gets the message that some important event has occurred. Only after the response has taken place can the egocentric mind react.

Of course, setting aside the negative baggage that accompanies the ego, the egocentric mind is a wonderful evolutionary advance. The egocentric mind has very important functions that enable growth, survival, and the evolution of identity (I am). For example, the egocentric mind can override an allocentric response—dampen it, learn from it, and alter responses going forward. The egocentric mind can also watch what is happening from a historical perspective—a delayed watching—it can then reflect and run future scenarios. The egocentric mind can learn, so it can develop a training program, a curriculum, for altering the knee-jerk behavior of the allocentric mind.

Each of our minds presents challenges and opportunities. The key idea is to remember that there are two different kinds of emotions that emerge from dual cognition. As we develop spiritually, we learn to perceive and manage our two kinds of emotions with ever greater sophistication.

Summary and Purpose

You are the soul
inside the soul
that is always traveling.
- "Hoofbeats," *A Year with Rumi*, 2006.

Although this book stands alone and can be read without reference to earlier publications, it is also an extension of my previous book *Consciousness: A New Slant on an Old Conundrum* (2017). I did not have the space in the previous book to elaborate on the consequences and responsibilities that arise from dual-process theory—there was much that had to be left unsaid and unsupported in the earlier book. Therefore, this is a completion for me, a final catharsis—it is no small matter that we have two minds in conflict. Indeed, our dual cognition is a very big deal that requires a great deal of future dialogue. I offer this book as a gift to you. My hope is that you will carry the dialogue forward.

Since you are joining me on this journey, I want to make it very clear where we are heading. My egocentric mind has crafted some fine bullet points below that state clearly the purpose of the book. But I know that this new perspective on dual-process theory is an evolving process. As I muse and consider, as I reread and edit, the pathway forward keeps changing—like a river flowing along natural channels. Many of the important insights in this book, and in my earlier two books, surfaced during the writing and editing process.

Articulating the purpose of the book hasn't been easy. My editor, a good friend (Karen Horwath), gently suggested that "making the world a more loving place" is honorable, and we might all agree it is a noble undertaking, but there has to be something new and interesting that convinces the reader to pack their bags and join the journey. Most of the content of the book is not common knowledge; there are many ports-of-call that will be new for you. I am pretty sure that you won't regret the journey because the evolution of consciousness is a process that no one escapes—in many ways, the book is an examination of *your* evolving mind and *your* fascinating life.

Anyway, you have come this far, so I assume you are getting comfortable with the oscillation between the two styles of expression. The paragraphs above are allocentric; they are relationship paragraphs that use caring language. Hold that good energy in your heart as I switch to the following egocentric mode of communication.

My purpose in crafting this document is to fill-in the details that were left unsaid in the earlier book. Specifically, my purpose for writing this book can be summarized as follows:

1. In *Consciousness: A New Slant on an Old Conundrum*, I made the case that human beings had an inherent mental duality because nature had

no choice except to craft a mutually exclusive, oscillating, whole-body sensorimotor system. I showed that there was anatomical, physiological, and evolutionary evidence for this theory. Once you begin to look for the duality, you discover it throughout our knowledge base. Therefore, a major goal in this book is *to give a few select examples* of our inherent duality as it manifests in poetry, religion, philosophy, psychology, science, gender, education, and music. I know that you do not feel like a dual creature, so you need to see this evidence. Then you need to take this mass of evidence seriously and reflect on the consequences and responsibilities that come with this new truth.

2. However, it is not enough to simply give examples of duality. A major purpose of this book is *to show that there is a repeating pattern that cuts across and connects all disciplines.* In every field of study, we find that researchers have discovered a general, overall system that manifests forms from a background—an alternation between excitation and inhibition. Unfortunately, each discipline has coined different words to explain this same phenomenon. The examples that I highlight demonstrate how each field has discovered duality and then gone about the task of trying to understand how this awareness fits within their professional domain. You probably are reading this book because you come from a domain of expertise that is interested in the evolution of consciousness—or you want to understand your own mind. Whatever professional or personal motivation, you have a responsibility to apply the dual-process perspective and consider how it reshapes your discipline and your personal behaviors.

3. Whenever a discipline—like philosophy or psychology—attempts to explain duality, an additional complexity is discovered: nothing is standing still, everything is evolving. The duality that we see within the human being is an ever-changing *process.* Evolution, maturation, and human development are unceasing. Therefore, *a major purpose of this book is to explain—as I currently understand this—how consciousness evolves over a lifespan and how dual processing has consequently evolved.* Dual-process theory is incomplete if it fails to consider human developmental cycles. Evolution, personal and universal, cannot be left out of the equation when we explore consciousness.

4. Humanity is at a crossroads. The puzzle piece called cognitive-duality is critically important for both self-understanding and for understanding

cultural conflicts. The planet is out of balance. Ego-bound human beings who have evolved with little to no access to their twin, the self—those who have weak and undeveloped souls—are busy crafting the future in their own egocentric image. Therefore, *a major purpose of this book is to hold up a mirror so that the ego can look at itself—I am also trying to show the ego its twin.* There is a poet (a creative soul) inside every human being. This book tries to show you, again and again, that all fields of study eventually unearth the soul. I am also suggesting that—just like me—you are on a journey; your spirituality is evolving.

5. Finally, *a major goal of the book is to motivate you to take action.* As the Dalai Lama said: *Karma implies action.* It is not enough to be smart and wise, not enough to be compassionate—we must each find a mission in life, a set of goals that drive our passions. We need to understand our duality and then we need to go on a shared mission to become saints and heroes. Remember that both the saint and the hero do their magic in their immediate communities—we don't have to change the whole world. We need to generate love and peace in the moment, in our communities—a ripple effect will then result that impacts the greater world.

Human beings are not cognitively equal. Some minds are more experienced (wise), or more educated (knowledgeable), or more compassionate (kind and empathetic). However, mental ability evolves over a lifetime; no one is permanently stuck at a level of consciousness. For example, a mind that is not so compassionate can become more compassionate. A mind that is not so wise can choose to have life-altering experiences. A mind that studies and reflects can become more knowledgeable. It is in the best interests of evolving cultures to cultivate as many wise, knowledgeable, and compassionate individuals as possible. Narrow-minded, mean-spirited, ignorant populations do not make a good democracy, or a profitable market economy, or nurturing communities.

We have a collective and personal responsibility to foster the development of human minds, starting with the awareness that there are relative degrees of compassion, wisdom, and knowledge—these three attributes in combination result in various levels of consciousness. Lower levels of consciousness display less kindness, less tolerance, fewer cooperative behaviors, and less charity than higher stages of consciousness. Lower levels of consciousness also display less wisdom and have less knowledge—less logical information about the world, fewer facts to work with. As we move upward around the spiral of spirituality, we become kinder, smarter, and wiser.

Buddhist philosophy acknowledges that suffering is part of the human condition; we are born into suffering and we endure degrees of suffering for a lifetime. Therefore, we have an obligation and responsibility during our sojourn on earth to ease suffering—using everything in our power. The first step on that journey is to realize that consciousness evolves; true human communication is much rarer than we suppose because people at differing levels of consciousness cannot comprehend each other. We need to elevate the cultural level of consciousness of individuals and cultures into ever higher stages of cognitive and spiritual evolution so that we are in the best situation for addressing personal and universal suffering.

If dual-process theory is correct, if kindness and wisdom arise from the allocentric mind and knowledge arises from the egocentric mind, then we have two kinds of consciousness that come with two sets of responsibilities. Dual consciousness needs to be acknowledged, and the consequences of dual cognition embraced if we are to foster greater wisdom, knowledge, and compassion—if the earth is going to evolve in a positive, healthy direction.

Notes

(1) **The poetry of Rumi.** The poet and mystic Rumi is a much loved and respected Sufi master. Because of the remarkable translation abilities of Dr. Coleman Barks, we are able to marvel at the wonderful magic of Rumi's words and insights. Rumi understood over 700 years ago that we have two minds, one of which is hidden. This hidden mind is where empathy resides and where we find compassion, loving-kindness, peace, joy, wisdom, and mindfulness. It's a shame that this allocentric mind is hidden because the best part of us, our spiritual nature, is inhibited and denied by the egocentric mind. It is strange that we can have vast amounts of knowledge, but not enough global empathy. We sit with the knowledge of our twin minds, and yet we don't take the key—a key that we hold in our hands—to unlock the door to release our allocentric twin from her prison cell.

In this book, I offer many lines or stanzas from Rumi poems that are quotations from a book called *A Year With Rumi, Daily Readings* (2006), by translator and poet Dr. Coleman Barks. You might find this book inspirational. I highly recommend it.

(2) The 2016 election in the United States demonstrated and exposed a frightening and **dangerously unevolved populous**. Mr. Trump's approach to debate—to name-call, to shout down, to be insulting and illogical—was a method that his followers admired, presumably because it mirrored and validated their own level of consciousness. On the other hand, ranting and name-calling coming from either political party does not help the polarized situation. Trying to shame others who have a different political habit is counterproductive and not very kind. We are all stuck on the spiral of consciousness. We are all at some level of ignorance, all of us are undeveloped, unevolved, and in need of help. It's time to go back to the meditation cushion.

(3) **Dual-process theory.** This was the main focus of my previous book *Consciousness: A New Slant on an Old Conundrum* (2017). What I present here and in that first book is a different understanding, an extension, of dual-process theory. The original formulation of dual-process theory did not have a navigational perspective, and it was championed more by the social sciences than the hard sciences. I have considerably expanded on the evolutionary and anatomical foundation of the theory, and here, in this book, I discuss the vast implications of being dual creatures.

(4) **The Akashic field:** From my perspective, this is another name for "background." Duality always seems to break down into background and foreground, and every field of endeavor unearths this basic concept. Ervin Laszlo took this ancient Hindu name and applied it to quantum theory in his popular book *Science and the Akashic Field: An Integral Theory of Everything* (2004). The core idea is that the universe itself is conscious—that consciousness is the basic building block of life. Consciousness originates from a universal fractal, which builds ever more evolved varieties of consciousness.

(5) **Our inner recording system can be used to receive insights.** Our brains can self-generate the words and images that constitute our inner dialogue. Outside influences (if they do, indeed, exist) could plausibly communicate through these images and words. If the Akashic field is real, if the universe is intelligent and conscious, then our evolved proprioceptive system is an avenue for communication. Spaces would be smart and spaces could inform us in an intelligent cosmos.

(6) Whether or not these **channeled messages** are actually coming from some unknown external support system—an intelligent, aware universe—I do not personally know. I have no experiences that convince me one way or the other. Like you, I am a seeker. I question, I dialogue, and I enjoy the mystery and the journey. I am also giving my scientific friends some wiggle room, some evidential space from which to reconsider the evidence for the reality of levels of consciousness. We need to evolve beyond our "right versus wrong" habits. As an alternative, we need to see humanity on a spiral of evolving consciousness. When we look at these levels of consciousness, both the wisdom and habits of science and religion are challenged to evolve as well.

(7) **We eventually become awash in love for ourselves and for others.** Of course, I am personally a long way from enlightenment, so this is speculation. I have little direct experience (wisdom) to back up my "logical" conclusions. Intuitively, it feels right that empathy, love, compassion, cooperation, and peace ought to become ascendant.

Two

Levels of Consciousness

Beat the drum.
Let the poets speak.
This is a day of purification for those
who are already mature and initiated
into what love is.
~ "No Room for Form," *A Year with Rumi*, 2006.

The Team Needs You, Dutch

"Good morning, Surge."

"Hey, Dutch. Welcome to the Third-Eye-Watching morning pep talk."

"No thank you. I'll just have breakfast."

"Breakfast comes with a pep talk this morning, nothing you can do about that. Here's our Emerging Dawn Special: Warmed-Over Conceptual-Eggs scrambled soft with Wry Toast. Are you ready to represent the team, Dutch?"

"What team? I came for breakfast. Fine, give me the eggs and rye toast."

"We don't have rye toast. We only have Wry Toast."

"Whatever nonsense you are up to, that's fine. Give me the not-rye toast with the wet eggs. And some coffee please, with vegan cream."

"The team needs you, Dutch. Tell it like it is. Go for the gold. Champion the oppressed. Strive to reduce suffering. We are running out of time, Dutch. Get your laptop and start writing."

"Can I eat breakfast first?"

"No. You eat after you save the planet."

"The planet is full of warriors and dimwits; it's not worth saving. Foul-mouthed imbeciles have seized power. The fools are torturing Buddhist monks and newborn kittens. Save yourself. I'm moving to Costa Rica."

"Shame on you, Dutch. Fleeing the scene is what chickens do when the fox shows up. Shouldn't the chickens stand and fight? Squawking and soiling the earth, that's the chicken way out."

"That's your suggestion? Fight like a chicken facing a pack of wolves? Courageous chickens rarely live to procreate, Surge. Those that run and hide survive to have sex later. Our ancestors were chicken-hearted and a good thing they were."

"Are you on the team or not?"

"Good luck with that team thing, Surge. I'll be fishing while the martyrs are being killed. Actually, what with modern technology, slaughter on a massive scale is our new glorious accomplishment. Look me up in the jungle."

"Sit down, Dutch. Chill for moment. I was just testing you. We need to know that our agents are ready for the battle."

"What battle? And I am nobody's agent."

"I am talking about the battle to raise consciousness. The battle that will find you no matter where you hide, Dutch. In the jungle or in the closet, you will be dragged out and enlisted. You can't escape or ignore this battle. It is already at the front door of the yurt, knocking to get in. Most of the planet operates in sleep mode. It is our task to wake the dead before the *humanity experiment* ends in disaster and embarrassment.

"You go first."

"This is a human challenge, Dutch. And I am not human. I come from an awareness that is alien to humanity. But trust me. I'm on your side."

"Yeah, sure you are. Except humans are wary of alien intentions; we can't know if you are a good guy or another devious salesman for consciousness."

"I know you think you are brilliant, Dutch, but you are just a pollywog in a murky swamp. You are stuck at a low level of consciousness. You need our help. There is no way you are going to move to a higher level of consciousness by talking about it. Look below you. See those starving masses? They need your help. Reach down and give them a hand. Here is my hand. Let's work together to raise the dead."

"I hardly have a clue what you are talking about, Surge."

"We will help you understand. You are a voice through which we can speak. Clear your mind so we can send you the words. Now get out there and do your duty."

"What duty?"

"Both your duties. Here's your Warmed-Over Conceptual-Eggs, scrambled soft. Chew slowly and taste your food for once."

A Geometric Perspective

This groggy time we live, this is what it is like:
A man goes to sleep in the town
where he has always lived, and he dreams
he is living in another town.
He believes the reality of the dream town.
~ "EVOLUTIONARY INTELLIGENCE," *A YEAR WITH RUMI*, 2006.

The Sufi poet Rumi knew very well that human beings had dual minds. He also knew that our dual minds evolve in parallel. Less evolved human beings live in a dream world, as an ego, and they believe that what this ego perceives is reality and truth. Many of Rumi's poems, like the one above, are about mankind's blindness to the self, the allocentric (poetic, spiritual) mind—the soul.

In their book *Frequency: the Power of Personal Vibration* (2012), authors Laural Merlington and Penney Peirce say we are moving from the Information Age into the Intuition Age. From my perspective, we are moving from an egocentric age—the hero's journey—to a new era in which the allocentric mind is gaining ascendancy. The power of the allocentric mind—an intuitive force, the journey of the saint—is being unleashed. Peirce and Merlington give us a nice geometric analogy to help us understand the transformation of consciousness into various developmental levels. I will explain their analogy below mixed with my own dual-process perspective.

Imagine that you are a single point in empty space. This dot-like existence is your level of consciousness. You think that egocentricity is all there is; that you are a spotlight creature, a shiny dot, like a distant star on a cosmic background of darkness. Everything revolves around you as you cast your spotlight and illuminate the objects within an environment. Now imagine that you wake up one day to find yourself joined by a series of compacted points in space, arranged so that a new reality appears: *the line*. This is a confusing dimension for a creature that assumed it held the only perspective. The community of points is a broader level of consciousness; indeed, the very idea that there can be more than one level of consciousness appears for the first time. This phase shift presents point consciousness with a new reality—emotions arise: shock, awe, disbelief, and defensiveness. *Point consciousness* now realizes that it is really part of a line, a community of points; that reality is linear. Spotlight egocentricity is still real, still there, but it now knows that it is also part of a larger dimension.

This goes on for a while and *line consciousness* takes to the lecture circuit to speak of the new perceptual miracle: we are not egocentric points drifting alone through empty space! There are more of us, we make a flat line in one dimensional space; we have entered the age of the line! We must help others who are stuck at the spotlight level to move their mentality into the highly evolved and greater level of consciousness called the Linear Age.

Except one day, years later, line consciousness has a bad day. While lecturing to a room filled with corporate executives, line consciousness has a heart attack. On the way to the hospital in the ambulance, line consciousness dies and its linear-soul floats to the top of the ambulance. Looking down on the dead body of line consciousness, an image appears: the line is joined by multiple other lines. An epiphany happens: We are not just a single line of points in space! If we make a series of lines we create *plane consciousness*!

While recovering from this near-death experience, plane consciousness writes a book called *Consciousness Explained!* and rushes the hardly-edited manuscript to the publishers—who have learned that sophisticated people, seekers, will purchase any book with consciousness in the title.[1]

You see where this is going, right? Plane consciousness eventually discovers that there is yet a higher level of consciousness where we can stack planes one on top of the other to create *cube consciousness*! That is where most of us are now, stuck in cube consciousness. And we are fairly arrogant about arriving at such a modern and highly evolved level of awareness—we are the pinnacle of nature's great work, the top of the consciousness food chain.

However, we have also learned to be cautious about what we take for granted. Someday we might discover that cube consciousness can be stacked in many innovative ways to build great cathedrals of consciousness. Then we might discover that the whole thing—points, lines, planes, cubes, sacred geometries, and conscious cathedrals—are all moving at the speed of light relative to each other. We think back to long ago when we were absolutely certain there was only one level of consciousness: egocentric spotlight attention. We were blind to all other dimensions of reality in our youth. However, these days we are not so quick to deny the possibility of ever more sophistication and complexity. Perhaps we are still asleep and there is more awakening ahead.

There are highly evolved people among us—those who have gone beyond cube consciousness, *beings of light* I call them. And yet, the world is still filled with people at all primitive levels of point, line, plane, and cube consciousness. They have divided themselves into political parties, religions, and nation states—they are busy being anxious and ill-tempered, each in their own way, and each with a different understanding of reality.

Penney Peirce and Laural Merlington come to this conclusion:

> The transformation of our world will become real when the next higher dimension . . . integrates with our "cube" reality. You might think [of the next] dimension as the *realm of the soul*, of spirit—*an experience* rather than a place, in which everything exists simultaneously in a *unified field* of energy and awareness, where everything contains everything else, all is known, and "love" is the primary substance out of which everything is made.
> ~ *FREQUENCY: THE POWER OF PERSONAL VIBRATION*, Laural Merlington and Penney Peirce, 2012.

The italics in the quote above are mine for "realm of the soul" and "an experience." These are allocentric terms. Basically, Peirce and Merlington are suggesting that human beings are becoming aware of the background field out of which everything manifests; allocentric consciousness is finally being let out of prison. Peirce and Merlington's geometric analogy gets us off to a nice start: if we live a restricted life, then our perception is narrow and self-centered, like a black dot on a white page. As we evolve, our consciousness expands to embrace wider and wider perspectives. Egos eventually disappear into a vast sea of mutual awareness. There arises a movement away from egocentricity into ever more sophisticated spatial dimensions, into ever more expansive allocentric levels of consciousness.

In the meantime, as we consider this notion that there are levels of consciousness, and as we consider that we are each somewhere on a spiritual path, we must realize that not all human beings are the same. We appear to be the same—we each have two eyes, two legs, a voice, we all have needs and desires—yet these similarities are deceptive. P.D. Ouspensky, in his book *In Search of the Miraculous* (1949), recalling discussions with his friend and mystic teacher G. I. Gurdjieff, put it this way:

> "At the moment it is not clear to you," Gurdjieff once said, "that people living on the earth can belong to very different levels, although in appearance they look exactly the same. Just as there are very different levels of men, so there are different levels of art. Only you do not realize at present that the difference between these levels is far greater than you might suppose. You take these different things on one level, far too near one another, and you think these different levels are accessible to you."
> ~ *IN SEARCH OF THE MIRACULOUS* P.D. OSPENSKY, 1949.

According to Gurdjieff and Ouspensky, each level of consciousness has its own form of art or expression. People at lower levels of consciousness cannot understand or appreciate art that is generated by more evolved human beings. Therefore, art and poetry often lead the way as examples for how consciousness should evolve. Artists are constantly showing us new realities that exist beyond our current perceptions.

But it is not only in artistic worlds where people of different levels of consciousness fail to understand each other. People also hold differing religious and political views because they are at different levels of consciousness—some are more spiritually evolved than others. Therefore, human beings at different levels of consciousness are unable to communicate with each other. As the quote above clearly states, the gaps between levels of consciousness are far greater than we realize. People who still exist as dots-on-a-page, the all-ego-no-compassion population, cannot comprehend or empathize with other humans who perceive life linearly, for example. A great irony is that individuals must have highly evolved cognitive skills to even conceive of levels of consciousness, yet alone to embrace the concept.

Laural Merlington and Penney Peirce are concerned with frequency differentials. They know intuitively that the universe is filled with frequencies (energy differentials). Humans are also filled with (and defined) by frequencies. Universal and specific frequencies interact and affect each other. Here is a useful analogy:

If you are a solid, if your individual frequencies are bound up, restricted, and contained, then you are not moving along the spiritual pathway—you are stuck, spiraling in place. If you are a liquid, if your frequencies are slow and heavy, you flow downward like water following the terrain—you are at the mercy of natural events. However, if your frequency state is vaporous, you rise up and follow the spiritual path. We can control our frequency state through mediation, intention, and through practice. We have a choice:

- We can stay frozen in place, immobile.
- We can drift sideways or flow downward.
- We can become ephemeral and rise upward toward an enlightened state.

Ironically, the higher up the spiritual spiral, the more an individual is aware of allocentric flow—the more awake they become and the more intention they generate. The lower on the spiral, the more asleep and unintentional an individual is, and the less they realize a need to evolve their mind.

What is Your Level of Consciousness?

We begin as mineral. We emerge into plant life
and into the animal state. Then into being human,
and always we have forgotten our former states,
except in early spring when we almost
remember being green again.
~ "EVOLUTIONARY INTELLIGENCE," *A YEAR WITH RUMI*, 2006.

There are different approaches to understanding how mental sophistication varies. One of the more elaborate and well-thought-out research approaches is the work of Dr. Susanne Cook-Greuter, an authority on human ego development. For our purposes, the importance of Dr. Cook-Grueter's work is that her research suggests that human beings evolve through levels of consciousness; consequently, some of us are fairly sophisticated, fairly evolved, while others of us are less so. On good days, we display a higher level of consciousness, while on other days our center of gravity is lower and we display a lower level of consciousness. These levels of consciousness are not inherent; they are developmental: we evolve through various levels as we mature and as we experience life.

There is a distinction between *levels (stages)* of mental development and *states* of consciousness. It is easy to get lost if these terms are not strictly defined. *States* of consciousness come and go, but *levels* of consciousness are more permanent. For example, we can have peak experiences, sudden feelings of clarity, or deep feelings of dread and despair that seem to appear out of no-where, and for no reason. We ride up and down on these *states* of consciousness; on sunny days we feel good, while on gloomy days we feel unhappy—without knowing why. However, *levels* of consciousness are stages that appear after hard work, after much practice and years of learning, and often after rites of passage—transformative experiences. Levels tend to remain dependable and become invariants, foundations for the leap up to the next level of consciousness.

Our understanding of levels of consciousness came from the pioneering work of developmental psychologists, especially Jane Loevinger who studied and wrote about the evolution of the ego. Susanne Cook-Greuter's research was heavily influenced by Loevinger's pioneering studies. *Spiral Dynamics* (2005), a book by Don Beck and Chris Cowan, is the application of ego-development theory to the world of business—to leadership and management. Developmental psychologist Roger Kegan is also well known for his "orders of consciousness," which are discussed in his two books *The Evolving Self* (1982) and *In Over Our Heads* (1994). Author Carol Gilligan, in her book *In a Different Voice* (1982), also identified stages of moral development that are now in popular use: pre-conventional, conventional, post-conventional, and integral.

Carol Gilligan understood that two forces were at work at different levels of consciousness. She, like many others, called these forces masculine and feminine. I would call the differentiation *egocentric* and *allocentric*. Her point is that masculine and feminine energies create different kinds of human beings, and so it is possible to postulate a level of consciousness schema that is—often unintentionally—overly masculine, failing to adequately understand the feminine mind. According to Gilligan, if you want to create a universal chart of consciousness that includes both masculine and feminine energies, you have to understand that the *female mind* is fundamentally different from the *male mind*. I saw this difference as I crafted the schema discussed below and I am well aware that my summary errs on the side of the masculine perspective.

However, the discussion about levels of consciousness needs to employ dual-process theory independent of gender. There are two ways human beings, regardless of gender, pay attention and remember. This was extensively discussed in my book *Consciousness, A New Slant on an Old Conundrum* (2017).

As I crafted my own composite of levels of consciousness (discussed below), I also used Steve McIntosh's excellent book *Integral Consciousness and the Future of Evolution*, (2007). McIntosh explains in his book how *levels of consciousness theories* can be applied to groups of people, including whole cultures. He draws heavily on Integral Theory which is the brainchild of popular author Ken Wilber. Wilber is a central figure in the understanding of cognitive development. I have listed many of his books in the bibliography. All of these approaches—of the writers mentioned above—look at both individual ego-development, as well as cultural evolution. For the discussion in this section, I will primarily (but not exclusively) use Susanne Cook-Greuter's research findings.

Terri O'Fallon wrote a thesis paper in 2010 that summarized the integration of developmental theory and spiritual guidance—a discipline that has grown in modern times to explore the evolution of spirituality. In O'Fallon's online thesis "The Evolution of the Human Soul: Developmental Practices in Spiritual Guidance," she provides an overview of developmental stages which includes examples from various historical models. O'Fallon's work is a good place to start if you are new to the evolution of developmental scales.

O'Fallon's essay begins with a discussion of James Fowler's developmental stages of faith. Fowler was one of the first researchers to join developmental theory with spiritual evolution. Not only do human beings evolve through physical and cognitive stages of awareness, they also evolve spiritually and morally. O'Fallon provides further historical background:

> James Fowler was one of the first people to bring adult developmental research into the spiritual traditions. He drew on the earlier works of Erikson, Piaget, and Kohlberg. Erickson's stages were related to the natural aging process. Piaget's work involved cognition from a developmental perspective, primarily for children, and Kohlberg delineated the stages of moral development in his research.
> ~ "THE EVOLUTION OF THE HUMAN SOUL: DEVELOPMENTAL PRACTICES IN SPIRITUAL GUIDANCE," 2010.

O'Fallon also mentions the work of pioneers Jane Loevinger and Elizabeth Liebert. Liebert, like many others, drew her inspiration from the pioneering insights of Loevinger:

> Liebert also drew on the work of Erickson, Piaget, and Kohlberg, as Fowler did, but added the research of Robert Kegan, whose work came from the same lineage as Fowler's. She identified three waves of structural theories from which spiritual direction drew. The first was Piaget's cognitive structures; the second was Kohlberg, who advanced the developmental research into the adult population and applied it to moral thinking. She also placed Robert Kegan's subject-object theory and Jane Loevinger's theory into the second generation of developmental models. Together they lend information to the third wave, which is noting the underlying assumptions that all developmental models seem to hold in common.
> ~ "The Evolution of the Human Soul: Developmental Practices in Spiritual Guidance," 2010.

I will discuss the underlying assumptions of developmental scales when I review Cook-Greuter's work below. Briefly, what Liebert found to be true for all developmental scales was that a person at a certain level of development could only perceive from within their stage (within a cognitive bubble)—they were blind to more sophisticated levels of consciousness. In other words, Liebert found that each stage had a worldview from which meaning was derived. Another assumption was that cognitive complexity increased with each higher stage—a threshold of knowledge and wisdom (experience) had to be gained before transformation to higher states was possible.

Before we explore each level in detail, it is important to get a general overview of Dr. Cook-Greuter's developmental stages. The overall developmental process can be imagined as an ever-widening upward spiral: we begin with a very restricted egocentric worldview—a single dot alone on a white background—and we evolve upward toward ever wider, more inclusive and embracing worldviews. Think of a funnel, the lower end is narrow, but the funnel is ever-wider as we move upward towards the lip. Cook-Greuter's first stages are near the narrow end of the funnel. As individuals (and cultures) become more spiritual and cognitively sophisticated, they move upward and are able to perceive from an overhead perspective—the higher up the funnel, the more spiritually evolved we become; the more our consciousness evolves. For me (and others), the word *spiritual* implies loving-kindness, peace, joy, wisdom, mindfulness, tolerance, and appreciation—all of which are more apparent at higher levels of developmental evolution. The more spiritual a person, the farther away they are from a purely egocentric (narcissistic) perspective.

Each higher level of ego evolution is more tolerant of differences and ambiguity, and less defensive when faced with beings of higher levels of consciousness—self-awareness expands with each spiral upward. This is obviously important for the evolution of humanity, since the lower stages of reality reside within individuals who are less tolerant, less appreciative of differences, and less able to adjust to change. It appears that higher levels of consciousness reside within individuals who are more compassionate, patient, and non-violent. However, this is a generalization; Dr. Cook-Greuter makes it clear that people within each level are capable of compassion and intelligence. It's not a good idea, she warns, to take her observations too literally or absolutely.

While we may transcend lower levels of perspective, all the previous levels of ego still reside within us; we are an amalgamation of all previous levels of unfolding—like Russian dolls, one inside the other. Each new level of ego sophistication is more complex, integral, and adaptable than earlier stages, but the primitive guys still hang out in our subconscious depths. This means that we have a center of spiritual gravity. We can be quite tolerant and loving one moment and not so much at another moment. It is not a steady and always smooth trek upward on our spiritual journey—it is a slow process most of the time, with many setbacks.

Mental evolution is *driven by experience*, through participation. However, no individual can be neatly stuffed into an ideal category. Human uniqueness and beauty are preserved—research shows that most individuals have scores that spread over at least three levels. There also is no happier or happiest level of ego attainment; every level is beset with the full range of human emotions—there is no escaping the Buddhist notion of suffering (Dukkha).

As I stated above, integral psychologists and philosophers differentiate between *states* of consciousness and *levels/stages* of consciousness. A *state* of consciousness can come and go, even moment-to-moment. Brain waves, like the restful alpha rhythms or the hyper-alert gamma rhythms, are examples of states of wakefulness that vary with the time of day and with nature's cyclical rhythms. However, *stages* (levels) of consciousness are more permanent. Once you get to the developmental level where you use language, for example, you tend to preserve the ability—memory becomes ever more reliable with repetition. Therefore, levels are pretty reliable once attained. However, it appears that we have a center of gravity around which we sink or rise depending on variables. Keep in mind this variability as we discuss developmental levels in the following sections.

Dr. Cook-Greuter identified six levels of ego development with four half steps—ten total—which I will paraphrase and summarize below. The first stage is total infant dependency with ego-potential but essentially no ego-manifestation. For this reason, Cook-Greuter ignores level one in her discussions of ego evolution and refers instead to *nine stages of ego development*. Refer to her work on the web for a more detailed and up-to-date discussion; see *Cook-Greuter and Associates*.

Dr. Cook-Greuter divided developmental stages into the categories discussed below. Each of these levels of consciousness is a separate universe. Concepts are defined differently at each level even if the words used are the same. For example, concepts like "God," "love," truth," "justice," "beauty," and so on, can mean something different at each level of ego evolution. A person at one level can speak with a person at another level, using the exact same words, and mean something entirely different. People can appear to agree or disagree, because they are using the same terminology, but they are not actually communicating. Each level makes sense of the world using a different foundational perspective. Each of the levels has a complete value system and each is defined by how they create meaning, what they see as the purpose of their life.

There is no indication that Dr. Cook-Greuter took dual-process theory into account; her research is ostensibly about egocentricity. However, her research shows an evolution from minds—levels two through four—that are *defined by knowledge* (a product of the egocentric mind), as opposed to levels 5, 5/6, and 6 that are *defined by wisdom* (a product of the allocentric mind). Cook-Greuter uses a phrase to describe post-conventional minds that is a very good definition of allocentric perspective: "A conscious sense of unity with the [back] ground."

The finding that we go from being totally egocentric and then gradually evolve into autonomous adults—with ever more complex allocentric abilities—is consistent throughout cognitive studies. The evolution of human cognition has the following characteristics:

- In the beginning as infants, we are immobile; infants are carried about because they cannot walk or crawl. Therefore, the evolution of cognition is a history and unfolding of *ever greater control over self-movement* as we evolve beyond infancy to adulthood.
- *The spatial domain over which a person moves gets ever larger* with developmental ability. The toddler wanders around the living spaces and slowly masters familiar indoor spaces. However, as the child grows and moves into new terrains, mastery of space widens outward.
- Mastery of space requires allocentric perception; *the child grows ever more capable of perceiving scenes and moving through them.* The more spaces the child experiences, the more spatial-temporal maps are stored in the brain.
- Egocentric processing also evolves as the child uses hand-and-eye and hand-and-foot perception to examine the world. *The more opportunity the child has to explore with hands and feet, the more neural networks are laid down in the egocentric brain.*
- *Egocentric perception develops from near-point and then extends outward.* The eyes especially "reach out" to touch objects across ever larger spaces.
- *As the child becomes an adult, there is a movement from ego-obsession to an awareness and appreciation of others,* and, eventually, to an ever wider understanding and appreciation of differences. A worldview gradually develops. This is followed by the attainment of a universal perspective. Then the ego-and-other perspective begins to dissolve into an awareness of the unity of existence. Adults can get stuck at lower levels of consciousness, but generally there is a flow toward unity.

Models like the Cook-Greuter scale of cognitive development observe a flow from egocentric mental dominance to ever more allocentric dominance. In other words, to be a self-actualized individual—approaching enlightenment—one must become ever more skilled at allocentric awareness. One must develop a highly evolved allocentric mind *forged by a lifetime of meaningful experiences.*

The various approaches for examining levels of consciousness, like the Cook-Greuter scale, should be reconsidered in light of dual-process theory. If we have two minds, then our goal is to have each mind be as highly evolved as we can manage. We want a highly evolved allocentric Christ-like (Buddha)

Consciousness *as well as* a highly evolved egocentric Einstein-like Consciousness. Furthermore, given the stark contrast and opposing goals of our two minds, there has to be a profound understanding of how these twins are to cooperate for the personal and global good. This is the great challenge of our generation: to rebalance our two minds so that they operate most efficiently.

I am well aware of the potential for self-righteous arrogance buried within ego-development scales,[2] so as I explore levels of consciousness, I will do so cautiously. Still, I think we are each curious where we fall on the various "sophistication scales." From a personal perspective, I was disappointed that my own mind didn't register at the far end of brilliance. I was "stuck" (as I began my research) at about level 4/5 on Cook-Grueter's scale: the evolving professor, the worried systems-thinker, the smart-guy-shallow-soul team. But that perspective was helpful for me on my journey, and I am grateful for Dr. Susanne Cook-Grueter's research and insights.

The Cook-Greuter Scale

As I said above, Dr. Cook-Greuter ignores stage one in her discussions of cognitive evolution because this is the infant stage where personality has yet to manifest. Therefore, she writes about nine stages of development. Refer to her work on the web for a more detailed discussion; the quotes below are taken from her website: *www.cook-greuter.com*.

Levels of consciousness 2 and 2/3 are called pre-conventional; *ten percent* of the population falls into this category. The general characteristic of this group is helplessness. Level 2/3 adults function with selfish attributes like the following:

> Rules are recognized, but mostly followed for immediate advantage or to avoid punishment . . . In a social context, people at this level of consciousness are often invested in a concrete visible world of "things" one owns—from tattoos, to clothing, to fancy boy toys. These are meant to demonstrate one's position of power.

Level 3 is called conventional consciousness. Along with stages 3/4 and 4, *seventy-five percent* of the population falls into this category. People at level 3 are conformists:

Their self-identity is defined by their relationship to a group . . . Being part of this larger entity allows one to be protected and share in its power. The price for inclusion is an unexamined demand for loyalty and obedience . . . The self is defined by and generated by the expectations and values of those others to whom one "belongs." Conformists tend to accept norms without inspection or questioning.

Stage 3/4 is called the self-conscious group:

They know all the answers . . . They enjoy oppositional battles with others who defend different positions. They live in a world where things are sure and clear, and they feel very much entitled to impose their views on others. . . . Self-conscious people intellectualize, rationalize and explain away what doesn't fit their expectations or set beliefs.

Stage 4 is called the conscientious group. They identify with others *who think as they do*. Examples include people with scientific or rational minds, like professors and highly educated individuals:

They treat reality as something preexistent and external to themselves made up of permanent, well-defined objects that can be analyzed, investigated, and controlled . . . This view is based on a maximal separation between subject and object, thinker and thought. It epitomizes the traditional scientific frame of mind that is concerned with measurement, prediction, and control. It also represents the goal of much of Western socialization and schooling.

Stage 4/5 individualists are systems-oriented; the stage is called post-conventional. Along with stage 5, *twelve percent* of the population falls into this category (Italics mine):

Stage 4/5 persons become interested in watching themselves trying to make sense of themselves. [They] abandon purely rational analysis in favor of a more holistic, organismic approach in which feelings, body sensations and context are taken into account . . . The need to reason and to explain everything by rational means lessens. *Experience itself becomes the new attraction.* Thus the process (or journey) of discovery is now more intriguing than reaching a goal or creating a product.

Stage 5 is the autonomous group, also called post-conventional. This group *identifies with others who have the same principles as they do.* They respect their own and others autonomy:

> Autonomous persons consciously commit to create a meaningful life for themselves and for others in the world through self-determination and self-actualization. [They] are often motivated and infused with a grand purpose and a vision of what could be . . . their enthusiasm is based on high ideals as well as on a more realistic view of what it takes to change old patterns in self, in organizations, and in society. [They want to be] instruments of change.

Stage 5/6 is the ego-aware group; this stage is also called post-conventional. This group *identifies with others who have the same principles* as they do, however (Italics mine):

> . . . an all-pervasive uncertainty is one of the characteristics of this stage. The suffering that can accompany this feeling of groundlessness is sometimes described as "the dark night of the soul" or in the existentialist French literature as "no way out." This level, like others, becomes the home base for some individuals and its very fluidity and uncertainty can become a source of élan vital and relish. *It is precisely in the tension between polar opposites that the most active engagement with life is experienced.*

Stage 6: This highest level of consciousness on the Cook-Greuter Scale is called the Unitive stage of ego development. This group *identifies with others who have the same "spiritual radiance" as they do. Less than one percent* of the population falls into this category. This is the level where an individual is aware and totally unified with a mental watcher, the witness.[3] The ego is transcended as this witnessing perspective develops:

. . . the Unitive stage presents an entirely new way of perceiving human existence and experience of consciousness. The term witnessing (rather than observing) can be used here to describe the capacity of people at this stage to metabolize experience without the conscious, willed focus and preoccupations of other adult stages. Truth . . . cannot be grasped by rational means and by making an effort. Unitive individuals therefore seem to transcend narrow ego-boundaries. They have open boundaries and are attuned to—rather than preoccupied with—whatever enters awareness.

This oversimplified look at one scale for understanding levels of consciousness, gives us a beginning foundation for dialogue. Later I will look at a few other attempts to assign levels of consciousness to our journey.

Levels of consciousness scales have several things in common; thus, these insights emerge:

- In the early levels of consciousness, we are totally ego-bound; we have little if any awareness of our allocentric twin. We even deny vehemently that such a twin could exist. Much of the world's cruelty arises from this primitive level.
- There is an evolution from the (sometimes) cruel, selfish tendencies of the ego toward ever greater compassion for self and others.
- There is a movement in "modern" cultures—which are predominantly patriarchal—away from materialism toward spiritualism. Belief in everyday reality, in objectivity, in sensory reality, is slowly eroded as we rise to higher levels of consciousness. We become less sure and more open to change and differences.
- There is an ever-greater awareness of a loving force that extends ever more outward. Souls get more peaceful, helpful, accepting, and blended-into-the-whole at each higher level of consciousness.
- A sophisticated ego is integrated with a sophisticated soul at the highest levels. These people are at peace with themselves even as they reach into the surround to make the world a more loving place.

When we move from the question of personal levels of consciousness to cultures and collective levels of consciousness, we enter the arena of inter-subjectivity. Objectivity is the outside world, subjectivity is our internal universe, and inter-subjectivity is communication between subjective minds. In 2007, Steve McIntosh published a very helpful book that contained an in-depth look at cultural levels of consciousness. Here, he defines inter-subjectivity (italics mine):

> When we begin to see the evolving reality of not only the objective external universe and the subjective interior of consciousness, but also the intersubjective realm of relationships, this constitutes a significant new way of seeing things. Just as Descartes' vision of the objectivity of external reality resulted in the opening of a new frontier of human progress, so too does integral philosophy provide a new way of seeing things by revealing how intersubjectivity (in concert with subjectivity) comprises the internal universe. This idea of intersubjective evolution emerges out of murky abstraction when we begin to see the presence of living systems within the intersubjective realm exhibited in the reality of human relationships. That is, *relationships* are the real, evolving, living systems of human culture.
> ~ INTEGRAL CONSCIOUSNESS AND THE FUTURE OF EVOLUTION: HOW THE INTEGRAL WORLDVIEW IS TRANSFORMING POLITICS, CULTURE, AND SPIRITUALITY, Steve McIntosh, 2007.

As Steve McIntosh makes clear in the quote above, human *relationships profoundly matter.* We need our fellow humans and they need us. Cultures have evolved around the degree and quality of human interaction. Developmental psychologists have been studying intersubjectivity along with individual ego-development for at least a hundred years. Below we will take a look at some of their findings concerning the evolution of cultural, intersubjective consciousness.

The German philosopher Georg Hegel (1770-1831) popularized the concept of Zeitgeist, a concept stating that society can be defined as a collective consciousness moving in a distinct direction and determining the actions of its members. Large groups of individuals, through their understanding of the meaning of life, determine the actions of the whole culture; that culture, in turn, influences individuals to believe a certain way—this is a self-organizing, self-generating mechanism.

The potential for greater levels of mental sophistication is a birthright of human beings. We can evolve ever-higher levels of consciousness. The trick is in knowing how to stimulate the mind to construct ever more sophisticated super-brain maps—witnesses of witnesses.

Piaget, Gebser, and Ken Wilber

People want you to be happy.
Don't keep serving them your pain.
~ "People Want You to Be Happy," *A Year with Rumi*, 2006.

The evolution of consciousness is mirrored by the developmental unfolding of human beings, from birth to death. For example, it is possible to use developmental psychologist Jean Piaget's (1896-1980) sequence for cognitive development in children as a model for the early stages of the evolution of human consciousness.

According to Swiss clinical psychologist Jean Piaget, children go through four stages as the mind develops:

- Sensorimotor (birth to 2 years)
- Pre-operational (ages 2 to 7)
- Concrete operational (ages 7-11)
- Formal operational (ages 11 and up)

This looks like the early history of ego development that Cook-Greuter explored:

- The infant has stage one consciousness on the Cook-Greuter scale.
- The toddler is at stage two. Both the infant and toddler are at Piaget's *sensorimotor* stage.
- Piaget's *pre-operational* stage is the birth of language and the explosion of egocentricity; it looks like Cook-Greuter's stage 2/3.
- Piaget's *concrete operational* stage of primitive reasoning and emerging allocentric awareness looks like Cook-Greuter's stage 3.
- Piaget's *formal operations stage*, the birth and development of abstract reasoning and sophisticated internal dialogue, seems to be the same as Cook-Greuter's stage 4.

Cook-Greuter was well aware of this correlation, of course—she used Piaget's early research—and that of others, in her own schema. However, her work went beyond Piaget's to look at various *adult* levels of conscious ability. She was also pointing out that adults can get stuck at unevolved levels like 2, 2/3, and 3.

We all go through physiological changes over a lifespan; we all age in the same sequence—there is not much argument about this. But it appears there is also a genetic recipe that repeats the evolution of consciousness within every individual. Inside of us is the complete history of the evolution of awareness. As we age, the potential exists within us to move through ever more evolved levels of consciousness. We can create greater awareness, beyond where we are now, individually, and as a species. The survival of the planet depends on having a population of highly evolved human beings. We have a long way to go, judging by the state of the planet as I write this in 2018.

Many people have contributed to the discussion of the evolution of cultural consciousness. One of the first was the European cultural philosopher Jean Gebser who crafted his outline of evolutionary consciousness during the mid-1900s. In Gebser's book *The Ever-present Origin* (1991), he offers a detailed look at a possible timeline for cultural evolution. Gebser was obsessed with the notion that we were entering a new, higher, and more integrated level of consciousness—he felt that something unknown in the evolution of mankind was emerging. I believe Gebser was correct—we are, indeed, moving to a more spiritual global stage.

Gebser was influenced by the Bohemian-Austrian poet and novelist Rainer Maria Rilke. I like these quotes from Rilke because they show his personal evolving consciousness:

I live my life in widening circles that reach out across the world.
~ RILKE'S BOOK OF HOURS: LOVE POEMS TO GOD, Rainer Maria Rilke, 1905.

Believe in a love that is being stored up for you like an inheritance, and have faith that in this love there is a strength and a blessing so large that you can travel as far as you wish without having to step outside it.
~ LETTERS TO A YOUNG POET, Rainer Maria Rilke, 1929.

The purpose of life is to be defeated by greater and greater things.
~ Rainer Maria Rilke. Popular quote on the internet; source unknown.

Poets like Rilke know how to live in the moment; they know that *experience is the thing*. Life is to be lived. Poets feel this to their core—but they also must write it down and muse over it. Poets need to nurse the words—they experience life, but they also seek to comprehend and share:

> Go into yourself. Find out the reason that commands you to write;
> see whether it has spread its roots into the very depths of your heart;
> confess to yourself whether you would have to die if you were forbidden to write.
> ~ LETTERS TO A YOUNG POET, Rainer Maria Rilke, 1929.

In the quotes above, Rilke is essentially saying that there is something loving and poetic that manifests out of an ever-present background, a well from which creativity is drawn. At a certain level of consciousness, we become aware that we are a child of this ever-emerging life force.

For Gebser, especially after he had a transformational experience in Sarnath, India, when he was traveling around the world as a young man, there is an *ever-present origin*, a background life-force that pushes us upward through ever more expansive levels of wakefulness. The *ever-present origin* means, for Gebser, that everything manifest must have come from a fundamental original background. Using my terminology, the background—that which is ever-present—is allocentric. A main theme in Gebser's magnum opus *The Ever-Present Origin* (1991) is that the new integral consciousness arising in our era *alters our relationship with time*.

As I stated earlier, the egocentric mind fabricates time—time is its business, its product. Gebser's new change in consciousness involves awareness that time is an illusion—that it is manufactured by the egocentric mind. Therefore, the new integral consciousness is atemporal, acausal, and aperspectival. In other words, the new mind that is arising in our era is a more sophisticated, more evolved allocentric consciousness—it is "aperspectival" meaning that it doesn't have an egocentric perspective—it doesn't use "perception" to understand the world. Instead, it uses an instantaneous translation from experience to action—there is no intermediary system of reflection for the allocentric mind.

Gebser divides the evolution of consciousness into five categories: archaic, magical, mythical, mental, and integral. Ken Wilber, who I will discuss later in the book, took Gebser's analysis, mixed it with many others, and blended all the various ideas with his own integral theory. Here is Ken Wilber discussing Gebser's theory and the evidence for stages of consciousness:

> So as we look at human history and see all the earlier ideas and beliefs . . . held by our ancestors . . . we can indeed trace their *genealogy*—actually look at history and look for any repeating patterns. When this is done carefully, as by, for example, somebody like the genius Jean Gebser, we find things like an unfolding sequence of genealogical patterns that remain as habits to this day—namely, the stages Gebser called archaic, magic, rational, pluralistic, and integral. Indeed, these are the repeating stages that developmental psychologists have empirically found . . . operating in over 40 cultures, including Amazon rain forest tribes, Aboriginal Australians, Indianapolis housewives, and Harvard professors—with no major exceptions in all of these. These stages were the basis of those hundred different developmental models that I mentioned, where you could see the same basic 6 to 8 stages repeated over and over again.
> ~ Trump and a Post-Truth World, Ken Wilber, an online document, 2017.

The stages of cognitive development that I outline below are based on considerable research by developmental psychologists and developmental philosophers. My contribution to the "levels of consciousness debate" is to introduce dual-process theory into the dialogue. We have a hard-wired sensorimotor duality that has been evolving: the egocentric mind and the allocentric mind are different at different levels of consciousness. Our two minds are less sophisticated at the lower stages of consciousness, and more sophisticated as development occurs.

I also find *blindness* at all scales of human existence. For example, there is a fundamental oscillation between awareness and non-awareness, perception and no-perception—there is a foundational on-off, yes-no aspect to our physical universe. We were created by this algorithmic invariant. What we perceive is balanced by what we do not perceive, by our *blindness*. Therefore, every act of human observation exists within a sea of blindness. If we apply this understanding to levels of consciousness, we can identify areas of blindness inherent at each stage. These invariant areas of blindness essentially define the stages—they reveal the dichotomy that is egocentricity and allocentricity. Furthermore, for each level of consciousness there exists blindness to more evolved levels of consciousness. Lower levels of consciousness are blind to higher levels.

If we use the geometric analogy that Laural Merlington and Penney Peirce gifted to us in their book *Frequency: the Power of Personal Vibration* (2009), we can clearly see how blindness is an integral part of every stage of human development. The spotlight creature, the single-dot-in-space level of consciousness, is blind to all other dimensions. This dot-on-the-highway entity cannot perceive that all of life is not about point consciousness. For years and years, the simpleminded dot-in-space felt alone, isolated—it was a level of consciousness so undeveloped and so unevolved that it was blind to higher spiritual dimensions. But then one day, through a transformational experience, single point perspective awakens—a phase shift occurs. Aha! There is more to life! There is linear consciousness! Suddenly, life is far more fascinating and far more magnificent than point consciousness ever could have imaged.

However, going from point consciousness to line consciousness is only the beginning of a series of awakenings. He who was blind (stuck at a low level of consciousness) is given the miracle of sight and insight over and over again as spiritual sophistication matures. But we can get stuck—or slowed way down—at every level of consciousness. Once we see this process and realize that there is a spiritual pathway, we then can go on a quest—a hero/saint journey—to discover "who I am" and "what this existence thing is all about." One thing is clear: the less we are concerned with our ego, the more we *actually care* about others.

Transformational experiences move us from line consciousness to plane consciousness to cube consciousness. That is where Laural Merlington and Penney Peirce concluded their thought experiment. They showed us through analogy how space becomes larger and more complex as consciousness evolves. Taking their analogy further, we can see that people with higher levels of consciousness—beyond cube consciousness—can conceive of endless space and endless time. Indeed, what higher spiritual beings come to know is that time and space are manmade illusions. The mind that was initially perceived to be confined to a body suddenly expands beyond the material.

As I said at the beginning of our journey, the egocentric mind does not just represent time, *it manufactures time.* The egocentric mind can conceive of eternity because it can always make more of this time-stuff. Likewise, the allocentric mind not only represents space, *it manufactures space*, as much of the space-stuff as it wants. Our allocentric and egocentric minds can also manufacture boundaries. We can conjure up beginnings and endings and put them wherever we want—we can imagine walls within walls.

Our infinity/eternity-building capacity and our wall/boundary-creating skills might turn out to have no relationship whatsoever to what is really out there in objectivity-land. Indeed, it is the main point of perennial philosophy, of the Wisdom Tradition, that space and time do not exist outside our human condition:

> Perennial philosophy is perennial not because it recurs in time, but because at its center is timelessness. That is what the term "perennial" means at heart. Nature in its magnitude and breathtaking beauty reminds us of timelessness and spacelessness, reminds us of the secret symbolic language that infuses everything in the cosmos . . . All that is reminds us of what cannot be described as "is" or "is not." That is the secret at the heart of nature's Language.
> ~ PERENNIAL PHILOSOPHY, Arthur Versluis, 2015.

The question that Arthur Versluis raises in the quote above concerns how we might reach this goal of esotericism: to go beyond space and beyond time, to find greater levels of consciousness, to understand secret symbolic languages. Others dwell in these higher realms—so they tell us—and we are not as skeptical as we once were, especially now that we can see an overview of stages of cultural cognitive development.

Besides my perspective on blindness, I have a second contribution to the "levels of consciousness debate." Years ago I became obsessed with brain frequencies. I eventually took the levels of brain wave frequencies—called delta, theta, alpha, beta, and gamma waves—and matched them with levels of consciousness (as many others have done). I then mixed in my understanding of brain wave frequencies with the levels of consciousness discussed below.

Intelligence and Wisdom

This is how a human being can change.
There is a worm
addicted to eating grape leaves.
Suddenly, he wakes up,
call it grace, whatever, something
wakes him, and he is no longer a worm.
He is the entire vineyard,
and the orchard too, the fruit, the trunks,
a growing wisdom and joy
that does not need to devour.
~ "THE WORMS WAKING," *A YEAR WITH RUMI*, 2006.

In the quote above, Rumi is talking about our two minds. The egocentric mind hunts for objects to devour, like grapes in a vineyard, but it is not capable of experience. Contrary to the egocentric perspective, the allocentric mind blends into the scene, participating rather than analyzing. In other words, egocentric attention devours the world one grape leaf at a time, but ignores the bigger picture. Allocentric awareness perceives the total gestalt—the whole vineyard is experienced.

Intelligence is the work of the egocentric mind. *Wisdom* is the work of the allocentric mind. Although they are often used as synonyms, these two concepts are very different. Intelligence is task specific. There are degrees of intelligence based on how well individual, task-based problems, are handled. The more primitive a creature in evolution, the fewer tasks it can do. An insect has inherent behavioral models for solving specific challenges within its domain. A rodent can solve more complex tasks than an ant. A dog has more intelligence yet because it can perform tasks that rodents and insects cannot. That is the pattern in evolutionary history; intelligence is defined by the number and complexity of tasks that can be addressed.

However, wisdom is a *sense of knowing*, which we call *faith*. Every organism is born with the ability to function in the world. Every organism comes equipped with a domain-specific wisdom suited for its survival. Therefore, wisdom depends in part on the genetic inheritance that an organism has been

77

blessed with. Somehow, there must exist a way for organisms to evolve ever greater wisdom and pass that on to the next generation. Knowing what to do—and quickly doing the right thing—is critical for moment-to-moment survival. Wisdom is based on experience. An organism *participates* as part of nature. When novel environments or situations arise, the organism learns and then passes this wisdom through to the next generation. This seems to be a different mode of transmission than the DNA system that passes intelligence from parent to offspring. We do not know at this point—mostly because we have not looked hard enough—what this allocentric generation-to-generation transmission might entail. However, the theories of Rupert Sheldrake—that postulate a Morphic Resonance—might be a good beginning.

Professor Sheldrake proposed his theory of Morphic Resonance in his 1981 book *A New Science of Life*, and elaborated on the hypothesis in *The Presence of the Past*, published in 1988. Morphic Resonance refers to a process used by self-organizing organic systems wherein memory is passed from generation to generation. Each individual inherits a collective memory from past members of the species, and then contributes to this collective memory, affecting new members of the species as they are born. Morphic Resonance postulates a new kind of holographic memory storage and a mode of biological/quantum inheritance that is independent of (or perhaps co-dependent with) DNA transmission.

Human beings at each of the levels of cognitive development discussed below have both a level of wisdom (the allocentric mind) and a level of intelligence (the egocentric mind) that define the character and skill set of that level. I will consider wisdom and intelligence as I address each stage of consciousness discussed below.

Crisis Mobility

Tears are worth more than money.
~ "A Dying Dog," *A Year with Rumi*, 2006.

As I taught blind children in special education, I developed a concept called *crisis mobility*. It was more an idea than an actual practice. Many of my blind students were fragile, and many of them were developmentally delayed—blindness had slowed their maturation. When they did reach a threshold level of competency, I would expose them to a gentle *navigational crisis*, a challenge to both their emotional stability and their rational mind.

I simply put these delicate children in a relatively more challenging environment, perhaps a little closer to heavy traffic than they were used to, or near train tracks—where trains rarely passed—or in a crowded urban setting with a cacophony of noises. I was careful to keep them safe, but the supposed dangers caused a spike in anxiety that sharpened their attention. I woke them up so they could evolve further skills. Crisis mobility was a small rite of passage, an inching up the spiral of conscious development.

Individuals and cultures (collections of people with the same average level of consciousness) go through periods of crisis. The world they are comfortable with and competent within reaches a tipping point. There is a crisis— a rite of passage—that results in (often) substantial change in their level of wakefulness. These crisis times seem necessary or inevitable if consciousness is to evolve. There seems also to be a threshold, a boiling point that must be reached before transformation happens.

Each of the levels of consciousness discussed below lasted culturally for a specific period of time before environmental or evolutionary pressures caused a radical change. Philosopher Ervin Laszlo has written about these moments in history and has called them *bifurcations in consciousness*. In the quote below by Laszlo he is speaking of a time of cognitive crisis in the early 21st millennium:

> We stand at the intersection of a system-level split of consciousness, facing either breakdown or breakthrough. The future is open to creation; to choosing something different from what you witness today. Future-creation is sourced in innovation, both profound and embracing. But first, there is a need to master the split in consciousness.
>
> A split in consciousness exposes a radical leap in the evolutionary trajectory of a system. A system evolves along certain lines, which can often be measured in terms of energy, size, complexity, or some other parameter. It evolves along its historic trajectory with minor fluctuations until the moment is reached when a linear evolution is no longer possible. It is at the point of the split that the system either transforms or collapses.

The concept comes from the theory of complex systems, and it applies equally to human societies and humanity as a whole. The split, known as bifurcation, is marked by a nonlinear system-level transformation. Bifurcation is one of the laws of complexity and it occurs in the evolutionary trajectory of nearly all complex systems.

The challenge is to select the path of transformation rather than that of extinction.

~ "The Challenge Before Us: Mastering the Split in Human Consciousness." Huffington Post Blog, Ervin Laszlo, 2016.

The last sentence of that quote (the italics are mine) is quite relevant to the magnitude of the problems that human societies face going forward. If we aren't very careful, we might inadvertently choose extinction rather than transformation.

Each of the levels of consciousness that I discuss below reached a crisis point in evolution that forced a shift in wakefulness; everything henceforth got more complex and interconnected. Everything also got more dangerous, especially as technology continued its relentless advance.

Another way to think about this need for a cathartic crisis-that-forces-change is to consider each of these levels of consciousness as paradigms. New paradigms generally don't evolve unless a shock occurs that exposes the limits of the pre-existing paradigm. From the ashes of the old perspective, a new paradigm arises. The old paradigm is incorporated into a broader understanding. Here is how Ken Wilber correlates levels of consciousness and paradigms:

. . . paradigms . . . whether functional or dysfunctional, are notoriously hard to get rid of. Max Planck (creator of the notion of a "quantum" of energy, thus ushering in the quantum mechanics revolution) is credited with being the first to notice that, paraphrasing, "Old paradigms die when the believers in old paradigms die"—which I summarized as, "The knowledge quest proceeds funeral by funeral." The point is that, put bluntly, boomeritis might die only when the Boomers die. But seeing the millennials adopting many of these notions, sometimes in even more extreme forms, it doesn't look like death is anywhere near strong enough to get rid of a really bad thing.

~ Trump and a Post-Truth World, Ken Wilber, an online document, 2017.

Here, Wilber is reacting to the 2016 presidential election, analyzing the electoral process from a developmental perspective. His primary point is that a culture's average level of consciousness determines voting patterns. In other words, the results of an election in a democratic society depend on the dominant level of cultural consciousness. Cultures are composed of individuals with varying degrees of sophistication. To move an individual to a higher level of consciousness requires a shock that propels that person into the next higher stage of development. The same is true for cultures as a whole—a shock is required to allow a new paradigm to transcend an old paradigm. But a prevailing cultural level of consciousness is usually so entrenched that only the dying of a generation will provide the necessary shock. Wilber is not seeing that the death of a generation of baby boomers will be sufficient to bring about the next paradigm shift, the next push into a higher level of collective consciousness.

My own take on the election of 2016 is that a moral crisis swept over the country causing a necessary shock wave. This shock wave caused all levels of consciousness to eventually ratchet-up a stage—I know that perspective is rather Pollyannaish, but I prefer hope to despair.

All levels of consciousness evolve because of a crisis. Something come along that causes a transformation that affects whole populations. In the early evolutionary stages change comes slowly, but as we consider the most recent transformations of consciousness we can discern exponential forces at work.

The discussion below was taken from multiple sources including (primarily) the research of Susanne-Cook-Greuter, Ken Wilber, Jean Gebser, Jean Piaget, and Steve McIntosh. My own views are blended with the thoughts of these brilliant thinkers. However, this is my interpretation and my prose—the views expressed should not be assumed to come directly from these developmental researchers, unless I reference them. I also do not wish to imply that there is agreement across the board. The discussion about levels of consciousness is complex and on-going.

Keep in mind that the eight levels of consciousness discussed below were initially proposed without an understanding of dual-process theory. The eight levels were crafted using a questionable scientific meme: that there is only one mind. The levels of consciousness discussed below come from a composite of *ego-development* scales. Early studies did not consider that human cognition is an oscillating, mutually-exclusive duality. That is one reason I called early approaches to stages of consciousness *masculine* (egocentric), because initial perspectives failed to adequately consider the *feminine* (allocentric) mind as an oppositional equal. I added in the allocentric perspective (to a small degree—not enough)—that is my small contribution to this important dialogue.

Notice that each category title is followed by either the term *egocentric* or *allocentric*. I am suggesting that each broad category can be considered to have emphasized or favored either one mind or the other. Notice also that there is an oscillation—for a while one of the minds is dominant but this is inevitably followed (rebalanced) by the following era when the oppositional mind is favored. This large-scale oscillation mimics (as above so below) the rapid and unconscious oscillating frequency state of individuals—our two minds must oscillate because they are mutually exclusive processes. If I label an era (level of consciousness) as egocentric, I mean that the era began as allocentric but was gradually rebalanced and then dominated by the egocentric mind. Each level of consciousness is a time of transition and transformation.

I have discussed this fundamental oscillation in my book *Consciousness: A New Slant on an Old Conundrum* (2017) and at the beginning of this book. However, dual-process theory is so non-intuitive that it is difficult to grasp. Here is another reiteration:

We *seem* to have one mind. However, we are *unaware* of the oscillation between opposite processing systems. This one composite mind operates by "blending" egocentric processing (the ego) with allocentric processing (the self). Egocentric processing *requires* duality—pattern or feature perception. Allocentric processing requires non-duality—perception of backgrounds or scenes. The ego and the self—the figure and the ground—are overlaid perceptually to give us the illusion of oneness, the gestalt. The categorization that follows was initially put forward—by Cook-Greuter, Wilber, and others—*without* an understanding of the importance of our inherent neural oscillation, and *without* consideration of a universal fundamental oscillation (as far as I know). This is not a criticism—I stand on the shoulders of the creative minds that initially formulated these stages; I simply offer a new slant on an old conundrum.

Archaic Man: Egocentric evolving towards Allocentric Life just Happens

Humankind is being led along
an evolving course through this migration
of intelligences,
and though we seem to be sleeping,
there is an inner wakefulness
that directs the dream.
It will eventually startle us back
to the truth of who we are.
~ "Evolutionary Intelligence," *A Year with Rumi*, 2006.

Jean Gebser says that within Archaic Man there is "complete non-differentiation between man and the universe." Gebser thinks there is not even an ego at this rudimentary stage. It is a proto-ego at best, not far from pure animal consciousness. For Gebser, Archaic Man has not fully manifested from the ever-present origin. In other words, Archaic Man has a dominant allocentric mind. The time period over which Archaic Man lived is a time in which a primitive egocentric mind began to ascend and rebalance evolving cognition. This is an era when the ego began to slowly assert itself.

Archaic Man navigated through life as if he was immersed in nature, part of a whole—this was his allocentric heritage, his connection with ancestral relatives who lived entire lifetimes without ego-awareness. He had a primitive proto-communication system based on animal instincts for territorial signaling and mate selection, but there was no inner voice. He could discern the differences between food sources or between male and female but his sensorimotor development was unevolved compared to later levels of consciousness. He lived hand to mouth. His consciousness was infantile. Archaic Man was a sensorimotor creature, stage 1 or 2 on the Cook-Greuter scale.

Archaic Man perceived from a first-person perspective. Perspective is not used here in the sense that it is commonly understood. In this context, I am speaking of a *bird's-eye-view*, an overhead spatial perspective. When we look down from above, we perceive relationships, networks of people and their

83

interactions. As we go from first-person to second-person, and so on upward, we are looking from ever higher viewpoints, taking in more of a spatial area and seeing ever more complexity in the interaction among sentient creatures within their specific domains. As we do this shifting of perspective, we become aware of an ever larger and more sophisticated network of lifeforms and complex environments. However, a first-person perspective is so unaware that no overhead view is possible. Personal needs dominated Archaic Man's perspective. In effect, Archaic Man was blind to any other viewpoint except his own. He could not comprehend bird's-eye perception. He did not understand *perspective*. He also had no sense of history or of future possibilities. His egocentric mind was rudimentary.

Suppose an Archaic Being came upon a newborn kitten. Archaic Man would kill the young animal and consume the beast raw—without fire. Archaic Man would eat quickly so that he didn't have to share. This level of consciousness arises from a primitive egocentric frame of reference. The Archaic Being is not much smarter than the cat.

Remember this graphic unsettling image of the newborn kitten. The young cat is a sentient being living on the earth alongside evolving human beings. Both egocentric and allocentric consciousness is evolving in the kitten, just as it is in human beings. Just like the human being, the cat is the pinnacle of nature's work at a moment in time. How we treat animals is different at each level of consciousness. I will return to this image, this insight, as I discuss each stage of consciousness.

Archaic Man is also blind to the pain of the kitten. Archaic Man has no empathy. As we spiral upward through levels of consciousness, we become increasingly more aware of other sentient creatures. We perceive them with ever more sophistication. In a way, our seeing them also manifests them. Seeing another living creature, and manifesting its essence, leads to ever more concern for their well-being. Archaic Man has only a rudimentary ability to perceive and manifest. Therefore, he has only a rudimentary, or absent, empathy. Manifestation seems to be linked with empathy.

Concern—attention, manifestation—eventually evolves into *compassion*. However, Archaic Man has no compassion and he has little concern for anyone else. He lacks any kind of awareness beyond the basic instincts. In addition, he is blind to the hunger of others. He has no awareness that if he shared his food, others might share their food with him. Archaic Man is totally blind to the world beyond his domain, his territory. He has no concept of time passing, no concept of spaces beyond his environment. Therefore, he is blind to space, to time, and to the needs of others. Both his allocentric and egocentric minds are essentially undeveloped possibilities at this primitive stage.

Archaic Man has an animal consciousness that cannot self-observe—he is not self-aware. His is a mind close to sleepwalking, not far removed from *delta-wave awareness*. Archaic Man is simply responding to events and needs as they arise. He has a genetic endowment which allows him to move upon the earth fluidly, but there is little intent beyond satisfying immediate basic bodily needs. In terms of levels of consciousness, Archaic Man is relatively asleep.

If God made man in his own image, then the Archaic God lives alone, naked in a cave. He has the close-set eyes of a predatory animal; he eats his food raw. Heaven soon fills with cavemen, cave-women, and lots of cave-children who died from all manner of illnesses and disasters. Heaven is also full of young souls because the life expectancy of the average cave dweller is between 30 and 35 years of age. None of these proto-people knew about religion or could conceive of religion. They are all in Hell unless there is some "caveman clause" that I don't know about.

The idea of a monotheistic judgmental God only arrived on the scene about 2500 years ago. Archaic Man, however, has been around for millions of years. If you died as a caveman, and you reincarnated, you came back as a caveman. The collective unconsciousness of human beings was strictly caveman consciousness. Initially, consciousness evolved at a caveman-pace.

My reference above to heaven and hell and to reincarnation—a pattern I will continue in the discussions ahead—is somewhat tongue-in-cheek and reflects my egocentric mind's skeptical perspective about traditional religion. However, from an allocentric perspective, there is something divine—a spark—in the essence of Archaic Man; a divine seed is present that holds the blueprint for the evolution of the sophisticated soul. I surmise that Archaic Man would not have been able to discern this evolving allocentric heartbeat. However, we have no way of knowing if a few poetic, saint-like humans did, indeed, walk through this primitive landscape.

Archaic Man knows how to kill to eat. He knows to seek shelter when the weather turns bad. He has a sexual appetite. He knows to drink water. He can sense fear and feel aggression. However, he doesn't say *please* or *thank you*, and he doesn't floss after meals. He doesn't have conversational language or good table manners. But he is the first creature with a modern human form, and it is from his rudimentary mind that all levels of consciousness eventually evolve. Within each of us is a genetic heritage gifted by Archaic Men and Archaic Women.

Ken Wilber uses a color scheme to highlight levels of consciousness. He labels the Archaic Period *Infrared*. He also states that this era began 250,000 years ago. I prefer a different date range. I believe that Archaic Man, the so-called caveman, has origins in Africa that date back millions of years. I reconsidered the date ranges for levels of consciousness because I believe there are exponential forces at work here—I will discuss this perspective later. Futurist and popular author Ray Kurzweil explains exponential phenomenon—as a basic consequence of technological advancement—in several of his books. I used Kurzweil's concepts to arrive at my speculations about the duration of each level of consciousness.

Tribal Man: Allocentric evolving toward Egocentric Life Happens to you

There are many guises for intelligence.
One part of you is gliding in a high wind-stream,
While your more ordinary notions
take little steps
and peck at the ground.
~ "BEWILDERMENT," *A YEAR WITH RUMI*, 2006.

Tribal Man (Gebser's *magical* stage) represents the next higher level of consciousness beyond the Archaic. Tribal Man evolved a hive-like mentality. He became a member of a small colony. This was a move away from the selfish egocentricity of Archaic Man, a swing toward a more sophisticated form of allocentricity, an awareness of a *self* contributing to a hive-culture.[4] Just as ants and termites work together for the good of a common mass organism, Tribal Man also did his duty as part of the whole. Tribal Man has a toddler, pre-operational consciousness similar to Cook-Grueter's stage 2/3.

Ken Wilber discusses three general stages of cultural evolution: egocentric, ethnocentric, and worldcentric. Archaic Man slowly evolved an egocentric, body-centered viewpoint. This eventually gave way to an ethnocentric perspective—the emergence of a creature with a mind that was more aware of the connection between *me* and *others like me*. Whereas Archaic Man is just concerned with *me*, Tribal Man becomes—for the first time—concerned with *us*. Yet, there is not a fully formed ethnocentric viewpoint in Tribal Man—that level of sophistication would evolve later in Traditional Man. World viewpoints would not evolve until Modern Man emerged on the scene thousands of years beyond the age of Tribal Man.

Tribal Man discovered sharing: give and you shall receive. He fills his belly first, of course, but then he offers the dregs to those around him, to other humans wandering about in his domain. The concepts of *concern* and *compassion* flicker on and off in Tribal Man, but they are rudimentary emotions present in a rare few. Tribal Man is aware that he lives an existence apart from others—his egocentric nature is evolving. His territory is expanding as well. He travels farther distances from familiar surroundings.

Like Archaic Man, Tribal Man is blind to the past and to the future. His egocentric mind is still at a primitive stage of development. He is also totally blind to empathy. His allocentric and egocentric minds are so rudimentary that he is locked into dream-like moments. He has a *theta-state consciousness*, so he perceives voices (sounds) in his head that seem not to be his own—these voices might actually be more clear and real than earthly voices. Sounds and proto-words in his mind seem like whispers from beyond. This is probably frightening for Tribal Man, but it is evidence that something or someone magical is ruling his life.

Movements are mostly routine and familiar. Tribal Man lives primarily within a small territory and spends his existence walking the same pathways. However, he is a nomad and has some understanding and control of the movement of men and animals. He is beginning to find rudimentary meaning in the world, although this is magical, theta-wave-meaning. His cave art is magnificent and suggests that he has a budding sense of time, space, and personal moments—he is becoming aware of events, of experience.

The roots of modern religions are, perhaps, found at this stage of consciousness. There is a primitive feeling that local gods sit on thrones in the firmament of the stars and manage the natural world. This pantheon of gods rules the sea, the weather, and the change of seasons. The world of Tribal Man is filled with things he does not understand that are outside his control. Consequently, he attributes unknown activities to magic and to magical beings.

Like modern religions, the roots of politics might also have arisen at this stage of consciousness. The isolated Tribal Man has to eventually cooperate within a primitive network of beings. Whenever two or more people gather, the potential for conflicts of interest arise. As the size of a network increases—as the population grows—conflicts increase. Power becomes an issue as individual interests clash. Some kind of governance becomes necessary, rules must evolve. Also, as the geographical size of the hive's territory expands, so also must the degree of governance expand. However, for Tribal Man, the population is small and so is the territory. Therefore, there is minimal need for rules and governance.

Tribal Man has a second-person perspective. He is aware of others and he tries to communicate with them, one-at-a-time. He is also aware of the first-person perspective that still resides within him, even though in a primitive way, he can perceptually rise above his body and look down as he relates to nature and to other sentient creatures one-on-one. He has a primitive ability to see from an overview perspective. However, he is still restricted to the present moment, the immediate past, and the immediate future. He cannot mentally project into different spaces or different times. His language is concrete; there are no abstractions to share or muse over.

If God made man in his own image, then this Tribal God is a good hunter who shares the kill after he eats his fill. Heaven and Hell have a new population hanging out with the primitive cave-guys. "Sophisticated" termite-like proto-humans walk the golden pathways of the soon-to-be-promised land.

Reincarnation has also evolved. Now human beings can return as hive-creatures or can revert to the good-old-days of caveman "freedom" before sharing and fire were invented. This hive culture existed for hundreds of thousands of years, but this is a much shorter period of time than the Archaic Age. There are far fewer tribal creatures existing in the Great Beyond.

We can surmise, especially as we gaze upon early artifacts and cave paintings, that Tribal Man's sense of the divine had emerged as a *flame* rather than an occasional *spark*. The community, the tribe, must have slowly evolved a sense of wonder and mystery as self-consciousness began to emerge as a product of the evolving allocentric mind—the soul as a collective sense of the divine—has emerged on the pages of history.

Within our genetic heritage, all human beings still hold the genes for Tribal Consciousness. If our collective unconscious is the sum total of all human cognition, then Tribal Man contributed a non-verbal caveman mentality, insect-like cooperative behaviors, and a single flame that was to become the soul.

In his Integral Evolutionary Theory, Ken Wilber color codes this level of consciousness as Magenta. Wilber also states that this era began 50,000 years ago. I suggest a different date range. I believe that Tribal Man was around for hundreds of thousands of years and lived intermingled with Archaic Man.

Warrior Man: Egocentric evolving toward Allocentric Life Happens by you

Forget safety.
Live where you fear to live.
Destroy your reputation.
Be notorious.
I have tried prudent planning long enough.
From now on, I'll be mad.
~ "BEWILDERMENT," *A YEAR WITH RUMI*, 2006.

The next level higher than Tribal Consciousness is Warrior Man (Gebser's *mythical* stage). This level of human being has evolved from a hive creature to become a member of a clan. The being of Warrior Consciousness would parade around the campfire holding the roasted "cat-on-a-stick" overhead, howling with savage pride over the kill. He would then make it clear to the clan that the coming of the cat is a sign from the gods that warriors must attack the neighboring village, burn it to the ground, kill all the males—young and old—and then reallocate the conquered women.

The fascist regimes in Germany, Italy, and Japan were perhaps at this level of consciousness when they climbed into their tanks to reallocate neighboring resources during WWII. The Warrior Man, especially the young man, can be whipped into such frenzy that he will march proudly off to war, holding the holy flag-on-a-stick overhead, howling with moronic pride. Too much of the world is still dominated by masculine creatures like this today, at war with other warrior males, because that is what Warrior Man does—that is as high as his consciousness has evolved.

How did peaceful men working in a collective hive turn into such stupid murderers? Perhaps it is because we are witnessing, for the first time in evolution, a fully formed personality-structure. Archaic Man was self-centered, a total ego, but that ego had no voice. Tribal Man developed language, rudely and incompletely, but there was probably no sophisticated internal voice forming whole sentences. Perhaps with Warrior Man we find the first internal (magical, frightening) human voices speaking inside the head. Or perhaps it was the coupling of the voice with emotions that made the difference. For males, emotions were created around the hunt, around danger, around murder, and around the satisfaction of the kill. An internal voice combined with aggressive emotions might very well have created this semi-articulate monster.

I suppose a geneticist might add something coldly rational about competition or gene migration. Perhaps, they might say that like an amoeba ingests food, colonies learned to absorb their neighbors. "Reallocating the women" is a crude and ugly metaphor for expanding the gene pool. Perhaps Warrior Man is about ever more groups coming into the colony and ever richer genetic diversity—better hunting grounds.

One thing is clear (if history books are to be trusted), males drive this Warrior Man insanity while females (historically) silently collaborate—or are powerless and soon dispensed with if they do rebel. This masculine war-like creature evolved when egocentric consciousness was still primitive, but on the rise. Warrior Man is an adolescent with concrete-operational consciousness. Warrior Man appears to be an early budding of Cook-Grueter's stage 3.

On a positive note, for the first time in evolution, the mind evolved the ability to construct patterns of space and time that enabled an emerging sense of future and past. Language evolved beyond the concrete, and the first simple abstractions appeared in the mind. Therefore, Warrior Man began to envision the consequences of his actions, and he could remember and repeat useful actions. He could now see the power in temporarily cooperating with other warriors.

Polarity appeared at this stage. Warrior Man can differentiate dualities. This creature has powerful uncontrolled emotions and he expresses them as they arise. The hidden (modern) allocentric mind flickers on at this level of consciousness as does the first hope that this primitive creature might someday experience empathy. But this War-Creature is very blind. There is little if any empathy. *Concern*, as a concept, is beginning to blossom, but *compas-*

sion has yet to enter consciousness. The allocentric mind is mostly hidden, unevolved, operating in silence while the moment-to-moment savageness of the unleashed ego goes about its cruel business. Warrior Man is blind to the ultimate nature of the brain mechanisms that generate endless time, endless space, and endless empathy.

But there is more to this Warrior-Man consciousness, something powerfully good. The Buddhists identify a level of consciousness called the warrior-bodhisattva. This type of person is *a warrior for the common good*. The same idealism that drives young men and women to enter into moral (and) actual combat against perceived evil comes out of this level of consciousness. There is a fierce fighting-spirit within human beings, a stubborn, defiant, action-oriented determination to make the world a better place. A sense of divine purpose has burst into flames and consumed hearts and minds. From a divine spark in Archaic Man, the soul became a single flame in Tribal Man, and has now become a raging fire in Warrior Man.

Modern fighters for social justice, for racial and gender equality, and for economic fairness draw on this ancient fighting spirit. The will to personally survive, to thrive, and to live an interesting and invigorating life also draws energy from this level of consciousness. Warrior Man, like Tribal Man, still resides in a *theta-wave state*, but he can at times move into a *primitive alpha-wave and low beta-wave* egocentricity. He has no clue and little interest in how his mind functions. However, he has learned how (often violently) to affect others and the world around. He can organize for a common purpose.

Warrior Man has a third-person perspective. He can rise above himself sufficiently to see not only his own body but also groups of creatures and the territories through which they roam. He has a territorial-view and is loyal to the geographical region where his group (his clan) lives.

If God made man in his own image, then God is sometimes cruel and aggressive. At other times God is determined to be a crusader for the common good—to craft His earth into a living sculpture worth admiring. God sees enemies everywhere. He can be ruthless in his effort to stomp down perceived evil. Killing men is no different from killing anything else that is hunted, so the Warrior God has limited empathy.

Warrior Men use God as an excuse to kill. There is no single agreed-upon God, of course, but the favorite of this age is the God of War. There are other gods in the Warrior family: gods of fertility; gods of the Sea; Earth and Sky gods; Father and Mother gods; gods of the past and future; as well as the Tooth Fairy and Tinkerbell goddesses.[5]

The halls of Hell and Heaven are now less safe and comforting because the souls of dead warriors rant and rave and find fault everywhere. These warriors are busy organizing the cave-people into labor unions, trying to make Heaven and Hell more responsive to the needs of the common deceased people.

As I said earlier, my comments about Heaven and Hell are tongue-in-cheek reflections on the primitive notions of early religious thought. The higher up the levels of consciousness, the more abstract the concepts of heaven and reincarnation become. If we accept a literal concept of heaven, we have to accept that this holy space is filled with people much less evolved than modern humanity, and that it is ruled over by a much less evolved God.

However, this is all good news for reincarnation since humans can now return with three choices: as caveman, hive-man, or clansman. The collective unconscious—an averaging of these three types of cognition—is now capable of hostility, weak collaboration, and basic survival algorithms. It is not a winning team, or an admirable team, but it is far more complex than anything evolution has so far hatched.

The age of the Warrior Man lasted for thousands of years. He is outnumbered in Heaven but he makes his presence known anyway. God constantly has to straighten out all the ideological messes these guys cause. But God is fond of them anyway.

The Warrior level of consciousness is appropriately labeled red. Wilber points out that this level is also called egocentric, pre-conventional, self-serving, self-promoting, and narcissistic. He also states that this era began 10,000 years ago. Using my own date range, I believe that Warrior Man was around for tens of thousands of years and lived side by side with Tribal and Archaic creatures.

Toward the end of the Warrior Age something miraculous happened all across the globe. A new kind of mind appeared during what is called the Axial Age—about 2500 years ago. This new mind existed alongside the warriors, the hive-creatures, and the cave folk. The Greeks appeared in the West, and Buddha appeared in the East. The Iliad and Odyssey were written at this time. The age of Traditional Man arrived.

Traditional Man: Allocentric evolving toward Egocentric Life Happens in you

Gamble everything for love,
if you are a true human being.
If not, leave this gathering.
~ "Half-Heartedness," *A Year with Rumi*, 2006.

Traditional Man (Gebser's mental stage) is a great leap forward in the evolution of consciousness, especially egocentric awareness, as complex language has arrived. Traditional Man has learned to employ powerful reasoning skills and to share what he has learned. A sophisticated level of consciousness came into existence during the Golden Age of Greece (500 to 300 BCE), and produced some of the finest art and deepest thought ever created by human beings in the Western half of the planet. The voice in Traditional Man's head evolved to be less magical and more rational— and Traditional Man became slowly accustomed to his internal dialogue. Here is author Gary Lachman's description of the transition from Gebser's mythical consciousness to mental consciousness:

> Gebser's next consciousness stage is "the mental." Like Julian Jaynes [author of *The Origin of Consciousness in the Breakdown of the Bicameral Mind*], Gebser believes the emergence—we might say *wrenching*—of the mental structure out of the mythical was violent . . . one myth stands out and seems to support Gebser's and Jayne's contention that the shift from the mythical mode—in Jayne's case, from "bicamerality"—was world-shaking. Gebser points to the myth of the birth of Athena as crucial: the god Hephaestus split Zeus' head with an axe and Athena, the goddess of wisdom and thought, was born fully grown. (The "head splitting" seems key; Jayne's book is full of accounts of psychic turmoil suffered by early post-bicameral men).
>
> ~ Classics from the Journal for Anthroposophy, Meditation and Spiritual Perception, "Rudolf Steiner, Jean Gebser and the Evolution of Consciousness," Gary Lachman, 2011.

As Lachman's vividly explains in the above quote, the head of Zeus—as a metaphor for all minds—was split into two by an evolutionary axe. Human beings suddenly had two minds, one to solve problems and the other to experience life.

Traditional Man developed abilities distinct enough to be recognized as the beginning of two kinds of consciousness. However, duality is only subconsciously sensed by Traditional Man. There is no articulation of the duality in such a way that would suggest to Traditional Man that he could imagine a world beyond his narrow spatial-temporal domain. Traditional Man is still blind to his ability to shape space and shape time. But finally, Traditional Man has evolved a primitive form of empathy. A genuine concern for others and a budding compassion—although primitive—have finally arrived on the evolutionary stage. In Traditional Man, we have the first glimmer of hope that this human creature can perhaps someday excel as a species worthy of a sophisticated God. Traditional Man is no longer totally blind to the essence and concerns of others.

Traditional Man would come to see the arrival of the cat—to continue that analogy—as evidence of Divine Creativity in the universe, proof of God's artistry. Traditional Man would then bring the cat into the house and give it milk, a bed, and cat toys.

Traditional Man, part of an extended agricultural family, has finally evolved a social conscience and a thirst for spiritual meaning. There is a caring for other life forms and a deeper mandate is felt, something nobler than just building bigger termite mounds and reallocating neighboring territories. The concepts of good and evil appear and there arises a determination to navigate towards a vague sense of *goodness*. If we are capable of goodness, then perhaps something greater than we are is also capable of Good. Perhaps *meaning* is cosmic and God-ordained.

Traditional Man put on his best tie and went routinely to the church, mosque, synagogue, or temple. He has an adult consciousness, the beginning of formal-operational cognition, a fully evolved stage 4 on the Cook-Greuter scale. Seventy-five percent of all humanity still resides at this level of consciousness. This is developmental psychologist Jean Piaget's stage of reason—a cognitive movement beyond concrete to ever greater abstractions. Importantly, Traditional Man has evolved a set of abstractions called "beliefs," which he uses to manage his life. The voice in the head has taken charge of the human body. Unfortunately, Traditional Man is still totally blind to "faith," which is a product of a sophisticated allocentric mind that had yet to appear on the evolutionary stage.

Traditional Man developed *low beta-wave consciousness* to a significant degree, and he was on the road to becoming articulate. Traditional Man came out of the age of agriculture. He could now alter (control) the environment in a primitive way to produce crops and to use animals domestically. He has gained control over and is immersed within a farming culture. He has discovered the concepts of *purpose* and *meaning* within himself, and thus he has begun to ponder what this life is about. For the first time in evolutionary history, human consciousness has become a distinct form, a fully shaped wave on the surface of the ever-present background origin.

Traditional Man has a 4th person viewpoint. He has all the other previous viewpoints embedded within his heredity. He can pull his perspective upward to view a vast territory with many types of groups and thousands of people networked in extremely complicated inner-relationships. He can conceive of government, leadership, and a crude form of local-citizenship. But Traditional Man is blind to higher levels of perspective. He can only rise so high in his effort to comprehend sophisticated networks of relationships.

If God made man in his own image, then God became a family man who kept animals in a pen in the backyard (some roamed around the house). He has a small garden and a few crude tools. He shares and trades, saves and organizes. God has begun to think and is confused by what he has created.

Traditional Man has been around for thousands of years. He is vastly outnumbered in The Great Beyond because it is mostly populated by creatures that procreated for a much longer period of evolutionary history. There are far more cavemen, hive-guys, and clansmen in the Netherland. Perhaps, we can conjecture that heaven was invented at this point in the evolution of consciousness as a place where men and women with a crude form of empathy can hang out for eternity without being hassled by non-caring, sleep-walking proto-humans. Or perhaps, having empathy is the portal to heaven.

Reincarnation also has gained greater complexity. Humans can regress (come back), if they want, to the good-old-days of aggression, or to caveman campfires under unpolluted skies, or to termite consciousness. Or, even better, human beings can come back as cutting edge Traditional humans. The earth has become interesting. The collective unconscious has added a fourth dimension. Walking the earth at the same time are now four kinds of human beings, each with different levels of consciousness.

Integral psychologist Ken Wilber color codes this level of consciousness as amber. This level is also referred to as ethnocentric. Traditional Man, despite the great mental leap forward, can be racist, sexist, xenophobic, homophobic, and hyper-patriotic. As I write this in 2018, it is still true that most of planet earth is now inhabited by this level of consciousness.

Wilber states that this era began 5,000 years ago. I believe this date range is too restrictive. I believe that Traditional Man has been around for *more than* 5,000 years, still walks the earth, and still votes for ethnocentric values.

Modern Man: Egocentric evolving toward Allocentric Life Happens for you

You set out to find God,
but then you keep
stopping for long periods
at mean-spirited roadhouses.
~ "Half-Heartedness," *A Year with Rumi*, 2006.

Modern Man is part of a nuclear family. He sees the existence of the kitten as evidence for the power of evolution, the end result of eons of materialistic creation. Consequently, with emerging empathy and a dull kind of awareness, Modern Man dissects the cat and labels all the body parts in the interest of science.

Life has now evolved sophisticated dual navigational goals. The species is hungry to examine and experience everything from the universal to the quantum. A deeper thirst arrives with each new discovery that brings ever greater mystery. Modern Man is ravenous to understand his own mind, to probe the depths of human evolution, to ponder where all this is going. However, Modern Man is busy dissecting abstractions and then recombining, pasting, and cutting, to evolve ever larger, ever more bewilderingly complex abstractions; he is self-absorbed in his projects and plans. Modern Man has discovered the tools in his head that enable him to imagine eternity and infinity. His point of view has exploded; a single point has experienced a big bang that has thrown open *everywhere and anytime*.

Meanwhile, the distance between our two minds, still essentially ignorant that they have a twin, is severe. A mental health crisis is at hand as Modern Men are drowning in their own egocentric purposeful soup. The internal mind has gone berserk as it embeds abstractions inside abstractions—as the ego evolves. Meaning itself has become abstract. The concrete origin of language has been forgotten; the linguistic ground that anchored language has been lost. Modern Man is simply mandated *to do*, to become, and this feels barren, empty, and seems to have no cosmic significance or any connection to Divinity. Indeed, Divinity is stored in mothballs in the basement of the mind.

Modern Man is still blind to faith, but not totally. He is a faith-impaired creature with glimmers that there is something deeper and more loving than his belief-riddled mind can reveal. Using only his egocentric mind, he is locked in the prison of his own belief structures. However, Modern Men are the first allocentric pioneers, the first wave of human beings to go on *spiritual quests*. Therefore, Modern Man is an evolving spiritual seeker looking for something vague but (maybe) wonderful. Modern Man senses a spirit arising within a physical body. It will take another leap of consciousness for Modern Man to transcend into the higher possibility that human beings are actually spiritual bodies having a physical experience.

Modern Man has discovered brain waves and has developed a deep thirst to understand how the mind works. He wonders about the relationship of frequencies to cognition and compassion. Meanwhile, he is busy dissecting the brain and body, searching for the secrets to this strange world of frequencies. He has seized hold of nature (by the neck) and is ripping her apart with a feverish intensity. He has unleashed his power over self, other, and the environment to an accelerating, unthoughtful, at times insane velocity. Modern Man has turned cultures into machines called societies. "Society" is a system for control over the self, others, and the environment to an extent beyond anything that has evolved previously. Society survives the death of individual men and woman—society continues to function in a machine-like way in total disregard to the purposes of the creatures caught in the cogs.

Ironically, George Orwell's Big Brother arrived long ago without much recognition or resistance. This awareness came to me as I sat in an airport trying to read about consciousness. I discovered that I could find no silence in this public space. In the airport, you cannot escape the opinions of Modern Men—they shout down their belief structure, their commandments, from overhead television screens. This level of Big Brother consciousness is firmly entrenched.

Indeed, Big Brother broadcasts from every wall. You cannot shut him off. He advertises in your face using sophisticated technologies that grab and hold your attention. He tells you what to wear, what to drive, what to eat, where to shop, what to buy, what to smell like, who your friends should be, what is good, and what is evil. He repeats and repeats. Big lies become embedded in the culture through the repetition of memes. The spaces created by Modern Man watch you and listen to you. You are recorded and remembered. Databases exist that track your opinions, habits, weaknesses, and, of course, your location.

However, George Orwell[6] was only half right when he crafted his terrifying totalitarian image of the future. Indeed, Big Brother actually *did* arrive; he is watching and remembering at this very moment—keystroke-by-keystroke. But Orwell missed something curious. Modern Man accepts, embraces, and even insists upon, the world constructed and controlled by Big Brother. To Modern Men, it is no big deal to live in this technological barrenness. After all, it is just a "gentle," mixed-bag, corporately-driven, robotic—no one is consistently in charge—set of authoritarian *habits*. What's the big deal? Pass the pizza and turn the volume up. This is the Super Bowl Age of evolution.

Big Brother speaks to you; he tells you how to live. When you get on an airplane, a Big Brother screen keeps asking for money so that Big Brother can tell you what to buy so you can live like a Modern Man—you must buy things; it is critical that you constantly make purchases. The society of Modern Man, especially in the Western world, requires life to be about making purchases and obeying the power structure and "laws" of the society.

Digital screens are in your face as you exercise at the YMCA or in grandma's living room. Indeed, we have found a way using our phones to voluntarily put the screens in front of our face all day, all the time. Big Brother doesn't have to covertly watch and record. We offer ourselves up for scrutiny.

However, Big Brother is also suffocating the soul—keeping the allocentric mind hidden. Modern Man is a deaf-blind corporate entity, ignorant of any level of consciousness beyond the Modern. Big Brother is often loud, rude, obnoxious, unsophisticated, and self-centered. Understandably, he has become an egocentric madman who is unaware of his egocentricity. Of course, this is to be expected since it is the logical conclusion for a mind that lacks discipline, intention, and a need to cultivate kindness.

Modern Man is Industrial Age Man, and he has moved from an ethnocentric perspective to a worldcentric perspective. He has found a way to replace his muscles with machinery. He lives in *beta-wave consciousness*, even at times moving into high beta-wave awareness. Yet he is blind to higher levels of being. He has not experienced gamma-wave consciousness. He vaguely feels that something is wrong, but he cannot articulate what it is—using language and reasoning is the wrong approach to finding the soul. He is still blind to his twin sister.

Modern Man sees himself as an organic machine that can only do and do and do, and for no greater purpose other than to enable ever more doing. Modern Man has an aging "mentally ill" ego-bound consciousness. However, he is beginning to feel the vagueness and isolation of cognition. He senses the uneasy, still vague arising of the "dark night of the soul." Something is very wrong. "You can feel it in your bones," he says. Modern Men sometimes stare into space with grief and dread, without knowing why. Modern Man has arrived at Cook-Grueter's stage 4/5.

Modern Man can look down from a great overhead perspective. He vaguely senses all the levels of consciousness that lie within his understanding—but they do not penetrate his waking consciousness. He has a 5th person perspective, a territorial viewpoint that includes states, provinces, countries, and empires. But he can rise no higher; he is stuck with patriarchal pride; he resists opening his heart and mind to a greater perspective. He has concern for others, especially his friends and associates, so the concept of "concern," and "caring" are firmly established. He can also define "empathy" and acknowledge its existence, but he cannot find it in his heart. Empathy for Modern Man is a concept, not a way of knowing and interacting.

If God made man in his own image, then God is a scientist who dissects small animals in a dimly lit laboratory at a small Midwestern college in central Ohio. He reads books on the toilet and calls his wife sweetheart. Meanwhile, the wife, Mrs. Sweetheart, wishes that a drummer from a rock band would steal her away from her spiritless mate. If she could, this wife would divorce God and escape from the technological convent that has been erected by Modern Priests.

Reincarnation is getting fun—finally. We can reincarnate at the cutting edge of a fast-paced evolutionary bio-machine. By the time we return to life after death, we will encounter something totally unrecognizable from the world we left. The collective unconscious has gotten complex and anxiety producing. It is a confusing and contrary mishmash of beliefs and fears.

I believe that what Buddha meant by reincarnation, and what Jesus meant by being born again, are not complex concepts. Every moment of our existence is a reincarnation, every second we are being born again. At any point in our personal life, we can become aware of our constant rebirth. We can jump in anytime we want with loving-kindness, compassion, wisdom, joy, and peace. We can find God-Consciousness at any instant. It is not about returning to earth after the final death of the body. It is about being reawakened at every moment so that we can feel the awe and mystery of life. It is about using the allocentric mind to experience life. Whatever this existence thing is ultimately about, it is well beyond what our minds can digest. We are just a few eons removed from our cave brothers.

Modern Men have only been around for decades. Something is speeding up, whipping consciousness into a growth-frenzy. There are very few Modern Men in Heaven compared to all the others with lower levels of consciousness. On the earth, Modern Men dominate in the industrialized developed nations. However, Third World humanity still exists at more primitive levels of consciousness, and many people in developed nations are still stuck at ethnocentric (Traditional) levels. Fewer still are the men and women of higher levels of consciousness beyond Modern Man.

<hr />

Here I reach a threshold. I am clearly, as I self-reflect, a Modern Man—I am 4/5 on the Cook-Greuter scale; there is no getting around the evidence. My mind was molded by scientific materialism—I cannot bring myself to disparage my beloved mythological belief that science is true and golden, and will deliver us from evil. I love my screens. I love technology. I am grateful to live in this fascinating time. Big Brother is a pain in the ass, but I am too small, too habit prone, too boring for Big Brother to care about. He ignores me; I ignore him.

However, I am also a minor league rebel, a mild social irritant. I am suspicious of all religious trappings—primarily because the history of religion is atrocious. I am disgusted with what men (mostly men) with more primitive levels of consciousness have done to their fellow man: burned them at the stake, buried them alive, beheaded them, drowned them as witches, stoned them to death, ripped them apart at inquisitions, machine-gunned them into mass graves, let them starve to death in concentration camps, dropped bombs on their grandchildren and aging parents—on and on marches *man's inhumanity to man.* I find this habitual pattern appalling. I blame the males, and yet ironically I am one of the guys. It's like yelling in the mirror: "You Moron!" And that doesn't seem quite right; something is missing from this emotion. I had a loving father and my sons are loving, empathetic souls. I also do my best.

However, I am a systems-oriented, anal-retentive, hyper-verbal male who can talk a good line about empathy but can't quite bring it on a daily basis. I have a long way to go on my spiritual journey, and I am not proud to be stalled at this level of consciousness.

When my friend Wayne O'Brien[7] read my notes for this book, he made a suggestion. He said I should use my imagination to pretend that I had transcended to a pure level 5 consciousness. Then he suggested I imagine further that I had evolved beyond level 5 to level 5/6. Finally, Wayne suggested that I imagine what it would be like to have evolved to a level 6 consciousness. Do a thought experiment, he suggested. Thanks to Wayne, that is where we go from here.

I can plainly see that I have a center of *conscious gravity.* Daily, I settle into my "high-beta knowing" as if that is reality. However, during my better moments, I drift upward and taste a greater wisdom—but that quickly fades as if it were a mere phantom.

On a rational level, I know that levels of consciousness are about whole-body frequencies. When I move from level 4 to level 4/5, there is a gap, a phase transition—a space wherein I switch from a slower wave frequency to the next higher set of frequencies. I suspect that allocentric meditation—sensing from greater and greater heights, higher bird's-eye-views—is the key to triggering phase transitions. Using the allocentric mind opens a poetic window through which spiritual messages come and go. I yield to our friend Rumi:

> *I was wrong. God has revealed to me*
> *that there are no rules for worship.*
> *Loosen your tongue.*
> *And don't worry what comes out.*
> *It is all the light of the spirit.*
> ~ "The Truest Devotion," *A Year with Rumi,* 2006.

I have been quoting Rumi, but he has a level of consciousness beyond mine, and he lived seven hundred years before my earthly role began. However, the background is the same: Mother Earth. Rumi *knew*, while I aspire to know. He was, I suspect, a person of level six consciousness, one of the first and one of the most eloquent. It must have been very strange to be enlightened and yet walk about the earth with men of such lesser awareness and appreciation.

Take a deep breath. Drop into the alpha brainwave state. If you, like me, are stuck at level 4/5—as many of us are—and not yet a solid 5 or 5/6 on the Cook-Greuter scale, then follow me as I imagine what these higher states might hold for us. This is a time for self-reflection and for asking how we would behave as a person of level 5/6 consciousness—as if we had a Post-Modern Mind. I am going to sit in meditation off-and-on for a few days and see what arises.

Besides sitting in meditation, Modern Man has several avenues open for altering personal states of consciousness. There are, I suggest, ways to get unstuck from 4/5 and be transported into the realm of 5/6, if only fleetingly, at first. We have, for example, access through the internet to frequency waves and animations. We can choose to entrain our mind to an alpha state of relaxation, or we can trigger higher frequency gamma waves with sound and images.

We could also travel to South America or Africa—this kind of travel is fast becoming a modern pilgrimage—to experience the spirit molecule (DMT) with the guidance of a shaman. Indeed, many children of the Post-Modern era are self-medicating, taking consciousness "into their own hands." For the first time in history, we can take control of our own transformation using hallucinogenic drugs. My son Noah went to Peru to experience Ayahuasca, a powerful plant-based mixture that contains DMT. After a terrifying beginning, Noah experienced a profound peace by the end of his spiritual journey. Noah came home with a message from the Earth Mother:

> She tells me wordlessly that . . . Love is not just a feeling but a real power that keeps us safe from the darkness. I am told to bring back home a single message: Keep honest love at the center.
> ~ From the blog HONEST LOVE, June 22, 2016, by dogsofperu.

Noah tells me that after his return to the United States, meditation became easier for him. He also felt less aggravated overall, and could interact with people from "a place of acceptance and love instead of competition and derision." He found a role as "a bridge between people." Noah also faced down death during his Ayahuasca experience, and after an intense struggle, he let his ego die. It was at the moment of egoic death that peace came and swept away the visions of terror.

In addition to using mind-altering plants (life altering experiences), seekers have used meditation, yoga (movement routines and postures), spiritual quests, and stubborn intention and discipline as tools for going beyond our current level of consciousness. I use meditation as my primary tool, although I do yoga, and I have a life mission. I also have a very small community of friends (and family) who understand what I am talking about and who are also on a spiritual journey. All of us realize that talk doesn't really help people spiritually evolve—only transformational experiences can alter one's level of consciousness.

Recently, I have been using two unusual meditation techniques to help with this transformation from 4/5 to 5/6. Look at yourself in the mirror, eyeball to eyeball. Practice loving-kindness and empathy toward the person in the mirror; accept what you look like and be kind to that reflection. The person reflected in your mirror is the main person on earth you are trying to understand and to help.

When we walk around our environment, when we communicate with others, we are not aware of our own face. We can't read our own facial expressions and our body language. In essence, we have no perception of our head. It is as if we have no head. This is a very strange and fascinating realization, but it also reminds us that we cannot perceive our own animation. The feedback system for understanding ourselves has a serious weak spot. Look at your reflection and consider who you really are.

The second technique I use involves seeing as if *transmitting* from the eyes—giving light rather than taking-in light. In other words, send photons *from the eyes* rather than passively receiving photons, as we habitually do. I doubt that we can actually send protons outwards from our retinas, but the meditation is very powerful. The feeling is an experience of allocentricity— you can't be an ego when you send photons in this manner. This is a way to taste level 5/6 consciousness. It is as if the act of *looking outward* triggers the opposite brain system than the egocentric inward-focused perception that is the norm.

❦

Okay, through the miracle of editing, the five days of meditation are up. Below are some thoughts that came to me during insight meditation. I suppose this is what life might be like as a Post-Modern Human:

- The foundation of level 5/6 consciousness is an unshakable spiritual practice—we work daily on a positive, loving attitude. We set a morning intention and live by it for the entire day. We have a determined practice to become empathetic, to meet every sentient creature with loving-kindness. Above all, we are conscious of food intake and eating habits, so we eat less often, we consume smaller portions, we eat slower, and we eat with consciousness—we taste our food. We eat only food that is fresh, organic, and balanced. Our practice also includes meditation and movement, including (perhaps) tai chi, yoga, qigong, riding bicycles, hiking, walking, and so on. We create positive intentions. We are mindful; we practice self-remembering, especially as we breathe and walk. We do not "fall asleep" and forget that a donut or another glass of wine is not good for our body—we remember to eat with awareness. *We have different habits* than we did at level 4/5, habits that are better for our mind and body.

    A spiritual journey is a personal exploration of how our mind works—how we *want* our mind to work. It is a process for cultivating a highly-structured empathetic mind. At level 5/6 we are on a full-force spiritual journey that we will follow for the rest of our life.

    Rumi has been helping us as I write this book and his helpful messages seem to arrive just in time. Here is Rumi talking about leaving behind the embedded level 4/5 habits—our addictions:

> *Tear down this house.*
> *A hundred thousand new houses can be built*
> *from the transparent yellow carnelian*
> *buried beneath it.*
> *The only way to get to that*
> *is to do the work of demolition;*

dig beneath the foundation.
If you wait and just let it happen,
You will bite your hand and say,
I did not do as I knew I should.
~ "The Pickaxe," *A Year with Rumi*, 2006.

- You and I, at level 4/5, have discovered and can call-up a mental witness. This is the ability to watch the self as we purposefully move our lips to talk, our arm and hand when we reach for a cup, etc. At level 4/5, we have a very rudimentary ability to call-forth this witnessing ability; yet we hardly ever witness—we remain on autopilot, half asleep. Awakening and strengthening the witness is the key to crafting and following new, healthier habits. At level 5/6 we practice witnessing every day and for longer and longer periods.

- As beings of level 5/6 consciousness we feel like we live in a mystery movie, a play—we move away from the "serious business" of living. We transcend routines, pettiness, and constant sleepwalking (living without the witness). There is a humorous, playful flow to being alive and we are surfing this wave. The stronger we become as a level 5/6 being, the more solidly we live in this magic world. Serendipity and synchronicity are common happenings for us now because we are open to receive the messages that manifest from the background; our mind stops filtering and inhibiting as much as it did in the past. Part of our transition to 5/6 is *our intention* to exist within this magic, serendipitous world. Others who observe us in this playful state realize that we are seekers; there is a glow that is attractive and fun—people want to share in this space. Being seen as a seeker is a catalyst for continuing the seeker's life—the inward and outward journey. As 5/6 beings, we enter the "seeker's existence," which is a monk-like, slow-paced, measured, aware, joyful, and purposeful life.

- At level 5, we take on greater purpose, greater intentions; we develop a mission for our life. We see each day as a gift. Level 5 is a self-actualized stage, a kind of pinnacle for a fading ego. Idealism is heavily tempered by awareness of the interconnectedness of all things, we perceive complexity; we see the dance and balance between chaos and order. However, as we head into 5/6, the old-style reality begins to unravel. The dark night of the soul arises as we witness our ego slowly evaporate. We realize that our grand mission is not our own—we are vessels, conduits through which the universe creates. We learn to allow and accept. As the ego disappears, so do the objects that the mind constructed; the seer and the seen start to blend into a unity. As the ego fades, so too does the sense of time, which now seems to speed up as we leave 4/5, transcend through level 5, and enter 5/6. The body seems less and less like an identity. Rumi tells us:

> *This is a rented house.*
> *You do not own the deed.*
> *You have a lease, and you have set up a little shop*
> *where you barely make a living*
> *sewing patches on torn clothing.*
> *Yet only a few feet underneath*
> *are two veins, pure red and bright gold carnelian.*
> *Quick. Take a pickaxe and pry the foundation.*
> *You have to quit this seamstress work.*
> ~ "THE PICKAXE," *A YEAR WITH RUMI*, 2006.

- You now surround yourself with the right "spiritual community," with others who also, in their own way, are reaching for a stable level 5/6. Everyday society—the level of consciousness of most people in any given "modern" culture—doesn't work for you at this level; it is not satisfying anymore. The people and the environment you surround yourself with shape your frequencies. Places and people nurture you, support you, guide you, and share with you. Each level of consciousness has a "friend-group" that fits well with the awareness level you seek. Finding the right people and the right spaces at levels 5 and 5/6 is more difficult than at all previous levels.

- Most early levels of consciousness are habit-heavy, mindlessly tied to routines. It will take willpower to overpower the habits. Habits are addictions, especially food habits; it takes extraordinary willpower, and perhaps external help, to go beyond these addictions. Therefore, as a level 5/6 person, you give up old habits of thought. The old stories that used to define who you are—for example, your career path, your lists of accomplishments, your material possessions, where you have been, and who you have met—will all be let go to make room for new experiences. The inner journey from 4/5 to 5/6 is a transformation of motivation. At 4/5 you were concerned with security and status. Now, at level 5/6 you let it go as you just flow from moment-to-moment. This is a move *beyond* self-effacement and self-deprecation, and *beyond* judging "self and other" as inadequate or unworthy.

- Old definitions and paradigms are let go. The world is no longer divided between conservatives and liberals, democrats and republicans, right wing and left wing. There is a new paradigm that can allow for levels of consciousness. *We are all on this spiral of consciousness*, all of us are on the journey together, and all of us need each other's assistance and compassion. Judgement has to be finally let go.

At level 4/5 (and to a lesser extent, even level 5) judgement is still with us, and we continue to categorize and ruminate about our differences. We are trying to comprehend this mishmash of levels of consciousness that exist simultaneously on our planet.

I write this in March, 2018. The Trump administration has taken the reins of government in the United States. Conservatives are elated. Liberals are horrified. The country is split down the middle. Anxiety and unrest predominate. In response to this global crisis, developmental theorist Ken Wilber wrote a small book to explore the conflict from a developmental perspective. I will use an excerpt from his book *Trump and a Post-Truth World*, (2017) to complete this discussion of Modern Man. I will then continue to use Wilber's insights in the discussion below about Post-Modernism. Wilber uses the color orange to represent Modernism, and green to represent Post-Modernism (which is discussed below):

Beginning in the 1960s, green began to emerge as a major cultural force and soon bypassed orange (which was the previous leading-edge stage, known in various models as modern, rational, reason, formal operational, achievement, accomplishment, merit, profit, progress, conscientious) as the dominant leading-edge. It started with a series of by-and-large healthy and very appropriate (and evolutionarily positive) forms—the massive civil rights movement, the worldwide environmental movement, the rise of personal and professional feminism, anti-hate crime, a heightened sensitivity to any

and all forms of social oppression of virtually any minority, and—centrally—the understanding of the crucial role of "context" in any knowledge claims and the desire to be as "inclusive" as possible. The entire revolution of the sixties was driven primarily by this stage of development (in 1959, 3 percent of the population was at green; in 1979, close to 20 percent of the population was)—and these events truly changed the world irrevocably. The Beatles (otherwise sacrosanct in my view) summarized the whole move (and movement) with one of their songs: "All you need is love" (total inclusion rules!).
~ TRUMP AND A POST-TRUTH WORLD, Ken Wilber, an online book, 2017.

Wilber is showing us that evolution is still marching forward, pushing the collective level of spiritual development to greater heights—whole cultures are struggling to evolve to higher levels of consciousness. Wrenching entire cultures from amber to orange, or from orange to green, is complex and disruptive. We are in the middle of a struggle—of a culture war—and it won't be an easy transition. On the other hand, consciousness will continue to evolve.

There is a level of consciousness—indeed many levels—beyond Modern, including Post-Modern which I (and others) call Post-Human. This transformation from Modern Man to Post-Modern Man is resulting in a crisis that dwarfs all that has come before. The debate we are having now about our cognitive differences pales in comparison with what looms on the horizon.

Wilber states that the orange modern level of consciousness began 500 years ago. I believe this date range is too restrictive. I believe that Modern Man has been around for more than 500 years.

Post-Modern Man: Allocentric evolving toward Egocentric Life Happens through you

> *Let them take you where they will.*
> *Listen to presences inside poems.*
> ~ "PRESENCES," *A YEAR WITH RUMI*, 2006.

Post-Modern Man, noticing that the kitten is an orphan, and recognizing that "orphaned cats" is a *category of cat,* would build animal shelters and cat rescue organizations. Post-Modern Man thinks in ever greater abstractions and seeks to heal the whole so that the individual might flourish. This level of awareness still has an overheated egocentric consciousness, but can intuitively feel the powerful pull of something hidden.

Meaning for this level of consciousness is becoming increasingly sophisticated. Post-Modern Man senses something greater than his small mind. He can feel this life-force as empathy grows. Healing from the barren emptiness and isolated angst of Modern Man's egocentric consciousness, Post-Modern Man is evolving genuine hope and a rising stubborn intention to save the planet. Without fully comprehending yet, this level of humanity has discovered the dual purpose that was there at the beginning of the evolutionary trip: to solve problems—the role of the *ego*—and to experience life—the role of the *self*.

Post-Modern Man is ready to accept that human beings have two minds. Post-Modern Man is no longer blind to this duality although the anatomy and physiology of the duality is yet to be fully grasped. Post-Modern Man is also ready to entertain the notion that belief is not faith, and that of the two, faith is by far the more powerful and beneficial. Post-Modern Man is no longer faith-impaired. Indeed, for the first time there is awareness that the reason Traditional Men and women allowed their sons to experience the horrors of war is because most everyone prior to this Post-Modern age were believers—and believers, by definition, are people who have yet to experience faith. That is why modern religions are going bankrupt and failing to save any souls; they are based on egocentric belief; they are still blind to allocentric faith. There is plenty of lip service for faith and compassion, and the words *faith* and *belief* are used incorrectly as synonymous, but this perspective is not supported with empathy and action. A famous quote by G. K. Chesterton is that Christianity has not failed; Christianity has just never been tried. Here is what Chesterton actually said:

> "The Christian ideal has not been tried and found wanting. It has been found difficult; and left untried."
> ~ "THE UNFINISHED TEMPLE," *WHAT'S WRONG WITH THE WORLD*, G. K. Chesterton, 1910.

Consequently, knowing about duality, about the split between faith and belief, Post-Modern Men are compelled to stand up at town hall meetings to say with conviction that people below Post-Modern consciousness, those who believe, are the reason we are stuck with so much ignorance and violence. The evidence is in. People who believe *can and will* kill you if you don't agree with them.

However, let's remember that belief—as an agent of the egocentric mind—is not a bad thing, despite what human beings have historically done with their rants and aggressive behaviors. Belief is the egocentric mind's knowledge-cataloging system. So long as emotion doesn't horn in, belief can add to our knowledge base and make the world a better place. The scientific method, as an example of an egocentric system, is a major advance for mankind. The basic tenet of the scientific method is that belief is always tentative. Absolutes are looked on with a wary eye. There is no scientific advancement—using the scientific method—unless we hold our current knowledge in abeyance. Post-Modern Man lives with this constant uncertainty, with nothing solid or ultimately reliable to hold on to. Quantum theory is a more realistic religion for Post-Modern Man, when compared to the failing traditional avenues for spiritual development.

Post-Modern Man has a new consciousness never before laid down in the genetic code of humanity. This new level of awareness is the beginning of a resonance between allocentric and egocentric perception. We are now at Cook-Grueter's stage 5, on the threshold to crossing into a budding 5/6 level of consciousness.

Post-Modern Man is fascinated with his brain and mind. He has discovered levels of consciousness and he is meditating, seeking, and exploring his inner self. He is refining his frantic, habitual drive to control everything. He sees a more sophisticated, subtle, gentler way to modulate self, other, and the external world. He is aware that his new mental powers can be used for healing or they can be used for destruction. He sees that he has an ethical choice and he has chosen the spiritual path toward intentional healing and compassion. He lives in high *beta-wave consciousness* but he can sense gamma-wave awareness and he is trying to get to this higher frequency domain.

However, there is still blindness within this level of consciousness. Post-Modern human beings are not yet integral, not yet spiritually complete. They cannot yet perceive higher levels of consciousness because they lack the necessary experiences. They are on the path, doing what they need to do, but they have not yet gone through needed transformations. They have the right intentions and practices, but they still have a distance to go to get beyond Post-Modern consciousness.

Post-Modern Man has a 6th person perspective, a truly global view, as if they are on the moon looking down at the earth. World government makes sense for Post-Modern Man, and if there were enough of them, there would, indeed, be a budding world-governing body with the power to actually govern—not the United Nations in its current evolution. Unfortunately, most of humanity is stuck at the traditional and modern levels of consciousness—people at these lower levels of spiritual evolution would vote down world government. The Brexit vote in Great Britain is an example of an ethnocentric populous voting down greater governmental cooperation.

If God made man in his own image, then God uses long abstract sentences and writes books on philosophy that nobody can comprehend. He is fascinated by fleas and galaxies and when he contemplates himself he is surprised to learn that there is more to his mind and powers than he could ever have imagined. However, heaven is still overrun by cave-folks, termite-humans, bad-ass Scottish clansmen, God-fearing farmers, and church-going family men. The Post-Modern few feel a little out of place in heaven.

Reincarnation has also become a tricky business. The evolution of consciousness is looking rather exponential. Whatever you reincarnate into will probably already be out-of-favor and out-of-date by the time you get back from the business of being dead. The collective unconscious is starting to feel mentally ill and panicked.

Here is how Ken Wilber discusses the Post-Modern level of consciousness:

> The leading-edge of cultural evolution is today—and has been for four or five decades—the green wave ("green" meaning the basic stage of human development known to various developmental models as pluralistic, postmodern, relativistic, individualistic, beginning self-actualization, human-bond, multicultural, etc.—and generically referred to as "postmodern"). The primary purpose of the leading-edge of evolution is to be just that: a LEADING edge of evolutionary unfolding, what Maslow called a "growing tip"—it seeks out . . . areas that are the most appropriate, most complex, most inclusive, and most conscious forms that are possible at that particular time and point of evolution.
> - Trump and a Post-Truth World, Ken Wilber, an online book, 2017.

That being said, according to Wilber, the Post-Modern age is now dead and past for the most advanced spiritual seekers. We have moved up a notch in our spiritual ascent:

> It's widely acknowledged that postmodernism as a philosophy is now dead; and books are everywhere starting to appear that are written about "What comes next?" (with no clear winner yet, but the trend is toward more evolutionary and more systemic—more integral—views). But in academia and the universities, it is a long, slow death, and most teachers still teach some version of postmodernism and its aperspectival madness even if they have many deep doubts themselves. (But it's telling that virtually every major developmental model in existence contains, beyond the stage generally known as "pluralistic," at least a stage or two variously called "integrated," "systemic," "integral," or some such, all of which overcome the limitations of a collapsed pluralism through a higher-level wholeness and unity, thus returning to a genuine "order out of chaos." Right now, only about 5 percent of the population is at any of these integral stages, but the evidence is that this is clearly where tomorrow's evolution eventually will go—if it can survive the present transition.)
> ~ TRUMP AND A POST-TRUTH WORLD, Ken Wilber, an online book, 2017.

The key phrase in the quote above is: "If we can survive the present condition." When a complex system reaches a transition point, two things can happen: either transformation to a more sophisticated state can evolve; or the whole thing self-destructs and divides back into more primitive fragments.

Wilber labels this level of consciousness Green. He also states that this era began 100 years ago. I believe this date range is correct.

Integral Man: Egocentric evolving toward Allocentric Life is you

> *Please, universal soul, practice*
> *some song, or something, through me.*
> ~ "SOME SONG OR SOMETHING." *A YEAR WITH RUMI*, 2006.

Integral Man is aware of all previous levels of consciousness that still reside within his essence. He knows that human beings are a mixture of levels of consciousness, capable of sinking to any of the lower levels of behavior if they become overtired, over-stressed, intoxicated, or mentally ill. But the fact that Integral Man can see ancient levels of consciousness residing within is a great awakening. He is no longer blind to the evolution of his mind, or to the evolution of a cultural mind. He is aware that there is a manmade planetary mind arising from the World Wide Web, and he senses that the earth itself is a living organism (Gaia). He is also open to speculation that there might be many levels of consciousness available for human beings through neuro-developmental evolution. He will even entertain a dialogue that the universe itself might be organic and intelligent—many integral humans are convinced that the universe itself is created by a level of consciousness. Integral human beings aren't offended when the word *God* is used to define this universal intelligence, although they avoid the anthropomorphic perspective.

Integral Man looks around at the people gathered for the town meeting and sees the full spectrum of consciousness in history. Over there on the right are the Neanderthal types who still want to eat their food raw and not share. Over there on the left are the Post-Moderns fighting for better cat shelters and universal empathy. In the middle are all the others, bewildered by why we cannot get along. Integral Man also sees a town hall meeting in his own head. He sees his own Neanderthal mind snarling next to his angst-filled Post-Modern cognition.

Integral Man knows that cultures and consciousness must evolve through stages and that multiple cognitive levels reside side-by-side, cohabitating the earth. Therefore, Integral Man treads lightly around the Archaic and Tribal morons, and while he admires the stupid spirit of Warrior Man, he runs like a crazed banshee from the gangster Neanderthals and street thugs who still dominate much of the planet's turf. He admires the gentleness and spiritual strength of Traditional Man. He honors the intelligence of the scientific method of Modern Man. He applauds the tempered idealism of Post-Modern Man.

Yet most significantly, Integral Man is a seeker after his next level of consciousness. He has seen that blindness is everywhere, thus he knows that there is more to the mind, something he has yet to discover, some experi-

113

ence waiting to transcend him to the next dimension of understanding. He is struggling to rise to a higher perspective, to an understanding of what a 7th person perspective might be about.

Unfortunately, and as noble as all this appears, Integral Man is busy sticking "Intelligent" nanochips into the body and mind of innocent cats, jacking the unsuspecting felines into the electronic mothernet. To be fair, Integral Man is also filling his own body with processing chips. He is jacking his own evolving consciousness into the mothernet.

Integral Man is also no longer waiting for evolution to do her slow, plodding, haphazard, and questionable work. Integral Man is crafting "consciousness modules," engineering new kinds of proto-navigators, taking matters into his own hands. Evolution has now given birth to a human organism that is re-engineering nature—co-evolving. Integral Man is exploring and experimenting with different genetic codes to create different kinds of consciousness. Integral Man is *integral* because he seeks to blend cognition (ego) with soul (self) now that he is aware of the duality. On the Cook-Greuter scale Integral Man is a secure 5/6.

Integral Man, however, is also in deep trouble and may be the cause of the final demise of humanity as we know it. If the egocentric part of this powerful new evolutionary creature spins out of control without proper balance from the allocentric twin, then unforeseen consequences can lead to disaster. One disaster on the horizon looms as Integral Man uses science to manufacture levels of consciousness, to soup-up the twin minds, to make new genetic variants of what used to be called humanity. Indeed, these new creatures, created by Integral Man and called Post-Integral by Ken Wilber and others, might better be called Post-Human.

There is blindness in this group that is as severe as in any other lower level of consciousness. Integral Man cannot see the results of a runaway science that goes on all the time in the background, seemingly unstoppable. There is no post-global oversight. One corporate lab is cloning animals—looking to clone humans so the CEO can live forever. Another lab in an obscure corner of a university is editing the genome and experimenting with hybrid creatures. The climate is hot and growing violent, and yet the craziness that is causing climate change goes on and goes forward because there are pockets of profit to be had and no global conscience, no global oversight, and no global government.

If God made man in his own image, then God has learned how to "fine tune" his creations. He is beginning to wonder if maybe all creatures might be better off with three or four heads rather than just one. Heaven is in revolt as strange human hybrids are showing up claiming that God also loves *them*. Reincarnated primitives now have the option of returning as half flesh/half metallic material/immaterialists. The collective unconscious is slipping into insanity. It is a mass of confusing contradictions. It hisses and snarls and runs in frantic circles, like a mentally deranged tiger in a small cage.

⌒────────⌒

Take another deep breath. You are at level 5/6 but you want to go higher; you want to be a fully autonomous, completely integral human. Let's use our imagination and feel what it might be like to be a person at level 6 on the consciousness scale. Unfortunately, Unitive Consciousness might only be achievable with help. Spiritual practices often insist that you cannot do this level of transformation without the help of someone who is already transcendent (good luck finding anyone). What are the techniques or technologies that might move you into a higher state? What chip implants might do the trick?

I wish I knew, but I don't. I do have a sense what a post-integral level might be, so let's look at that next. However, I am hanging over the cliff here with a weak hold on crumbling rock. I am in my favorite zone: *flaky speculation*.

Wilber labels this level Teal and states that the era began 30 years ago. I believe this is correct.

Post-Integral Man, Post Human-Man: Allocentric evolving toward Egocentric Life is a Cyborg

Watch your thoughts; they become words.
Watch your words; they become actions.
Watch your habits; they become character.
Watch your character; it becomes your destiny.
Watch characters gathered together; that is your culture.
Watch the actions of your culture toward others;

that determines the destiny of your culture.
Watch the actions of the cultures in this world;
that determines the destiny of your planet.
It all begins with the thoughts in individual minds.
~ Unknown author

The above quote is an overhead view of the evolution of thought, but it leaves out the impact of technology. Therefore, let me add one more line to the above poem: *Watch as technology feeds you thoughts. Pay attention to where thoughts originate.* Social media, alongside standard media, is filling minds with platitudes and memes. The destiny of the planet is in peril because the quality of accumulating thoughts is deteriorating. We are currently blending with technologies that are disruptive and divisive.

We will not make wise, compassionate decisions for the good of all humanity if the leaders of the world are stuck at low levels of consciousness. We need sophisticated leadership that has climbed the mountain and can see the biggest of the pictures. Those who have climbed the mountain are called Post-Integral beings. We can only speculate about the qualities of individuals at this level of development because there are so few of them. I used my science fiction skills to arrive at the outline below.

Post-Integral Man can move at will, with the help of quantum biology, between levels of consciousness—moving considerably beyond the *gamma-wave range* where monks and gurus go to attain enlightenment. Post-Integral Man can also switch back and forth between his allocentric mind and his egocentric mind. He knows when he is on a hero's quest and when, awhile later, he switches to the quest of the saint. We are now part technology but mostly human. Soon, we will be part human but mostly technology; human-beings-as-cyborgs are the future of cognitive evolution.

Post-Integral Man can employ different varieties of *manmade quantum-minds*, minds that are a blend of technology and molecular science. These new humanoid entities are complex cyborgs. *Natural man* will gradually disappear from the evolutionary stage. There will be no more *He this* and *She that*. Male consciousness will be transcended, blended with the feminine. Androgyny unfolds and then dominates. The human being as we know it is fast disappearing, blending with *quantum machinery.* This sounds really bad, except that we are already quantum machinery. We are just going to add "photon-apps" to our essence.

116

Meanwhile, there is a slight hope for retaining some degree of the old human creature if we want to stay at that level, but only if our dual minds re-confirm their wedding vows. The marriage of the Egocentric and the Allocentric, living with each other with love and respect, making plans for a future filled with compassion and wisdom, can brighten the outlook for mankind. If mindfulness can manifest in ever more human beings, then more people will walk about awake, witnessing their own essence, moment-to-moment.

Quantum awareness is arising from a completely integrated mind. If we do this journey correctly, we will become ever more spiritual creatures with deep empathy for our common sufferers. However, this will only happen if we nurture and unleash the powers hidden within our allocentric inheritance. Our goal should be to pull humanity upward through the various lower levels of consciousness into the higher, more spiritual planes of being. Many people see this need and are working on it.

We are struggling to climb higher up the mountain of consciousness, to 7th, 8th, and 9th levels of perspective. Beyond global is a universal or cosmic viewpoint, a big-bang perspective. Beyond that is the perspective of Gebser's Ever-Present Origin, perhaps the Everest of Perspectives.

If God made man in his own image, then God has become aware of other gods in other dimensions, other supreme beings who also live forever. There are gods nested inside gods who are living inside glass bottles inside other glass bottles. There are many ever-present origins. God has discovered angst and overcrowding. People in heaven are relocating to other heavens (you can only tolerate the cave-neighbors for so long). If you reincarnate at this stage of evolution, you might come back inside a bottle of formaldehyde on the other side of No-where. The collective unconscious will become programmable. There will be many software "solutions" to post-human angst.

The color code for Integral Man is teal. According to Wilber in his book *Trump and a Post-Truth World*, the election of Donald Trump to the presidency of the United States, was a door that opened, an opportunity for a significant portion of the culture to move toward Integral Consciousness. Most Americans who voted for Trump were at Red, Amber, and Orange levels of consciousness, the lower rung where egocentricity dominates.

Ken Wilber's integral evolutionary movement does not seem to consider technology as the primary force shaping modern evolution. This could simply be my ignorance of what has been written (it probably is, knowing Wilber's energy and scope of understanding), but I don't see any alarm being sounded about the impact of technology on what it means to be human. Many technologies will exponentially impact the human body and mind (turning us into digital cyborgs), including cloning, molecular implants, DNA manipulation, artificial intelligence, quantum computers, DNA computers, brain stimulants, and virtual realities. I also don't see an understanding of allocentricity versus egocentricity in any culture, anywhere—which is understandable at this stage of the debate, but we are running out of time.

The integral evolutionary movement uses different terminology at this level of the discussion. After Post-Modern, are three more divisions that I have uncovered: Emergent Integral (teal), Mature Integral (turquoise), and Post Integral (Ultra-violet). Wilber states that this is the era we are in now—post-human. I believe this to be true.

Or Not

On the other hand, maybe none of this *level of consciousness rant* has validity. Space and time are foibles found in this dimension. Perhaps they exist nowhere else. We can only know about our reality within the confines of space and time—we are ultimately limited by our sensory apparatus. We have no means to perceive or understand any other kind of non-perceptual universe. We are stuck in the Hotel California, going from room to room, floor to floor, and we can never check out unless we die (and maybe not even then).

Worldviews and Conflicts

I love this world,
even as I hear the great wind
of leaving it
rising.
for there is a grainy taste I prefer
to every idea of heaven:
human friendship.
~ "Grainy Taste," *A Year with Rumi*, 2006.

As I write this, the United States population is split down the middle, polarized, and traumatized by the presidential election of 2016. However, the year and circumstances are not what is important. What is relevant is the duality that manifests during times of crisis. Why do we become polarized? How did behavioral duality evolve? What can we do about this harmful divide that shows up again and again in cultures across the globe? How can an understanding of levels of consciousness and dual-process theory help us see the causes and the solutions to our shared dilemma?

Let's explore each level of cognitive staging and see if we can shed some light on the problems facing us. A good beginning to this discussion is the Buddhist concept (meditation) called the Four Immeasurables. Whereas science makes generalizations and predictions based on quantitative measurements, Eastern religious principles center on experiences, values, and well-being. Science is concerned primarily with the material world, while spiritual practices center on hard-to-measure human values and emotions.

Here are the Four Immeasurables that Buddhists evoke during meditation:
1. May all sentient beings find happiness and the causes of happiness.
2. May all sentient beings be free from suffering and the causes of suffering.
3. May all sentient beings never be separated from the happiness that knows no suffering.
4. May all sentient beings live in equanimity, free from attachment and free of aversion.

To reword the above, I suggest that these four attributes refer to love, compassion, joy, and balance:
1. May all sentient beings find love.
2. May all sentient beings find compassion.
3. May all sentient beings find joy.
4. May all sentient beings balance their egocentric mind with their allocentric mind.

The Four Immeasurables are valuable for two reasons. First, they can be used to better understand what is happening at various levels of consciousness. They help us see why we are polarized. Second, the Four Immeasurables offer a common ground to soothe conflict and move toward peaceful solutions. No matter what level of consciousness we explore, love, compassion, joy, and equanimity exist in some form—these are common values that link all of us together.

As we look at each level of consciousness, we see that each stage has a different worldview—a different comprehension of reality—and a different perspective on the meaning of life. Each worldview is built around different definitions for love, compassion, joy, and equanimity. Furthermore, each level of consciousness has a different overhead perspective, so that complexity and relationships are seen differently within each stage. There is also a built-in blindness—or perceptual impairment—that defines each level of consciousness. Blindness is balanced by mindfulness, which also varies at each level. Likewise, wisdom and intelligence are understood differently at each stage, and there are different levels of silence (and mental noise) at each stage. Finally, it should be understood that individual human beings go through each of the levels as they mature. There are necessary benefits that accrue at each cognitive waystation.

The spiritual path is long and curved (a spiral) and no two individuals are at the exact same location at any given time. However, we can *average* worldviews to arrive at current levels of cultural consciousness. Also, remember that there are levels of gravity, or density, for individuals and cultures. This means that an individual's worldview can vary somewhat from day to day and moment to moment, and the sophistication of a culture can flow up and down like a roller coaster.

Most of the population of the United States falls within the red, amber, orange, green, and turquoise levels of consciousness—these are the five levels we will consider below. Very few members of our society fall at the extreme poles, in the red or turquoise levels—perhaps 5 percent in each. Most of the population falls in the amber (25%), orange (40%), and green (25%) levels. The mess we are in, as a country, comes about because of the different worldviews and the varieties of perceptual impairment (degrees of blindness) that are inherent at each level.

In the pages that follow, I have taken eight attributes and discussed how each of the levels of consciousness might perceive or interpret these attributes. The eight attributes are:
1. Equanimity and Blindness
2. Love

3. Compassion and Joy
4. Relationships and Complexity
5. Benefits Gained and Lessons Learned
6. Wisdom and Intelligence
7. Silence versus Noise
8. Intolerance to Tolerance

Equanimity, Mindfulness, and Blindness

Red, the Warrior Level of Consciousness: The lower we go on the scale of consciousness, the more the ego dominates. The allocentric mind is hidden so deeply from view at this stage that it is essentially non-existent. There is a total blindness to allocentricity. Warrior Man is so cocooned within the ego that there is little to no sense of belonging and contributing to groups. There is an extreme imbalance within this population, and no understanding of worldviews. This is the level of pure egocentricity. From the perspective of mindfulness, this group is deeply asleep and severely perceptually impaired (blind).

Amber, the Traditional Level of Consciousness: Ego still dominates this level, but now there exists a collective of egos gathered together in exclusive groups. Religious and political affiliations are powerful exclusive attachments, as are race and ethnic origin. Sports teams also are favorite attachments at this stage of evolution, and often the favorite sports team trumps politics, religion, and family. The amber stage of consciousness identifies with such groups, and is attracted to the group's language, history, uniform, and aspirations. Contrary to this, all groups and individuals outside the narrow confines of the select group are seen as threats, causing a collective adverse reaction. The amber level of consciousness can see the red level below, but it cannot perceive or understand any level above. The amber population is blind to the orange, green, and turquois stages. Amber is the level of ethnocentricity. There is a high degree of perceptual blindness at this level and very little mindfulness. The amber population is deeply asleep.

This group started to use a phrase called *Fake News* in response to viewpoints coming from other levels of consciousness. Notice that *Fake News* happens for all levels. Everything below a level of consciousness is seen as unsophisticated and false information. Everything coming from above is incomprehensible and is, therefore, dismissed as gibberish, as if the political power structure was imposing their will on masses of innocent people. The intellect of Modern Man is *Fake News* to the Amber level of consciousness because Amber populations don't use intellect to manage their lives—they use a habitual belief structure already laid down by the group to which they belong. The Amber group is meme-driven; they are not guided by wisdom, compassion and intellect.

Orange, the Modern Level of Consciousness: Ego is carefully examined at this level of consciousness, and the notion that ego is a separate entity and amenable to study is accepted. The possibility that the ego has a twin is also entertained at this level, but generally not accepted or understood. Consequently, the ego still dominates at this stage of development. Science and rationality dominate at this stage, and mega-organizations have arisen to support this level of consciousness—including political think tanks and mega-churches. Modern Man can perceive and understand the amber and red levels of consciousness—having gone through these stages previously—but there is blindness to any level above. The modern stage of consciousness upholds a worldcentric viewpoint. Mindfulness is arising as a concept at this stage, but relative sleep is still the primary mode of everyday operation.

Green, the Post-Modern Level of Consciousness: At this stage of development, individuals are busy trying to transcend or tame the ego using meditation, yoga, contemplative prayer, hallucinogens, exercise, diet, therapy, and discipline. There is a beginning understanding of dual-consciousness, *an acceptance of duality in a world where all is actually one.* Post-Modern individuals understand this paradox; they have a deep feeling of symbiosis and connection with all other life-forms. The greens are well aware of the other levels of consciousness below them on the spiritual path, but they are blind to what the levels above have to offer. This level has a firm worldcentric view, with glimpses of something wider and more sophisticated. Using their various mind-control techniques, this group is balanced between perceptual blindness (being asleep) and the awakening that evolves as mindfulness is cultivated.

Turquois, the Integrated Level of Consciousness: At this high end of the scale, where few individuals reside, there is an overview that encompasses all that is below. There is a clear comprehension of the egocentric/allocentric duality (using my terminology). This level can also dwell in the allocentric mind and can switch knowingly between the minds. The ego is under control, harnessed to do necessary work. Even at this high level, however, there is blindness to what is above. This group is not enlightened. They can discuss and glimpse enlightenment (God-Consciousness), but they do not dwell in this realm. The worldview of this group is cosmic or universal, allowing for the possibility of something unknown beyond space and beyond time. The Turquois level is characterized by wakefulness. The alternation between sleep and mindfulness continues, but now mindfulness overpowers sleepiness and perceptual blindness. This group is more awake than any other stage. This group seeks ever more mindfulness.

Love

Red, the Warrior Level of Consciousness: There is love at every stage: love for children, love for a mate, and self-love. However, this red stage is so dominated by ego that self-love can be weak, and the challenges of earning a living and receiving due respect can hamper the ability to love others. Love is often understood to be about attachment, rather than mutual support and admiration. Empathy, an allocentric skill, is confused with sympathy, which is an egocentric skill. There is paternal love and *a need to be loved* at this stage, but the ego struggles to feel or express love. Hate can easily well up, and violence is potential wherever the ego is unrestrained.

Amber, the Traditional Level of Consciousness: The family group especially, and the religious connection to a church, synagogue, temple, or mosque are paramount at this stage. Love expands spatially and includes not only self and family, but it embraces the group—political, religious, ethnic, and race—acceptance and recognition (love) are valued most when it comes from the group.

Orange, the Modern Level of Consciousness: Love expands to include the entire planet and includes all the groups that populate the earth, regardless of race, political persuasion, ethnicity, or religious preference. This advanced form of love is mostly cognitively understood at this level, especially at the lower ends, and the orange population cannot necessarily feel the empathy and compassion that it can articulately express.

Green, the Post-Modern Level of Consciousness: The green population is trying to actually feel (live with) love and compassion. They are trying to be empathetic. They are trying to stretch their loving energy beyond the body. They are doing whatever they can, through meditation, education, spiritual practices, and through discipline in mind and body, to sense and express a sophisticated kind of expansive, all-inclusive love.

Turquois, the Integrated Level of Consciousness: At this level, individuals can sense the propagation of life, and they know this is a cosmic kind of love. The unfolding of the life force that infuses all creation is sensed, and an effort is made through consistent discipline to maintain the connection to cosmic love. There is an awareness of the various eddies of love—empathy, sympathy, fraternal and paternal love, maternal love, the love of a mate, a child, and of self—all of it is sensed, honored, and integrated.

Compassion and Joy

Red, the Warrior Level of Consciousness: Compassion is an awareness of individual and universal suffering. There is an inverse correlation between ego and compassion—the more ego, the less an individual can feel compassion. Therefore, it is not surprising to find that the red level of consciousness is almost devoid of compassion. There is no cognitive understanding of the concept, nor is there emotional awareness. Joy comes out of love and compassion; this group is not in touch with joy or with bliss.

Amber, the Traditional Level of Consciousness: The amber population can articulate compassion, but the ego still restricts this group from understanding and feeling compassion. Compassion is a narrow expression at the amber level of consciousness, confined within the strictures of the organizations that the individuals identify as their family—religious, political, racial, and ethnic. There is a restricted sense of joy within this group—it is a shared joy that is restricted to small groups.

Orange, the Modern Level of Consciousness: The orange level of consciousness expands to a worldview—there is global compassion and a sense of responsibility to the planet. This compassion is more cognitive then it is spiritual. The orange population has a scientific concern for the network of relationships on the earth, but the depth of emotion and resolve necessary to push this group to action, is not as evolved as the green level of consciousness. There is more joy at this level than at the red and amber stages, but it is refined and restricted by a materialistic perspective and an overactive ego.

Green, the Post-Modern Level of Consciousness: The green level of consciousness strives through a discipline—like meditation, for example—to actually feel compassion, and to allow the joy that naturally arises when love and compassion are evolved. Ego is still present at this stage and it restricts a higher unfolding of compassion and joy.

Turquois, the Integrated Level of Consciousness: This level of consciousness has experienced all the levels before. Individuals at this blue end of the spectrum understand and feel all the levels of compassion and joy that make up the lower levels of development. They do not live with their high level of compassion and joy, because they slip backwards into lower levels from time to time as stress and circumstances impede their awareness. They have a discipline that helps them rebalance and re-allow love, compassion, and joy, but they have not reached enlightenment—they are not constantly connected to the source that pumps out love, compassion, joy, and equanimity. Despite their high level of attainment, they have more evolving to do.

Relationships and Complexity

Red, the Warrior Level of Consciousness: The higher up the spiral of spirituality, the more complexity can be perceived. For example, from the top of a hill, a person can see the whole valley below, including the relationships taking place. It becomes obvious, from a high vantage point, that everyone and everything is part of a vast interdependent network. However, from the vantage point of a single ego, at the red stage, complexity and interrelationships cannot be perceived. At the red level of consciousness, which is completely egocentric, there is no comprehension of complexity. The world is simple and relationships are few and unsophisticated.

Amber, the Traditional Level of Consciousness: At the amber stage of consciousness, the web of interrelationships within and between groups is perceived, but the greater complexes of the world are blind to this group. They cannot see the interdependence of groups within groups. The amber level of consciousness has a restrictive worldview, a higher perspective than red, but below all higher vantage points.

Orange, the Modern Level of Consciousness: Standing on a hilltop that is higher than the perspectives of red and amber, the orange population can see a whole world of interactivity. Science thrives at the orange level.

Interrelationships that are inherent within the natural and physical realms are quite obvious to the scientific mind. However, the orange population cannot see the spiritual pathway that leads to a mountain that stands above.

Green, the Post-Modern Level of Consciousness: Standing on a mountain-top, the green level of consciousness can look down on the hills below. Post-Modern individuals can see whole populations of people standing on lower hilltops proclaiming that they—at red, amber, and orange stages— have attained the highest peaks. However, the green population also does not look up to see Mount Everest. The green stage is composed of individuals who glimpse realms beyond the physical. These realms have an atmosphere of love, compassion, joy, equanimity, complexity, and interrelationship.

Turquois, the Integrated Level of Consciousness: The turquois level contains individuals who stand in rarified air, as if at a base camp on Mount Everest. They know they have not attained the highest peak, even though they can imagine it above as they stare into the clouds that stand between them and the summit. Levels of consciousness below on the foothills far away seem tiny. The interconnectedness of all things, all sentient creatures, is obvious from this height. Complexity is awesome, nothing is simple, and the propagation of the life force is spawning ever greater complexity—in an exponential march.

Benefits Gained and Lessons Learned

Red, the Warrior Level of Consciousness: There has to be a way to move from one level of consciousness to another. This can occur through deliberate rites of passage, but most often it is caused by natural shocks. A death, a divorce, the loss of a job, some kind of dramatic event, perhaps even a stage play or movie, or even a passage in a book can be the seed for change. The sleeping conscious state is jolted awake by unforeseen circumstances, and a new clarity arises. There is also slow cognitive growth that moves individuals along the spiral of mental development—this mental evolution roughly parallels physical maturation. The red level is the opportunity to experience the full blown ego. It is a showcase time, in which a person is totally self-absorbed and oblivious to the perspective of others. This baseline selfishness, when shocked by ego-threatening experiences, is forced to glimpse second-person awareness. The needs and suffering of others eventually moves the individual—in the Red level of consciousness—off the frozen tundra of greediness and narcissism.

Amber, the Traditional Level of Consciousness: Group identity is a powerful high. The reinforcement that comes from the attention, admiration, and solidarity of fellow travelers is addictive. It is this addiction that eventually, over time, moves the amber folks into the orange population (where the group of supporters and admirers is even larger). The lesson at the amber level is that other egos exist and have an impact, for good or ill. There is power in collective action, a thrill and comfort from being in a group. Amber populations are especially open to memes—these one sentence mind-viruses are effective propaganda tools that enable people in the orange level (and above) to manipulate the amber mind. Amber-level populations marched for German, Italy, and Japan in World War Two. Amber minds make good soldiers. The amber mind can be molded by *big-lie memes*, which can cost them and their victims untold misery. Knowing the joy of belonging to a group and feeling the positive and negative attributes of this level of consciousness is the lesson that this stage provides.

Orange, the Modern Level of Consciousness: The lesson at the orange level of consciousness is that the ego has a twin, an equal twin. Rationality is not the whole story, and the physical universe is not necessarily real. Doubt is sowed at this level by the very scientific method that brought so much technological progress. The quantum world is eroding the bio-chemical perspective. This group watches as the greens meditate and do yoga. They watch others go to therapy and take hallucinogens. They know the science behind change, but they have not changed themselves. Professors and thinkers are saturated at this level of consciousness by their own frozen perspectives—which they cannot yet perceive. They can think abstractly, but they are not so good at empathy and feeling compassion. This is a barren universe and the experience of this barrenness is the lesson to be learned.

Green, the Post-Modern Level of Consciousness: Greens want to change the world for the better. They want to save the earth from pollution. They are social justice warriors. They see the levels of consciousness below them and they are dismayed—they blame the people with lesser (but evolving) consciousness for the mess the planet is in. The greens have gone in and out of power (influence), and when they are in power they regulate the game in the name of love and compassion. The greens know that something must be done, but they do not fully grasp the idea of levels of consciousness. They can see the levels, but they do not have a strategy based on cognitive evolution. Their lesson is that ideals cannot be shoved down the throats of others who have no idea what is happening or why. A calm strategy that uses love as the key tool for transformation is still missing from this stage, but it is their lesson to learn.

Turquois, the Integrated Level of Consciousness: The turquois group has learned all the lessons from the past, but they have one understanding that is paramount—they know that love and compassion are the tools for transformation. They have a calm strategy that can result in positive change. They know how to help each stage evolve. They know the language of the different groups, and can speak that language. Their lesson has to do with consistency and reliability. They have a discipline but it is relatively weak—not yet strong enough to impact whole populations. They also have a secret. They know that we are all part of a single mind. If a large part of that mind has a collective intention, and states that intention in unison, then collective consciousness can be moved and altered.

Wisdom and Intelligence

Red, the Warrior Level of Consciousness: Wisdom and intelligence are not the same. Wisdom is based on accumulated experiences—wisdom means *to taste* existence. However, intelligence is a pattern recognition ability of the neo-cortex of the brain. The capacity for ever-more intelligence unfolds with physical maturation. In addition, education, practice, and discipline provide the avenue for intellectual development. Living a life, event-to-event, and having experiences, develops wisdom. People at a red level of consciousness lack the life-altering experiences that cause the ego to transcend. However, at every stage, there can be exceptionally intelligent individuals. Therefore, the determining factor is not how smart you are, but rather how wise you are. What has experience taught you? Too often, harsh experiences have taught the warriors to fight and distrust.

Amber, the Traditional Level of Consciousness: Knowledge gained through the intellect is passed down mostly through the groups that the amber members belong to—the church, the family, and the ethnic belief structure. Indeed, at the amber level of consciousness *belief* rules the intellect.

There is no understanding of wisdom as separate from intellect. What you believe is who you are at this stage of development.

Orange, the Modern Level of Consciousness: Knowledge, science, and reductive thinking rule this level—facts are king. Belief is tempered by the scientific method, and some exceptionally intelligent people find a center of gravity in the orange level of consciousness. Wisdom is understood as separate from the intellect, and diverse and exceptional life experiences define this group. However, a certain kind of wisdom is missing from the life story of individuals at this stage. They have not meditated, used contemplative prayer, or practiced yoga—they don't have a spiritual discipline or foundation—so they lack the awareness (the experience) that comes from being in touch with the allocentric mind.

Green, the Post-Modern Level of Consciousness: In general, there is greater knowledge acquisition and greater wisdom as we go up the spiritual spiral. Developmental age is definitely a factor, since older individuals have had more experiences and a longer time to acquire knowledge. The greens have more wisdom than the groups below. That is the chief defining difference of this population. They know things through experiencing them. Communicating non-verbal experiences, explaining the allocentric mind, are near impossible tasks, which is why it is so difficult for levels of consciousness to interact and exchange views.

Turquois, the Integrated Level of Consciousness: Again, at this stage we find highly sophisticated individuals with vast amounts of knowledge and many transformative experiences. This group is both wise and intelligent. This level of consciousness also knows that wisdom and intelligence are held in a polarized tension by a force called *emotion*. There is a balance of three mental centers at this stage: there are equal parts compassion (positive emotions); knowledge and the intelligence to use knowledge, and wisdom that is gained through self-initiated intentions and discipline.

Silence versus Noise

Red, the Warrior Level of Consciousness: The lower on the spiral of consciousness, the more noise there is. The monkey mind is very active, and silence is not a concept that is dear at the red level of consciousness. There are no quiet gaps as life races forward. There is no pausing to consider.

Amber, the Traditional Level of Consciousness: The noise lessens at this level, but not by much. The monkey mind repeats the memes that drive it to rant and rave, review and stress. Mental loops go round and round and leave deep grooves in brain networks—all of which are linked to the emotional center. There is very little effort to comprehend the mind. There is an acceptance that the internal voice is who you are. The egocentric mind rules and the allocentric mind remains hidden.

Orange, the Modern Level of Consciousness: At the orange stage, quieter reflection enters the mind. The rants and raves are watched, or noticed, and there is a budding awareness that the ego is somehow not all there is to life. There is dismay as the inner voice rules daily life and causes undue stress. The orange population is not happy with the incessant voices in the head and in the street. The egocentric mind rules, but the allocentric mind is glimpsed.

Green, the Post-Modern Level of Consciousness: The green population understands that stress is coming from the inner voice, and from an unbalanced ego. Therefore, greens fight the ego-demon by using various strategies for quieting the inane chatter of the monkey mind.

Turquois, the Integrated Level of Consciousness: At this level, long practiced routines begin to bring peace to the mind. There is a measured and calming flow to the body and brain that is brought on by disciplined practice. The monkey mind can be shut down at will. There is clarity, a clearness, and a beneficial emptiness inside the brain and body—a deep peace can be called up from within.

Intolerance to Tolerance

Red, the Warrior Level of Consciousness: This level is so ego-bound that the only response available to even mild conflict is anger and sometimes violent confrontation. When faced with the suggestion that there are levels of consciousness, levels of experiential sophistication, and degrees of knowledge (that become levels of intelligence), the person of red consciousness goes into extreme defensive and/or attack mode.

Amber, the Traditional Level of Consciousness: This ethnocentric level of consciousness is also ego-bound, but group identity influences and guides individual responses to conflict. Rejection and exclusion define this population's response to conflict. They are capable of pity, derision, and scorn.

They are much less drawn to violent conflict, although they will participate and support harsh responses to those they disagree with. This is the level of consciousness that keeps war and violence alive on planet earth.

Orange, the Modern Level of Consciousness: The intellectually-bound orange level of consciousness will study, consider, and reflect on differences. They logically reject violence and intolerance. They reject ethnocentric viewpoints as being too narrow, and they adopt a worldview that allows for global differences. They tend to be stoic and passive in the face of anger. They mirror back stability and calm, although this population can go on rants as they slip back into ethnocentric language and emotion—they use reason to combat the illogical stances of the red and amber populations. This group is capable of tolerating and supporting violent decisions—like the decision to go to war to settle conflicts. They can provide logic to support violence.

Green, the Post-Modern Level of Consciousness: The green population is composed of people who have learned to deeply listen, to use-loving-kindness during discourse. They notice the differing responses that arise from the red, amber, and orange groups, and while they are tolerant of the positions and emotions arising from these lower levels of consciousness, they confront and reject many of the conclusions and practices of the less-evolved groups. Although the green population has evolved a sophisticated degree of tolerance and compassion, they still see the world as a division between insiders—those who get it—and outsiders (those without a clue). This group is composed of pacifists—they reject violence as a means to resolve conflicts. They will not support or participate in war.

Turquois, the Integrated Level of Consciousness: This level of consciousness has evolved sophistication as they genuinely use deep-listening, loving-kindness, and sincere tolerance for differences. They perceive all levels of consciousness and seek to converse in the language used by each. They are peace-keepers and peace makers. They have faith in love and in the power of loving. They forgive. They seek to heal. They strive toward a higher level of kindness and helpfulness.

Addressing the Polarization

As I write this, the United States is emotionally divided. My proposition (theory/suggestion) is that this divide is between levels of consciousness.

On the lower end of the scale, the red, amber, and early stage of the orange level, are ego-bound and lack empathy. These lower levels of consciousness do not understand the high end of orange, the green, and the turquoise levels of development. Perceptual blindness, lack of mindfulness, dominates at the lower levels; even the need for dialogue and love are genuinely questioned at the earlier stages of development. How then might we go about healing the divide? How can we help those at lower levels of consciousness move upward on the spiritual pathway? How do we *cure* developmental cultural blindness? Here are some suggestions:

- Use the language of love. Use positive and respectful communication. Use loving-kindness to bridge gaps. The same words can be cautiously used, even if there are differing shades of meaning. For example, words like God, compassion, love, and justice mean something different to each stage of consciousness, but they don't have to be avoided.

- Focus on issues, like health care, climate change, and job creation. These are problems common to all levels. Groups can agree that *something* needs to be done regardless if they comprehend each other.

- Meditation and contemplative prayer are common to all levels. Use meditation and intention to generate harmony. For example, in some schools, meditation is used as a form of time-out for struggling students. Rather than isolate, or expel, or condemn the actions of youngsters in a school setting, quiet contemplation is used to calm and restore balance. Meditation and contemplative prayer can give a group of people a bonding experience—moments of silence, for example, are powerful ways we traditionally pause to consider and remember.

- Create forums where collective experience can affect spiritual development, and where cultural levels of consciousness can interact. In the electronic age, the levels have been given an isolated collective voice. For example, the Red stage and Amber stage have their own talk radio hosts. They have Fox News on TV. Websites and Facebook pages reinforce the Red and Amber stages of development. These media bubbles solidify the stages rather than offer an opportunity to evolve. The same is true for each level of consciousness—there are media bubbles where individuals can go and not be bothered by others who see the world differently. We need to see this problem and find ways to transcend the isolated bubbles.

- A campaign to discuss and debate levels of consciousness might be helpful, especially for individuals at the higher levels of development. Such an effort would be expected to meet with resistance from lower stages of consciousness, but at least the debate would be helpful to some.

- The highest levels of consciousness, the blue end of the spectrum, have a responsibility to use their awareness of the collective mind to influence the evolution of mindfulness. Contemplative prayer and meditation, when done in synchronicity with others and when accompanied by a specific intention, have been shown to have influence on communities and individuals. The major theme of this book is that individuals at the higher stages of consciousness have a responsibility to bring about beneficial change. These individuals have an obligation to do as much good as they can during their sojourn on the planet.

The Turing Test of Artificial Intelligence

The Turing Test is named after Alan Turing who first proposed it in 1950. Turning suggested that a machine might be considered intellectually equal to a human being if it could converse in such a way that a human would be unable to tell it was a computer. The test is set up so that a human being in one room communicates with "someone" in another room—perhaps a computer with a highly-evolved artificial intelligence—via text, emails, or natural-language voice. The goal of the human is to determine if they are speaking with another human being or a machine. The goal of the designer of artificial intelligence is to create a machine intelligence that is indistinguishable from human intelligence.

As we look at the levels of consciousness, we see that at low levels of cognitive development human beings are just bio-machines. This is the conclusion of mystical thinking—Gurdjieff (who I will discuss below) said straight out that almost all human beings are sleepwalking *machines*. Therefore, the Turning Test measures the judgment of a bio-machine relating to a silicon-machine.

It didn't take long for the computers to fool most of the population. Most people these days cannot determine if they are talking with a machine or another human. Why would that be? There are at least two reasons why human beings are so easily fooled by artificial intelligence—or at least why they are fooled by tests such as the Turing Test:

1. The Turning Test only measures the egocentric mind. There is very little understanding and less acceptance of dual-process theory, so the allocentric mind is typically ignored. There are two kinds of minds and two kinds of consciousness. What scientists typically try to do is duplicate the egocentric mind which grows in sophistication through knowledge acquisition. However, the allocentric mind is based on experience, compassion, and on relationships. Wisdom and emotional development result from experience, not from knowledge-based intelligence.
2. If both our minds evolve upward through a spiral of evolution—if there are levels of consciousness—then the Turning Test would first have to determine the level of consciousness of the human. Presumably, a person at a lower level of consciousness would be more easily fooled. The test is invalid without an understanding of both the human being's and the machine's level of spiritual evolution.

I point this out because science is marching along under the assumption that they are duplicating the human mind during AI development—especially when designing robots. It is very dangerous to assume that there is only one mind and one level of consciousness, and yet that is what is occurring across the board and around the world.

Additional Thoughts and Caveats

Build a ship, and there will be water to float it.
The tender-throated infant cries,
and milk drips from the mother's breast.
Be thirsty for the ultimate water.
Then be ready for what comes
pouring from the spring.
-"Having Nothing," *A Year with Rumi*, 2006.

After I wrote the above sections on levels of consciousness, I had insights and afterthoughts, which needed to be included. The following is a collection of these insights. There is some repetition—ideas mentioned previously—but I felt these insights needed further clarification and emphasis.

There is an exponential component to the evolution of consciousness. This perspective differs from that of Evolutionary Integral Theory. For example, Integral Theory states that the Archaic Period started 250,000 years ago. This is one perspective and it is backed up with historical evidence. However, from another perspective—using Raymond Kurzweil's evidence of exponential doubling (applied to technology)—the early stages of cognitive evolution can be surmised as lasting much longer. From this second perspective, the exponential nature of the evolution of consciousness becomes plausible. Archaic Man, if we include our proto-human ancestors, was around for millions of years. Tribal Man was around for hundreds of thousands of years. Warrior Man was around for thousands of years. Traditional Man was around for hundreds of years. Modern Man emerged in the late 1800s during the Industrial Age and has been around for decades. Post-Modern Man came into prominence in the mid-1900s. Integral Man rose up in the latter half of the 1900s. Post-Integral life forms will soon explode all over the landscape as humanity moves beyond the year 2020. This is an exponential rate of change in human consciousness—something extraordinary is happening here that is truly amazing and unnerving.

Exponential doubling seems to parallel the rise of new technologies, to be linked to the evolution of new tools. Raymond Kurzweil,[8] who was called "Google's chief futurist" when he worked for that corporation, has written about this exponential phenomenon and has made many correct predictions about the future based on the understanding of the velocity of change.

There is also a singularity at work here. The time frame for the arrival of each newly evolved level of consciousness is drastically shortening. If the pattern holds, new kinds of consciousness will emerge monthly, weekly, daily, hourly, minute-by-minute, and then second-by-second. This seems obviously absurd and impossible (and it is), but we are staring at an exponential unfolding—cognitive change is coming like a runaway bullet train. The evidence is clear: the law of change is working as it has always worked: exponentially. What derails and redirects exponential unfolding are half-step intervals in the octave of change—called the Law of Seven. This is a mystical (esoteric) understanding of how frequency plays out. I don't have the space to elaborate on the Law of Seven (I don't fully understand it) although I will return to it below when I discuss The Fourth Way philosophy popularized by G. I. Gurdjieff.

It could very well be that what we are observing when we study levels of consciousness is driven by—and will be limited and influenced by—technological changes that are essentially destructive to the status quo. However, technological change, as influential as it is, may only be part of a greater understanding.

In my book *Consciousness, A New Slant on an Old Conundrum*, I discussed another kind of exponential doubling[9] inherent in our genetic nature and related to a concept called *syntropy*.[10] Therefore, it may also be true that exponential change is more fundamental than our technology—changes in levels of consciousness and advances in the sophistication of technologies may be following a more basic universal law.

Levels of consciousness can be depicted as a spiral, with egocentricity at the bottom of a funnel. As we spiral upward within the funnel, we are moving around as well as up—in other words we can picture an ascending spiral. If we now lay the spiral of spiritual development on its side and view it in two dimensions we see a sine wave. This perspective allows us to see more clearly the exponential nature of levels of consciousness. The Archaic Period was a slow, low amplitude (low frequency) rise and fall, which took place over millions of years. Then it faded out as the Tribal Age began to slowly dominate. As time moved forward, the frequency of the wave increased. In other words, the sine waves get more compact and the amplitude of waves is higher as we approach the present.

By the time we reach the Post-Integral age, the frequency rate is relatively high. This matches how brain waves look when we do EEGs. Slow, lazy, flat-like sine waves are the states of sleep—delta waves. However, as wakefulness grows, the waves increase their frequency—they have a higher energy state—and the peaks and troughs of the sine wave come faster and faster, approaching the gamma frequencies associated with highly awake states of mind that monks reach during meditation. It is as if Post-Integral people have gamma-wave states of consciousness available to them. Individuals and cultures seem to be approaching—moving in the direction of—enlightenment. As long as the exponential rate of change keeps going, we can expect to witness more and more evidence of enlightenment[11] in people around us.

But what is enlightenment? It seems to be an ultimate arrival, the highest state possible for human awakening. I cannot speak from experience; I do not know. However, during a meditation session one morning, I realized that when an individual (or culture) jumps a quantum level—say from the amber ethnocentric stage to the orange modern stage—it is a *form of* enlightenment (a relative awakening). The person who was at the amber level of consciousness becomes amazed as awareness expands into the modern mind—a knowledge explosion occurs; new wonders and revelations are revealed. Each quantum leap into a higher spiritual stage is a leap to a kind of enlightenment. Perhaps the ultimate awareness that gurus and monks reach—the gamma state, the nirvana level, heaven on earth—is just the highest stage that these early spiritual pioneers perceived. Perhaps there are endless levels of awakening. Perhaps enlightenment is a process and not an ultimate arrival. Whatever the enlightenment process, it is approaching the earth at an exponential speed.

Of course, there is extensive overlap; one age slowly fades out while another age slowly arises. Older levels of cultural consciousness don't go away as more advanced versions slowly gain ground. Today, because of this overlap, many kinds of human beings with various levels of consciousness walk the earth together. Multiple levels of consciousness have existed simultaneously throughout modern history and continue to do so.

French philosopher and Jesuit priest Pierre Teilhard de Chardin (1881-1955) studied evolution (as a paleontologist and geologist) and concluded that there was a planetary bubble within which consciousness evolved. He called this *atmosphere of consciousness* the noosphere. Chardin anticipated our recent discoveries about consciousness: that consciousness is pervasive and necessary for human evolution just as air is pervasive and necessary for human survival. Chardin would not at all be surprised to find out that the noosphere was evolving at an exponential rate.

❧⎯⎯⎯⎯⎯❧

Each level of consciousness has a different concept of meaning. There is a gradual change in meaning from the Archaic to the Post-Integral. At the lower ends of the spectrum, individuals and cultures seek to *find meaning.* They seek to *understand.* This is what the egocentric mind does—it is the job of the ego to attach to "objects" in the objective world, actually *to create* an objective world. However, at the higher end of cognitive evolution, individuals and cultures discover that they are *agents of change*—they realize that their own minds are manufacturing space, time and meaning. At the highest levels of consciousness, *meaning is made.* Language also gets more sophisticated, more complex, and is used in more nuanced ways, as we go up the evolutionary scale of consciousness.

It is no wonder that human beings struggle to explain themselves to others. Each new level of consciousness adds more abstract and complex terminology into the cultural milieu. Existing concepts like beauty, truth, goodness, justice, love, and God, for example, take on new shades of meaning with each spiral up the ladder of consciousness. Misunderstanding, conflict, and bewilderment are natural consequences that occur because we cannot accurately and consistently communicate with each other. I like poet Billy Collins' sentiment in his poem "Early Morning" because it captures our bewilderment when we try to communicate with another level of consciousness:

> *I don't know which cat is responsible*
> *for destroying my Voter Registration card*
> *so I decide to lecture the two of them*
> *on the sanctity of private property,*
> *the rules of nighttime comportment in general,*
> *and while I'm at it, the importance of voting to an enlightened Citizenship.*
> *Of course, it's hard to get them to stay*
> *in one place let alone hold their attention for more than two seconds.*
> ~ "EARLY MORNING," *THE RAIN IN PORTUGAL*, Billy Collins, 2016.

I can really relate, quite humorously to the cat problem. My cat Napoleon routinely marches every morning to the backdoor—before I can get my morning coffee—demanding to be let out. I carefully explain that it is the dead of winter, two degrees and a wind chill of minus ten, and that his cat fur will instantly freeze if I even open the door a crack. He doesn't care. I open the door and he says (in cat language). "What the f....!" And he turns around and heads to his cat dish for breakfast. We do this routine every morning, all winter. I saw a documentary on the human gut biome last night—evidently we have as many neurons in our stomach and intestines as cats have in their entire brain. Enough cat science, back to the egoic dialogue.

As I was musing about the above poem, it occurred to me that the ability to *sustain attention* egocentrically and to *sustain awareness* allocentrically probably varies with levels of consciousness. At the red and amber levels, we would expect low scores on a scale of allocentric awareness and higher scores on egoic attention. As the level of sophistication advances, in the green and blues stages, we would expect to see greater ability both to attend and to be aware. A main task of mysticism and esotericism is to awaken the mind to the difference between attention and awareness. A foundation of spiritual work involves practices that strengthen awareness. Also notice that for each level there are two kinds of meaning: an egocentric meaning and an allocentric meaning.

❧━━━━━❧

Notice that ***human beings, individually, as part of their personal cognitive development, evolve through these same levels of consciousness***. It is as if we are not capable of grasping certain concepts until our minds have sufficiently matured. It also suggests that as one level gives way to another, there is *internal conflict*, just as there is external conflict between cultures with different worldviews. Consequently, different worldviews are at war *within us* as we are wrenched from a lower energy (consciousness) level to a higher level. We might plateau for a while but evidently we are fully capable of evolving into these different energy levels as we mature. There is as well, I believe, some kind of cultural heredity (Morphic resonance? Epigenetics?[12]) available to each new generation, as if our children are born with potentials that we did not have at birth.

Cosmic rhythms (physiological, cyclical, and quantum) affect both our *states and levels* of consciousness. In the morning we might be fully integrated, but by evening, after a long exhausting day, our level of consciousness might have sunk to the warrior state, or worse. The fact that we know we are acting like a Neanderthal tells us that the level of consciousness we think we have attained is not fully formed—we are actually in transition. *States* of consciousness fluctuate, while *levels* of consciousness are more stable, but *each has a center of gravity.*

❧━━━━━❧

Overlapping levels of consciousness are naturally in conflict. This is because each level of consciousness has a fully formed worldview supporting their understanding of what reality is. The meaning of life is different at each level. There is no way that the overlapping groups are going to understand each other or be able to fully cooperate. Only at the Integral level do we finally arrive at a cognitive stage where individuals have *consciously* incorporated levels of consciousness into their worldview, where they can grasp the big picture and have tolerance and compassion for each group's perspective. Understand that each level of spiritual development is an integration of all previous levels. The newly evolved cultural level contains the latest integrated minds. However, it wasn't until the level called Integral Man arrived on the scene—very recently—that there was conscious awareness and appreciation for *levels of consciousness* as a concept. How will conflict play out, we wonder, in light of exponential change, as more and more levels of sophistication overlap and clash?

As we go from a waking state to a state of deep sleep, we also go through levels of consciousness. At deep levels of sleep we shed all our hard-earned ego-development, and we return and dwell within the archaic mind—we are again cavemen living with an unevolved cognition and without sophisticated language. The fascinating theta-state (between sleep and waking) is an opportunity to once again become tribal and warrior beings, as if we were alive in that long-ago era. ***Each evening as we fall asleep, and each morning as we slowly awake, we relive the evolution of consciousness in humanity.***

Furthermore, if all human beings have the history of consciousness embedded in their DNA, if human maturity from infancy to death is a recapitulation of physiological history, then for a time each of us is archaic, then tribal, then warrior and so on. This means ***that those being born now come equipped to take humankind to the next level of consciousness***. It also means that the earth is full of people going through their varying levels of consciousness. This is a rather significant understanding for educators to comprehend and employ, especially as they nurture the young (and old).[13]

The levels of consciousness outlined above, starting with the archaic and ending at the post-integral, are usually depicted as spiraling upward and outward, a funnel standing on end. Starting at "point consciousness," the spiral widens—it explodes from the big bang beginning of egocentricity toward the ever-expanding allocentric sky. This spatial widening is significant because allocentricity *is* spatial awareness; the greater the spatial scope of a person, the more they are awake. At the point where it all began, our consciousness level was purely egocentric. ***Therefore, the road to enlightenment is a steady movement away from point consciousness, away from saturated egocentricity toward ever more allocentricity.*** Indeed, the spiritual journey is the evolution of ever-expanding spatial perception. Actually, it is more complicated than that, as I will explain in the following paragraph.

Envision the spiral in its upright position—a funnel widening outward from a point. If we envision the spiral curving upward to the right before it turns and goes upward to the left, we notice that the right-hand levels of consciousness are all stages of egocentricity. The archaic, warrior, modern, and integral levels are progressively more advanced versions of egocentricity. On the left-hand side of the spiral, all the levels are allocentric. Tribal, traditional, post-modern, and post-integral are ever higher levels of allocentricity. This image underscores my major point that ***human minds and the cultures they bring about are dual, and the two minds are equal in value for human existence***. Although it appears that we should be developing ever greater allocentric spiritual minds, the reality is that ***nature has decided to oscillate as it creates ever more sophisticated levels of both allocentric and egocentric consciousness***. Furthermore, it looks—as we study the graphic—as if our two minds reach down and pull their twin upward, as if they are cooperating to raise individual and cultural levels of consciousness.

Another way to see this alternation between allocentric eras and egocentric eras, within individuals as well as cultures as they develop, is through the

141

philosophy of Georg Hegel. According to Hegel there is an oscillation that flows up and down like a sine wave; periods of dialogue are followed by periods of conflict, which are followed by resolutions and a return to dialogue. In his words, thesis and anti-thesis lead to synthesis, but the synthesis is just a new thesis and comes with the baggage of an anti-thesis—ad infinitum. Every egocentric period is a time of conflict, of unease (decoding, assimilating, and remembering patterns). There is an evolution away from the current traditional, habitual, comfortable way of being. Allocentric eras are peaceful, cooperative and yet get boring and are, therefore, challenged—they eventually get passé and uninteresting. Conflict arises and egocentricity returns—although in a more sophisticated and evolved form.

Still looking at the spiral, notice that the **egocentric right-hand side represents** *levels of consciousness that honor the hero's journey*. These are patriarchal societies where individualism wins out over sainthood. The Greeks, for example, were tribal; Socrates is the father of the hero's myth. However, on **the left-hand side of the spiral all the** *levels of consciousness honor the journey of the saint*. These are matriarchal societies and they embrace compassion for the group over adoration of the individual. Looking at our modern day struggles, it seems that Modern Men (egocentric, patriarchal men) are being wrenched into the Post-Modern allocentric matriarchal era—but not without a fight.

How can the oscillation between ego and allo change with time? For example, we are able to say that a whole culture is stuck at an egocentric or allocentric stage for decades or centuries, while for individuals there is an oscillation between ego and allo (as they move through levels of consciousness) that can only last over a lifetime. The cells in the body also have an allocentric-egocentric rhythm, but this lasts only days before the cell dies. The molecules that make up the cell oscillate but only for a few minutes. The atoms within molecules oscillate but only for microseconds. Quarks within atoms oscillate but only for nanoseconds. *Vibrations are oscillat-*

ing inside vibrations. Frequencies are embedded within frequencies. Let me see if I can provide an image to display this pattern.

Think of a stone thrown into the water that generates waves around it. In a third dimension, the propagated waves are spherical, radiating outward, expanding. Notice that as the spherical wave (the leading edge) moves ever further from the initial point, the distance between two random points on the leading edge increases exponentially. Think of a balloon expanding. If we put a red dot on the balloon and then place a blue dot directly opposite—on the other side of the balloon, we can watch as the red and blue dots are separated by ever greater distances over time. If we now look down from an overhead perspective at this process—from the moment of impact when the propagation began—we will observe our funnel (the spiral) as it evolves. Hold that image for a moment.

If we envision energy jumping (oscillating) between the blue dot and the red dot, we will notice that the oscillation is very fast right after impact. However, as the balloon expands and the distance between the dots increases, the speed of oscillation slows down.

I make this observation because I am trying to explain scales (as above so below). As we pull back and look from overhead, the oscillations appear to slow down over time. This is why we can have a slow alternation between ego-centric cultures and allocentric cultures yet have faster alternations between ego and allo at the level of individual evolution. Inside a person, the ego-allo swings are much faster inside cells. Deeper in, the oscillations at quantum levels are extremely fast.

There also seems to be a correlation between levels of consciousness and the architecture of the human nervous system. It is as if Archaic Man, for example, could not have become a Modern Man—even with the right effort and opportunity—because the neural architecture of Archaic Man had not evolved sufficiently. Traditional Man had within his history, his neural architecture, ingredients that created Tribal Man, Warrior Man, and Archaic Man. However, Traditional Man could not become a Modern Man because some kind of change within the human nervous system had yet to occur. During the Axial Age—about 2500 years ago—a major neural adaptation must have occurred, if this speculation is correct. Roughly between about the 7th century BC until the 1500s AD, history records the emergence of Buddhism, the Greeks, Lao Tzu and Confucius in China, the Upanishads in India, Zarathustra in Iran, and the Christian prophets in Palestine. "Modern" philosophical perspectives were defined and debated in this short span of history, including materialism, skepticism, nihilism, and the conundrum of human duality. Karl Jaspers was the German philosopher and psychiatrist who defined this pivotal age. In his book *The Origin and Goal of History* (1953) he wrote:

> What is new about this age, in all . . . areas of the world is that man becomes conscious of Being as a whole, of himself and his limitations . . . Measured against the lucid humanity of the Axial Period, a strange veil seems to lie over the most ancient cultures preceding it, as though man had not yet really come to himself.
> ~ THE ORIGIN AND GOAL OF HISTORY, Karl Jaspers, 1953.

What changed in the human nervous system, across the globe, 2500 years ago? We can only speculate, but whatever it was, the impact on the evolution of mankind was profound.

***

Color coding levels of consciousness is not random. Philologists, anthropologists, and evolutionary biologists have studied the evolution of color and have arrived at similar conclusions. First, color perception evolved. It is wrong to assume that human beings who lived a million years ago, for example, had the same kind of vision system we have today. To the contrary, we now can speculate that they did not.

Second, there are correspondences between the frequencies of colors and levels of consciousness. The color spectrum that we see in a rainbow follows the same pattern used to label levels of consciousness: on the lower end we find red, then orange, then yellow; in the middle we find green; and at the other extreme we find blue, indigo, and violet. Notice that this is a scale of frequency: red is a low, slow sine wave. Orange is faster than red with higher amplitude. Yellow is even faster. As we go from red to indigo the sine wave is ever faster with greater amplitude. These amplitudes and frequencies correspond to the degree of wakefulness in human beings. When we are asleep, or nearly so, we register low frequency brain waves similar to the red end of the spectrum. As we wake up further our brain waves register in the orange, then yellow, then green range, then—at a high level of wakefulness—we register blue-end brain frequencies. The implication is that people on the red, orange, and yellow end of the spiral of consciousness are not as awake as people above them on the spiral—their brains process at slower speeds.

Third, the evolution of color perception cannot be separated out from the evolution of our two minds. In 1858, William Gladstone (scholar and former Prime Minister of England) wrote a book called *Studies on Homer and the Homeric Age*, in which he speculated that the Greeks might have been color blind. Interestingly, when he was challenged about this speculation, Gladstone (1809-1898) responded by saying that Homer operated on a quantitative scale, contrasting white and black, or light and dark, instead of a qualitative scale that would have taken into consideration color perception. In other words, Gladstone really said (using my terminology) that Homer knew our two minds and chose to write from the perspective of the egocentric (quantitative) mind rather than the allocentric (qualitative) mind. Therefore, the debate around the evolution of color perception may be better framed as a debate between the *evolution of the allocentric* mind versus the *evolution of the egocentric mind*.

Personal states of wakefulness are related to cultural levels of consciousness. It has to do with the percentage of the population residing in any cognitive state. If almost the whole population walks around in a theta-state consciousness, the result is a very low level of cultural wakefulness—a culture ruled by the limbic system. These are hand-to-mouth cultures that are self-serving. Archaic, tribal, and warrior cultures are made (so it seems) mostly of individuals with theta (perhaps also alpha) states as their highest ability. Theta is low energy, dream-like mindfulness. To go beyond the theta state takes increased cognitive energy. To arrive at alpha and then beta consciousness—to use language, to project beyond the ego—the brain (the whole body) has to kick into a higher gear, to use higher frequencies. When most of the population is in a beta-state, we get more advanced cultures: the traditional, modern, and post-modern levels of consciousness evolve. Integral and post-integral cultures don't exist yet because not enough of the population can generate the brain/body frequencies sufficient to process at this high level of mindfulness.

Movement to higher levels of consciousness is accompanied by an ever expanding circular horizon of knowledge. Therefore, each level of consciousness is beyond reach unless a threshold of knowledge is breached. As the level of consciousness rises, so does the knowledge base. The two go together: ***the higher the level of consciousness, the greater is the base of knowledge***. It is also true, as I mentioned earlier, that "jumping to the next higher level of consciousness" requires some kind of shock (a rite of passage). This is analogous to electrons around a nucleus that stay in stable orbit unless increased energy shocks the electron to another discrete energy state. A "shock," in human terms, is another name for "a threshold experience." Wisdom is gained through tasting life, through participation, through experience. Therefore, the spiral widens outward and upward as wisdom is gained. ***The higher the level of consciousness, the greater is the base of wisdom.*** Remember that wisdom is allocentric and is related to discipline and experience, while intelligence is egocentric and related to discipline and knowledge acquisition.

The above paragraph has a symbolic, metaphorical correspondence with the Christian image of Christ on the cross. Each level of consciousness must die and be reborn (be resurrected) before there can be a transformation into a higher spiritual realm—a higher level of consciousness. The cross that human beings hang from is the paradox of our two minds. The horizontal part of the cross that spreads outward in a single plane represents the knowledge base of the egocentric mind. The vertical dimension of the cross represents the allocentric mind as it gains wisdom from experience. The important image to carry forward is that Christ—symbolizing mankind on the cross—is rising upward toward higher dimensions of awareness. *The whole cross*, with mankind attached, is rising through levels of consciousness. We cannot shed our duality as both our egocentric and allocentric minds evolve—as our attention and our awareness evolve. Here is the Hindu guru Osho (Bhagwan Shree Rajneesh) talking about the dual evolution of humanity and this analogy of Christ on the cross:

> Horizontal [on the cross] you are Jesus. Vertical [on the cross] you become Christ. . . . If you move from one thought to another, you remain in the world of time. If you move into the moment—not into thought—you move into eternity. You are not static; nothing is static in this world, nothing can be static—but a new movement arises, a movement without motivation. Remember these words. On the horizontal line you move because of motivation [intention]. You have to achieve something. A motivation is there.
>
> A motivated movement means sleep. An unmotivated movement means awareness—you move because to move is sheer joy, you move because movement is life, you move because life is energy and energy is movement. You move because energy is delight—not for anything else. There is no goal to it; you are not after some achievement. In fact, you are not going anywhere, you are not going at all—you are simply delighting in the energy. There is no goal outside the movement itself; movement has its own intrinsic value, no extrinsic value.
> –*AWARENESS: THE KEY TO LIVING IN BALANCE*, 2001.

There are many insights in the above quote by Osho. As I explained in *Consciousness: A New Slant on an Old Conundrum*, a fundamental movement fractal is exponentially building the universe. Movement balanced with no-movement is the prime creator. In the quote above, Osho is also talking about the allocentric mind and the egocentric mind, although he does not use this terminology. He is telling us that the ego runs on intention and motivation, on achievement and striving toward goals. To the contrary, the allocentric mind rejoices in the sensation of movement, without judgment or thought. The egocentric mind *inhibits* joyful animation. The ego dwells in (creates) a world of time, while the allocentric mind is timeless. I have discussed all these ideas before. The key understanding in this context is that the egocentric mind expands as the cross-with-Jesus (humanity) rises toward heaven, while the concept of Christ is the allocentric evolution of awareness—moving through levels of consciousness. **The egocentric and allocentric minds both evolve in parallel, but they do so differently.**

Breathing in and out is a frequency cycle—breathing must occur for the lifetime of an individual. The in-breath is egocentric and the out-breath is allocentric. Think of smelling a fragrant flower. The in-breath is a search for meaning: What flower is this? What does this fragrance remind me of? The out-breath is not a search for meaning; it is a whole-body experience. This analogy of breathing in and out works across the spectrum: everything is breathing in and out, oscillating between meaning extraction and experience. However, because everything is in relative motion, we don't get a simple sine wave graphic depiction when we attempt to picture what is happening, instead, we get an elliptical spiral. The point I am reaching for in this paragraph (which I discussed above in another context) is that *frequencies are embedded within frequencies*. Breathing in and out is happening inside the cycles of the moon, and inside the rotation of the earth. The cycles of the moon and the seasonal and diurnal cycles of the earth also have egocentric and allocentric oscillations—a pulse, a spin, a push-pull, expression and suppression, and an alternation with a dual *purpose*.

The higher the level of consciousness, the more the mind is silent. At the level of enlightenment, the level somewhere beyond integral consciousness, there is total mental silence—no need to speak or to think in words. As we observe lower levels of consciousness, we see ever more noise and internal banter. At the red warrior stage, there is no separation between the ego and self. The warrior *is* the emotional voice in the head—all ego, with no awareness of the allocentric mind. There is no empathy at the warrior stage, no tolerance, and no compassion—no caring about the suffering of others. The amber stage is the level of the incessant internal voice—the ranting, meme-obsessed, repetitious, internal dialogue that self-justifies, judges, blames, and separates all humanity into groups—into good guys and bad guys. There is little to no awareness of allocentricity at the amber level, and not much tolerance or empathy beyond that which is directed at chosen groups. At the orange level, there is a vague awareness that the internal voice is not who we are. What the mind ruminates over is for the first time suspect. Empathy begins to arise as a concept. Tolerance is considered. Group loyalty is questioned, as is the language of groups. At the green level, tolerance, empathy, and spiritual values become political tools as well as personal intentions. The voice is quieter; deep-listening is championed. Meditation is a way of life. Measured speech and a quieter mind are goals. At the integral level, the self and ego are merging. The witness is becoming stronger as it looks down and watches over all lower levels of consciousness—within oneself, in others, and collectively in cultures. Silence is prized at the integral level. Beyond integral is ever more silence.

From an egocentric perspective, as we spiral upward we gain greater and greater control over attention, intention, and will-power. As we become more egocentrically sophisticated, we become more able to control the spotlight beam of attention, to hold it on a subject for ever greater periods of time. In other words, our ability to concentrate increases as our level of consciousness elevates. With greater *attentional* power comes greater *intentional* power. Attention must combine with intention to move up the spiral of consciousness; however, there can be no actual movement without will-power (discipline). The notion that mankind sleepwalks through existence is a relative understanding. On the lower rungs of consciousness, human beings sleepwalk through life—reacting to events rather than steering their own essence. The higher up the spiral of consciousness, the less human beings sleepwalk, the more awake they become. Notice that attention, intention, and will-power compose a familiar triad. In Christianity it is called *the Father* (attention), *the Son* (intention), and the *Holy Spirit* (the source of will-power). In the famous equations of quantum physics we see the same triad: E (movement) = M (no movement) times the speed of light squared (the power source, the mover).

The evolution of consciousness can be viewed as a law of three forces. There is creative, upward pressure that must be opposed by an equal downward counter-pressure. It is only through the oscillation between balance and imbalance, that evolution unfolds as the third and resultant force. A balance results as the egocentric mind pushes against the allocentric mind. However, imbalance—caused by the third force—naturally causes these two minds to vibrate and rebalance. The evolution of consciousness is molded by this invariant pattern. The esoteric Wisdom Tradition has called this the *Law of Three*. Understanding this necessary balance of forces has important consequences for the evolution of cultures, as Cynthia Bourgeault explains in the blog post below:

> One can only imagine how greatly the political and religious culture wars of our era could be eased by this simple courtesy of the Law of Three: (1) the enemy is never the problem but the opportunity; (2) the problem will never be solved through eliminating or silencing the opposition but by learning to hold the tension of the opposites and launch them in a new direction. Imagine what a different world it would be if these two simple precepts were internalized and enacted.
> - Richard Rohr's Daily Blog; insights by Cynthia Bourgeault, March 22, 2017.

Cynthia Bourgeault is saying that the feeling and concept we call *enemy* (our perceived adversary) is a sign of imbalance within cultures and within the minds that make up specific cultures. If we see the enemy as an opportunity for cultural and individual growth, we can use the moment to create rather than destroy.

∽————∽

Before I leave this discussion, let me point out that within any democratic political landscape, **all levels of consciousness line up at the ballot box.** In the democratic congresses of modern nations, we have Warrior Men and Integral Men debating the future of humanity alongside Traditional Men and Post-Modern Men. There are relatively few women in this debate— more about that later—and very few Integral or Post-Integral spokespersons.

If beings of lower consciousness win the debates, expect a bloody, barren, stupid future for all lifeforms. Meanwhile, we are evolving so fast, so out of control, so lacking in fundamental overall perspective that hope seems to be fading. However, if we trust to the spiral it appears there is a grand plan that we need not fret about. If we let the process run its natural course, it will all work out, if arriving at ever higher levels of attention (egocentricity) and awareness (allocentricity) is the goal of evolution.

That being said, the rise to higher levels of collective consciousness *requires work* on the part of individuals and small groups. Our task is to pull the majority of mankind into the Integral Age where they can witness and accept all levels of consciousness. Dual-process thinking is also a higher perspective that we need to understand if we are to survive and thrive. We need to be able to make both wise (allocentric) and intelligent (egocentric) decisions. The Law of Three adds that *compassion* is the third center that must be combined with intellect and wisdom—all three must be in balance.

One final warning: we have to be very careful as we move forward because the egocentric mind will always try to control and take credit as it continues to categorize and label individuals. ***We delude ourselves when we assume that we are on a high rung of the spiritual path.*** If we are comparing, then we are judging, and that is the ego at work. We can influence each other through example, by our animation, our mission, our genuine compassion, and through dialogue. However, as soon as we get that "holier than thou" feeling, know that our ego has taken charge and our level of consciousness has sunk.

If the concept of *levels of consciousness* catches on in cultures, we will find that the ego will inevitably create conflict and comparisons. Of course, the claim will be put forward that people at higher levels of consciousness are more knowledgeable, have more meaningful experiences informing their wisdom, and are more spiritually sophisticated. There is validity to this statement. However, if we say that people at higher levels are more intelligent, wiser, and more compassionate, we will only foster division and defensive reactions. ***There is intelligence, wisdom, and compassion at all the levels***—of different character and degree, and at different and varied stages of evolution. If the ego needs to categorize and make comparisons—which is the ego's job—it is best to compare levels of consciousness and not to compare intelligence, wisdom, and compassion. Our goal is to bring people together and to make the world a safer and happier domain.

Personal Reflections

I am so small
I can barely be seen.
How can this great love be inside me?
Look at your eyes. They are small,
but they see enormous things.
~ "YOUR EYES," *A YEAR WITH RUMI*, 2006.

On the Cook-Greuter scale, I personally vacillate between levels 4 and 5. On a bad day, I sink below 4—those are the days when I am not worthy of my social security check. I should mail it back to the government with an apology letter.[14]

On the developmental scales described above, I drift between Modern and Post-Modern with occasional good days when I bump my head against the Integral level. Still, I am light years from enlightenment, and I am aging fast.

On the Pierce scale, I have transcended dot-on-the-highway egocentricity, moved beyond line and plane consciousness, and I am now firmly cemented in cube mind, kind of a 3-D rectangular-geek mind. I am not sure how to get out of cubidity.

What about you? Are you a happy camper, coasting along in bliss and blindness on your chosen level of consciousness? Or are you, like me, rather disgruntled at our poor showing among the planet's elite minds?

How exactly are we normal folks expected to step out of one level of consciousness and rise like triumphant ghosts into the next wonder-level of wakefulness? And if there are two minds, should there not also be a second ghost arising triumphantly and in synchronicity? The answer, as far as I can tell, is that *only experience causes change.* The ancients have known this for centuries (we seem to have forgotten), and they developed elaborate, time-tested rites of passage to change the consciousness levels of participants. In the book *The Consciousness Revolution*, Grof, Russell, and Laszlo discuss the profound impact of these cultural rites of passage:

Rites of passage are conducted in native cultures at the times of critical biological and social transitions, such as the birth of a child, circumcision, puberty, marriage, menopause, aging, and dying. In these rituals, the natives have used similar methods ("technologies of the sacred") for inducing non-ordinary states as the shamans—drumming, rattling, dancing, chanting, social and sensory isolation, fasting, sleep deprivation, physical pain, and psychedelic plants. Typically, the initiates have profound experiences of psychospiritual death and rebirth.

~ THE CONSCIOUSNESS REVOLUTION, Stanislav Grof, Peter Russell, Ervin Laszlo, 2003.

In other words, we have to be shocked into change; we have to go through profound experiences. Ordinary life doesn't provide the necessary catalyst. Good shock, or bad shock, it doesn't matter; they both produce a reaction that lasts. *Telling* someone how to be carefree, or joyful, or successful, or evolved, or loving (and so on) doesn't work. Talking doesn't work because we don't listen well and because we don't trust the ideas of another mind. Mostly, however, talking doesn't work because knowledge exchange comes with little to no shock.

We can wait around for shock and awe to happen in our daily lives—good luck with that—or we can get outside our comfort zone. Travel (somewhat) does it for me. Of course, lots of people "go on vacation" and come back with the same prejudices and mental habits. The difference is in the intention of travel. You have to expose yourself to situations that challenge your habits; you have to have powerful *experiences*.

Since we have two minds, we need to be searching for two kinds of shock and awe. Egocentrically, we need problems to solve, accomplishments to check off, mountains to climb, birds to identify, languages to learn, and so on. We need a mission in life, inequalities to challenge, environments to clean up, and causes to champion. On the other hand, allocentrically we have to practice being awake in our moments. We need to stop trying to accomplish things. Instead, we have to get off our mental horse and watch the sun go up and down, the clouds pass overhead, and the children laugh and play. We need to *actually* smell the flowers.

One further thought: consciousness shrinks to fit the box. If we live inside a house and never get out of that box, we will die there without ever gifting ourselves with transforming experiences. I like this poem by Rainer

Maria Rilke because it captures for me this need to leave spaces that confine our consciousness:

> *Sometimes a man stands up during supper and walks outdoors,*
> *and keeps on walking,*
> *because of a church that stands somewhere in the East.*
> *And his children say blessings on him as if he were dead*
> *And another man,*
> *who remains inside his own house,*
> *dies there,*
> *inside the dishes and in the glasses,*
> *so that his children have to go far out into the world*
> *towards that same church,*
> *which he forgot.*
> ~ Rainer Maria Rilke

I write, in part, so that my children and grandchildren will go beyond—carefully and mindfully—the spaces that confined me during my visit to the earth. I have not remained in the dishes and glasses, I did go in search of spirituality, but I was a child of my time—only able to go so far toward that church in the metaphorical east.

Gurdjieff, the Enneagram, and Levels of Consciousness

> *The enneagram is the fundamental hieroglyph of a universal language which has as many different meanings as there are levels of men.*
> *The enneagram is a schematic diagram of perpetual motion, that is, a machine of eternal movement . . . it is the philosopher's stone of the alchemists.*
> ~ IN SEARCH OF THE MIRACULOUS, P. D. Ouspensky, 1949.

As a student of human consciousness, I became interested in the writings of G. I. Gurdjieff and his student P. D. Ouspensky. These two remarkable men—somewhat independently—created a philosophical and esoteric school in the early nineteen hundreds called the *Fourth Way*. Gurdjieff was the chief architect of the *Fourth Way*, while Ouspensky, a journalist by trade, was the primary path through which the *Fourth Way* reached the public.

According to *Fourth Way* teaching, human beings can reach higher levels of consciousness by various routes. For example, we could use a *First Way* of transcendence, educating and disciplining our bodies (the way of Hatha Yoga). Or we could use a *Second Way* of transcendence, educating and disciplining our emotions (Bhakti Yoga). Or we could follow a *Third Way* of transformation by educating and disciplining our intellect. (Jnana Yoga). Gurdjieff's plan was to combine all three and call the combination the *Fourth Way*. As such, Gurdjieff's approach was unique and in a way independent of what had come before. All four ways have this in common: they are all methods used to sustain attention. They are each a way to study and control the process called *mind*.

Gurdjieff taught that human beings have three minds—he called them *centers*—that must be in balance. There is a *head-mind* where knowledge is processed; a *heart-mind* where compassion flows, and a *gut-mind*, where wisdom resides. Modern men are almost always out of balance; usually one of the minds is dominant and inhibits the others. For example, intellectuals can be unbalanced if they lack intuition and compassion; emotionally intelligent people can be unbalanced if they lack practical experience; and people with great wisdom can be unbalanced if they lack sufficient knowledge. The problem that faced Gurdjieff was that most human beings believe they understand themselves. They have a story about their life that they repeat until it becomes a personality. This personality is actually false—all of us, according to the *Fourth Way*, are more or less unaware of our real self.

We can take any abstraction, any cognitive concept, and examine it from three perspectives. For example, if we explore the concept/emotion called *love*, we find that there are three kinds of love: instinctual (biological) love, emotional (compassionate) love, and egocentric (head-bound) love—which needs to attach (identify) with the loved one. Gurdjieff taught that true love was a balance of all three—unbalanced love is weak and cannot endure.

In his youth, Gurdjieff traveled (with a group called *Seekers after Truth*) from esoteric school to esoteric school throughout Asia. He gathered knowledge and practiced various disciplines until he felt that he understood the objectives of these mystical schools. Ouspensky, who was on his own spiritual journey when he met Gurdjieff, translated Gurdjieff's ideas into books, which became popular among people studying the evolution of consciousness.

It is important to see the historical context in which these two souls lived. Both men lived through WWI, then the Russian Revolution, and then

WWII. Chaos, cruelty, and madness constantly overshadowed their young lives. It is little wonder that Gurdjieff felt that men would eventually destroy the earth if something wasn't done. Therefore, Gurdjieff became a man on a mission. He believed that the planet would be destroyed if human beings did not develop higher levels of consciousness, and he felt that he knew how to help souls transcend to a higher spiritual plane where peace and love overpowered chaos and hate.

One of the esoteric disciplines that Gurdjieff found and used in his own teachings was the Enneagram—as it is called today. I felt that my discussion of the mind, especially my focus on dual-process theory, would not be complete unless I provided this small section about the Enneagram, a very complex subject, which I can only introduce here.

On a simplistic level, the Enneagram in its present incarnation can be understood as an ego-personality scale. It is based on an ancient diagram—a geometrical configuration called the Enneagram. Various questionnaires are used to determine where on this diagram your personality (a combination of three numbers) is depicted. A primary number is assigned (arrived at through a process) which corresponds to a personality type. For example, I am a *seven* on the personality scale—a happy, adventurous person who avoids conflict. This is an incomplete and yet strangely true depiction of my behaviors. However, analyzation quickly gets complex as the many levels of sophistication of the Enneagram are employed for self-understanding. A uniquely personal picture eventually emerges as we apply the logic of the personality Enneagram. On the other hand, there is another way to think about this so-called personality tool.

The Enneagram, besides being an ego-personality scale, also represents a process; it is a *symbolic depiction of movement and evolution*. All human beings are somewhere along the path of physical and spiritual development, caught in the unceasing march of evolution. G. I. Gurdjieff and P. D. Ouspensky felt that the Enneagram could only be understood through participation in movement experiences. The diagram that is the foundation of Enneagram wisdom cannot be adequately grasped without comprehending the significance of relative motion. This idea of the Enneagram as a method for transformation and as a process for self-understanding came long before its use as a personality scale.

The idea that the historic Enneagram[15] (as a process related to movement) could also be used as a tool for understanding inner experience and personal evolution came only in the 1970s through the work of Bolivian Oscar Ichazo (born 1931) and Claudio Naranjo (born 1932), a Chilean-born psychiatrist. Ichazo's esoteric journey began in his youth. As a teenager he used the hallucinogenic plant Ayahuasca to counter epileptic seizures (catalepsy). He got more than a remedy for seizures from the plant; Ichazo also got out-of-body experiences, a sense of unity with nature, and a thirst for esotericism—he went on a spiritual journey for the rest of his life. Along the way, he studied Gurdjieff, Ouspensky, Rudolf Steiner's theosophy, Pythagoras, the Kabbala, hypnosis, and spiritual practices like yoga and meditation. Ichazo's creation of the personality Enneagram came after a long period of seeking and studying.

Claudio Naranjo studied Ichazo's new interpretation of the Enneagram, eventually bringing his (Naranjo's) own insights into the dialogue. Naranjo studied at Harvard and was also a Fulbright scholar and a Guggenheim fellow. He was also an expert on hallucinogenic plants, like Ayahuasca; he probably experimented with the drug (my speculation). Naranjo's career centered on human typology, which is the study of human differences—ways to classify and differentiate personalities. Therefore, Ichazo's use of the Enneagram as a personality tool was especially relevant to Naranjo's academic interests.

Ichazo and Naranjo were well suited to create the revolution that became the Enneagram movement which is still sweeping around the world. In other words, these two pioneers understood at a deep level the esoteric and mystical roots from which the Enneagram emerged. They felt that their interpretation of the Enneagram was a natural outgrowth of the ancient Wisdom Tradition. Unfortunately, but not very surprising, the early pioneers, Ichazo, Naranjo, and a few other early proponents, ended up embroiled in personality conflicts over who should get credit (fame and financial reward) for the emergence of the Personality Enneagram. The worst features of the egocentric mind ended up dominating.

Whatever the complex history as a personality analysis tool, the Enneagram became a worldwide phenomenon and is now a sophisticated instrument for studying the human mind. However, from my perspective, it is in connection with movement and with levels of consciousness that the Process Enneagram holds a key for understanding the operation of the dual human mind.

In his lectures, Gurdjieff essentially stated that a person's level of consciousness determined how the Enneagram was understood. A person with a low level of consciousness would have a simplistic comprehension of this complex set of interwoven archetypal symbols, while those with higher levels of cognitive sophistication *would* be able to see the complexity inherent in the diagram.

It is fascinating to realize that the Enneagram is used in these two ways: as a process for transformation—moving through levels of consciousness— and as a personality scale that looks at character types and their motivations. Although it is not stated in articles and books, the Enneagram has become a set of methods for observing and studying our two minds. The allocentric mind is concerned with flow, with the awareness of transformational experiences, with expanding awareness, and increasing interconnectivity. This is a match for Gurdjieff's Process Enneagram. However, the egocentric mind freezes motion (and people) into entities. It is not concerned with process. It is concerned with labeling, categorizing, comparing, and analyzing—with fixed patterns and with sequential events. Therefore, the Enneagram as a personality scale—based on Oscar Ichazo's school of thought—correlates with the evolution of the egocentric mind.

As I mentioned above, Gurdjieff traveled through Asia as a young man searching for the sacred knowledge hidden in Eastern esoteric traditions.[16] Essentially, he was looking for the fundamental practices of the Wisdom Tradition as it was understood and practiced in Eastern religions. On one of his adventures in a remote mountainous region of Turkey, Gurdjieff discovered the Sarmoung brotherhood. In the film *Meetings with Remarkable Men*, based on a book by Gurdjieff, the Sarmoung brotherhood is called the most ancient of the secret schools—the order was founded in the year 2500 BC. It was thought to have disappeared in the 6th century, but Gurdjieff found it still alive and still passing on the oldest secrets. While staying with the brotherhood, he was taught sacred dances—sacred movements—used to raise an individual's level of consciousness.[17]

Sacred dance uses a movement language that is thousands of years old. Each movement sequence is a letter in an alphabet of movements. The sacred movements can then be used to spell out emotions, or truths, or non-verbal stories. This was a way, like oral history before the advent of writing, to pass down wisdom from one generation to the next. The performer is required to balance two opposing forces that flow within the body. When these two forces are brought into harmony, a force is released that contains tremendous power. That power can be used to control allocentric awareness and egocentric attention. Whereas oral history could pass on the verbal knowledge of a culture—the egocentric heritage of a time past—movement language could pass on experiential wisdom, the knowledge-base of the allocentric mind. Sacred dance gains power when groups perform; it is the gestalt and the animated relationships between the dancers that becomes the story. Gurdjieff experienced and learned this sacred language and then he taught others to "speak" this unusual language in his various schools.

After a series of life-altering experiences as a seeker, Gurdjieff established his first school in Russia in 1912. At this school, he taught the *Fourth Way*, which was the study or science of *being*. This exploration of *being* used a variety of methods to *wake up* the initiates. Gurdjieff also spoke of another kind of intelligence—unlike how the ego defined higher cognition. This new form of intelligence (what I call *wisdom*) is a whole-body knowing, not an egoic head-bound pattern-categorization system. In other words, what Gurdjieff discovered—what the Wisdom Tradition taught—was that we had two minds, two kinds of consciousness, one of which was hidden. Gurdjieff spoke of "i am" (the egocentric mind) and "I Am" (the allocentric mind). Of course, "egocentric and allocentric" are modern terms, not used by Gurdjieff or Ouspensky.

The Russian Revolution drove Gurdjieff out of the country and he eventually ended up in Paris, France. In Fontainebleau near Paris, he set up a school called the *Institute for the Harmonious Development of Man*. People came to the school from all over the world to learn how to awaken the hidden mind—the "other kind of intelligence." The primary method used for awakening the hidden mind was called *gymnastics* by Gurdjieff. This terminology had to be changed when the school was moved to France. The French language did not have the right connotations for what Gurdjieff intended.

Gurdjieff taught his students movement routines, which were essentially an allocentric language expressed through sacred dances. Gurdjieff taught that human beings moved (danced) the same way they lived. There were those who could flow with grace and rhythm and there were others who were flat-footed and stilted. The body restrains (imprisons) and defines a certain kind of adaptive personality. Gurdjieff could presumably "read" people by observing their movement habits; he learned to categorize people into variously "imprisoned types." Movement patterns of individuals have been studied by modern Enneagram researchers and they find that specific energy signatures correspond to certain personality types—research is catching up to Gurdjieff's insights.

As I pondered Gurdjieff's contributions to the study of mind, I began to see that what he found in the ancient traditions, and what he passed down through his students, was what I am trying to tell you here, in this book: the animation that comes from proprioception holds the key to understanding dual consciousness. Movements that we make become habitual and obsessive. Eventually, these movements (gestures, postures, flow-patterns), define who we appear to be when others observe us—*we are the way we move*, the way we animate. Personality types can be "read" from these repetitive, automatic, deeply ingrained movement behaviors. Gurdjieff could look at a person and get a pretty good read on their character. Understanding proprioception is a key to using the Enneagram.

Human bodies are shaped by the stresses and restraints of modern society. We are twisted and frozen, our faces set like masks by the mechanical cultures we are born into. In the language of dual-process theory, our proprioceptive system has become impaired by the artificial, unnatural forces of modern societies. In other words, our allocentric mind—the source of animated flow and interrelatedness—has been crippled by modern society, driven into hiding. Learning about sacred movements and experiencing scared dance is healing and freeing.

Gurdjieff also saw that the world of music—the beats, harmonies, melodies, rests, and accents—matched the rhythms of sacred movements. With Russian composer Thomas Alexandrovich de Hartmann, Gurdjieff created music with the same frequencies as the sacred movements. Music written for the sacred dances was calculated (purported) to help transform individuals from a lower to a higher state of consciousness. This music is available on the web at: https://lightintheattic.net/releases/2905-the-music-of-gurdjieff-de-hartmann. Here is a quote from this website that provides more detail about the relationship between De Hartmann and Gurdjieff:

> The music of Gurdjieff/ de Hartmann is the result of an extraordinary collaboration between the Greek-Armenian spiritual teacher, G. I. Gurdjieff and Russian composer, Thomas de Hartmann. Gurdjieff traveled for twenty years in the Middle East and Central Asia to discover and develop the teaching which now bears his name. Medi-

tative and mindful, Gurdjieff's music stems from Eastern melodies and music he heard in remote monasteries.

From 1923 to 1929, Thomas de Hartmann worked closely with Gurdjieff at his *Institute for the Harmonious Development of Man* outside Paris, translating into European notation the music Gurdjieff composed from his travels. The original tapes containing these tracks were recorded in the 1950s under informal circumstances with rudimentary equipment, never intended to be heard by the public. As for the instrumentation, the performance is stripped down to nothing more than a single piano (played by de Hartmann) but lacks absolutely nothing; rather de Hartmann uses the negative space between notes to revel in resonance, in turn capturing remarkable depth and meaning.

~ *LIGHT IN THE ATTIC*, Thomas de Hartmann, The Music of Gurdjieff /de Hartmann

Thomas De Hartmann and his wife Olga collaborated on a book about their relationship with Gurdjieff called *Our Life with Mr. Gurdjieff* (1964). The musical partnership explained in the above quote came about after a decade long journey that the De Hartmanns undertook with Gurdjieff and his wife Julia Osipovna Ostrowska. A small group of friends and followers of Gurdjieff, including the Ouspenskys and De Salzmanns, moved slowly, over a period of years, escaping from the revolution in Russia, eventually ending up in France. All the while that they traveled, Gurdjieff worked with the group on the connections between movement (sacred dances), music, and the evolution of the mind.

Ouspensky and Gurdjieff both emphasized that human beings are asleep on their feet. The level of consciousness of most humans is so low they are dominated by habits, going from task to task like mechanical robots. In other words, the movements of sleepwalkers are automatic, robotic, habit-driven, not fluid and not under self-control. In this sleep state, they cannot self-observe—their true self is hidden from them. Importantly, for our discussion of dual-process theory, *the Fourth Way is a set of methods used to develop and control attention and awareness*—a method for waking up.

The key is attention. Gurdjieff is asking: Are you a head-bound creature, or are you embodied? How are you paying attention? The scared movements (very much like tai chi, qigong, and yoga) expose our duality. Precisely-controlled movements and unusual postures inhibit the egocentric mind so that the allocentric mind can be experienced and validated. In their book *Our Life with Mr. Gurdjieff*, Thomas and Olga De Hartmann provide this insight into the importance of attention:

> One feature which all the exercises have in common is that they require all our attention and so avoid the flow of uncontrolled associations which waste our life energy through very stupid, sometimes very painful, sometimes fantastic, sometime erotic thoughts, feelings and sensations, which we more or less experience. Mr. Gurdjieff frequently said that "intentional suffering" and "conscious labour" by reducing this unconscious flow of associations could prolong life. For those who work on attention and use it in the struggle with associations, who do not forget to "remember themselves"—for those people attention begins to be not only the center of life, but also the factor which lengthens it.
> - *OUR LIFE WITH MR. GURDJIEFF*, Thomas and Olga De Hartmann, 1964.

There is much wisdom embedded in this quote. To *self-remember*—a favorite phrase of both Gurdjieff and Ouspensky—means to use allocentric awareness rather than egocentric cognition. In other words, we should remember moment-to-moment that we are not just an ego; we are also a self. The ego is lost in a sea of anxiety, grief, dread, anger, and hope, an emotional stew that causes the mind to flow from one association to the next. While this internal monkey mind is doing its thing, there is a complete shut down in awareness of the environment, including the body and other sentient beings in the surround. A person who is self-remembering is aware of personal body movements (animation), the body movements/animations of others, and what is happening in the surround. Evidently, those who are able to control animation, as well as the flood of emotions and the resulting stress, lead happier, healthier, and longer lives than the rest of us stressed out types.

Gurdjieff taught what he called "attention without tension." In my book *Consciousness: A New Slant on an Old Conundrum*, I differentiate *awareness* from *attention*. *Awareness* is "embodied experience in the moment;" it is the system used by the allocentric mind to know and react to a relational world. *Attention* is the egocentric mind's way to perceptually study, categorize, and recall invariant patterns. "Attention without tension" is really *awareness*. The Wisdom Tradition that Gurdjieff drew his knowledge from is (in modern terminology) just the contrast between, and the study of, our egocentric and allocentric frames of spatial reference.

It is also helpful to think of attention and awareness as ways to move energy around. Attention draws energy to a location. If that location is within the body, then energy can be focused where it can reduce stress, heal wounds, and soothe emotions. This ability to move energy around is the basis of Qigong and the martial arts. As we move energy around, we are shaping, controlling, and fine-tuning animation (the proprioceptive system). In other words, our levels of consciousness can be altered through energy manipulation.

The Enneagram, in the *Fourth Way* tradition, first appeared in print in P. D. Ouspensky's 1949 book *In Search of the Miraculous*. "Enneagram" comes from the Greek words ennea (nine) and gramma (written or drawn).

The Enneagram was first introduced, as a method of self-study, by Gurdjieff to his students in St. Petersburg and Moscow in 1916. Gurdjieff told his students that the Enneagram was an ancient secret, the truth behind the Philosopher's Stone. Ouspensky quotes Gurdjieff as saying; "In order to understand the enneagram it must be thought of as in motion, as moving. A motionless enneagram is a dead symbol." In other words, process—unceasing and relative motion—is at the heart of the Enneagram. "Unceasing and relative motion" is another way to say "animation."

In the lore of alchemy, the philosopher's stone is a chemical substance capable of turning base minerals into valuable metals like gold or silver. It was also said that this alchemical substance could extend one's life and so was called the elixir of life. This powerful substance could rejuvenate and open the doors that led to enlightenment. Another way to think about the philosopher's stone is that it was a set of methods for transforming a base level of consciousness into higher forms of consciousness. It was a way to reveal the allocentric mind to the egocentric mind so that a balanced spiritual (internal) dance could be experienced.

Two fundamental processes are depicted in the Enneagram diagram according to Gurdjieff: *The Law of Seven*, also called the Law of Octaves," and *The Law of Three*.

The Law of Seven is a relational process. Everything is moving relative to everything else in invariant patterns—that is the Law of Seven. We assign different words to this relative movement. For example, words like *frequency*, *vibration*, and *energy* often substitute for "relative patterned-movements." There is a movement fractal at the heart of existence. Everything is created from this fractal. The Enneagram symbol depicts the flow of this fractal as it creates.

The Law of Three states that all phenomena are produced by three forces. Catholics call the three forces "The Father, The Son, and the Holy Spirit." In dual-process terminology, there is a background potential out of which manifestation (creation) occurs in the form of figures or patterns. The third force is what causes manifestation and which also causes a return to the background. Human beings can conceive of backgrounds and foregrounds, but they are blind to the third force—that is why the Holy Spirit is sometimes referred to as a (Holy) ghost. You can feel its presence but your senses (vision and hearing especially) cannot verify anything solid. Gurdjieff used a simple analogy to explain the Law of Three. He said you need three things to make bread. The basic ingredients must be correct; you need flour and water. However, a third force is needed for the creation of bread. That force is heat.

The Enneagram symbol incorporates and unifies the Law of Seven and the Law of Three. The Enneagram symbol can be used to describe (understand, study) any process. For our purposes, the Enneagram is a pictorial, symbolic way to envision patterns of animation that result in levels of consciousness.

Before I conclude this brief look at the ideas of G. I. Gurdjieff, I want to make sure you understand the key starting point for Gurdjieff's philosophy. In his book *Beelzebub's Tales to His Grandson* (1973), he is only a few pages into his dialogue when he asserts the importance of cognitive duality:

> In the entirety of every man, irrespective of his heredity and education, there are formed two independent consciousnesses which in their functioning as well as in their manifestations have almost nothing in common.
> ~ BEELZEBUB'S TALES TO HIS GRANDSON (1973)

After years of studying and practicing esoteric teachings, Gurdjieff concluded that we have two mutually exclusive minds. He uses the terminology of his time to explain these two minds, calling the subconscious our real mind and the conscious ego-state our small mind. We think we are the egoic mind, Gurdjieff tells us, and so we miss our real essence—we don't even look for it because the ego is so sure that it is the only act in town.

The best books I found about the *Fourth Way* and about Gurdjieff (I used them to craft this section) were authored by William Patrick Patterson, who spent his life studying and promoting the *Fourth Way*. I listed Patterson's books in the bibliography and recommend them to you. I also enjoyed the books of C. S. Nott, Thomas and Olga De Hartmann, Rodney Collin, James Webb, Jeanne De Salzmann, and Kenneth Walker, former students of Gurdjieff and Ouspensky.

Gurdjieff died in October, 1949. His friend Thomas De Hartmann wrote the eulogy and gave it to the Russian Priest who was performing the ceremony. De Hartmann deliberately concluded the eulogy with words from Gurdjieff's ballet *The Struggle of the Magicians*, which Gurdjieff wrote in 1914 and staged many times in various cities in Russia, Europe, and the United States:

> *God and all his angels*
> *keep us from doing evil*
> *by helping us always and everywhere*
> *to remember our Selves.*

Consciousness is Evolving

> *The place that Solomon made to worship in,*
> *called the far Mosque,*
> *is not built of earth*
> *and water and stone,*
> *but of intention and wisdom*
> *and mystical conversation*
> *and compassionate action.*
> ~ "THE FAR MOSQUE," *A YEAR WITH RUMI*, 2006.

It is not surprising that consciousness, as a subject of study, has such a short history. This is because consciousness is evolving; people in earlier cultures had lower levels of consciousness—levels that did not lend themselves to self-reflection. Indeed, the very awareness that there are personal and cultural levels of consciousness requires a sophisticated mind that did not exist until very recently on the evolutionary journey of mind. We also did not have the technology or research expertise to probe levels of consciousness until quite recently—and these technologies are still rudimentary.

The notion that we have two minds is very old and can be seen in mythology, folklore, philosophy, and religion. However, the physiological and anatomical evidence supporting duality is only now arriving. Furthermore, few experts in the field of consciousness studies have taken a look at dual-process theory, and there has been no evolutionary cause put forward for our duality except for hemispheric specialization, which is a manifestation rather than an explanation.

In a general way, we can say that those who have pondered consciousness fall into two camps. Given our two minds, this is no surprise. One camp argues that consciousness can be understood through reductionism and science: for this group of materialists, it all comes down to synapses, genetic codes, molecular bonds, particles and waves, frequencies and probabilities, all of which, when properly assembled, explain the whole thing. The other camp, using experience as a guide, argues that consciousness is about qualia, experiential meaning, ineffable forces, and an ultimate wisdom behind evolution.

Some thinkers believe that consciousness is mostly about language, others that the evolution of social interactions crafted our minds. A few, including myself, think that navigation—purposeful movement—is a key understanding. The ego can be reduced and dissected while the spiritual mind can only be experienced. Each is true. The paradox has to be accepted. We have a physical right-side body and a physical left-side body seamed together at the midline. We can argue the left-side body is the more real of the two or that the right-side body is the more real of the two. The truth, of course, is that they are both equally real. We are crafted by, and in the image of, a universal duality.

Keeping in mind that consciousness is an evolving duality, let's next explore the confusion caused by being your own twin as it is discussed in poetry, religion, philosophy, psychology, science, education, and culture.

Notes

(1) People will purchase any book with "consciousness" in the title. This interest in consciousness is wonderful, hopeful, and is evidence of both individual and cultural progress. Individual minds are slowly evolving. This will raise the level of our collective consciousness. The world will get wiser, more intelligent, more tolerant, more cooperative, and more compassionate as decades pass. Or so I suppose. We shall see.

(2) I am well aware of the potential for self-righteous arrogance buried within this perspective. I imagine myself in a biker bar explaining to some guy with tattoos on his bulging muscles that he has a low level of consciousness. Before I have time to go over the ten pages of caveats—it's not about intelligence and self-worth, for example—I get punched in the face and lose a few of my favorite teeth. Like any classification system, we have to be careful that it doesn't turn into an excuse for the egocentric mind to display itself—like a male peacock showing off its feathers. However, levels of consciousness are amenable to change; we can each evolve through personal effort.

(3) Unified with a mental watcher, the witness: In *Consciousness, A New Slant on an Old Conundrum* (2017), I proposed that consciousness was a proprioceptive phenomenon, a memory of memories. In that discussion, I suggested that the ability to witness your own thoughts was a memory of internal dialogue. I also proposed that eventually there would have to evolve a watcher of this witness—and perhaps an infinite regress of watchers of watchers as consciousness becomes ever more complex and awake.

(4) Hive-culture: Designating the Tribal level as a hive-culture and Warrior Man as a clan-culture simply felt right to me. I did not do research to determine if these speculations were entertained in the developmental literature. I am having fun as I let my allocentric mind loose on the page. I leave it to the academics to deny or defend my allocentricity. I am riding the waves of a theory.

(5) Tooth Fairy and Tinkerbell Goddesses: These are my just-for-fun deities. Feminine forces arise from *magic musing* mixed with self-effacing humor. Such musings bewilder the egocentric mind. The allocentric half of this book is fine with musing and playfulness. The egocentric half is trying to make this a textbook—and so the ego is impatient and critical.

(6) George Orwell was a pen name for English writer Eric Arthur Blair (1903-1950). Blair was an English novelist, a journalist, and most importantly he was an articulate social critic. His novels, especially *Animal Farm* (1945) and *Nineteen Eighty-Four* (1949) were biting satires that described an authoritarian future driven by impersonal and invasive ideologies. After the 2016 elections in the United States, Orwell's novels again became best sellers. *Nineteen Eighty-Four*, about a totalitarian state run by *Big Brother*, surged to the top of Amazon's best-seller list on the sixth day after Donald Trump became president. Orwell's dystopian vision highlighted the madness (and strategies) of authoritarian regimes, and several phrases from the book entered the common lexicon: *double think, newspeak, thought crime*, and *memory hole* are famous examples. Sophisticated modern readers could see the parallels between these phrases and the language of extreme politics. Political extremists, during the run-up to the 2016 elections, used new technological tools to craft a deliberate strategy that looked and felt dystopian. These extremists used propaganda, fake news (truth decay), manipulation of social media, deliberately placed lies, orchestrated events, the discrediting of journalists, anti-intellectual rhetoric, distrust of innovative and creative people (artists of all kinds), and a meme-driven strategy to manipulate the minds of people on the lower rungs of consciousness. Then, of course, it was all helped along by the Soviet Union's (Vladimir Putin's) hatred for Hillary Clinton.

(7) Wayne O'Brien is a friend and professional confidant. Wayne founded the esoteric study group at Saginaw Valley State University. This group of like-minded seekers has nurtured my soul for many years. I am deeply grateful to Wayne for his intellectual leadership and for his gentle and loving personality.

(8) Exponential doubling appears to be a fundamental fractal of our universe. Ray Kurzweil has written about exponential doubling in his many books, including *The Age of Intelligent Machines* (1990) and *The Singularity is Near* (2005).

(9) In my book Consciousness, A New Slant on an Old Conundrum, I also discussed another kind of exponential doubling inherent in our very genetic nature. See Chapter Seven in *Consciousness, A New Slant on an Old Conundrum*: "Quantum Science, Duality, and Consciousness" to review this detailed discussion.

(10) Syntropy is the life force that runs counter to entropy. Entropy is the universal force that constantly deconstructs, disassembles, and chills creation. Syntropy constructs, assembles, and warms creation. Entropy brings death and syntropy brings life. It is as if syntropy is forward motion—the evolution of consciousness—while entropy is backward motion, the evolution of the "unconscious," to use Jung's terminology.

(11) As long as the exponential rate of change keeps going, we can expect more and more enlightenment. Of course, my center of gravity is between 4 and 5 on Cook Grueter's scale—enlightenment is a vague concept in my worldview. In other words, be careful and don't take my speculations as solid fact. I am just following a theory as I explore logical consequences of a line of thought.

(12) There is some kind of cultural heredity (**epigenetics**) available to each new generation, as if our kids are born with potentials that we did not have at birth. I arrive at this observation by looking back through recent history. The cultural level of consciousness in each century is relatively shallow compared to where we are today—something is driving the mind of human beings to evolve at an exponential rate. Normal genetic heredity and the mechanisms of DNA cannot account for this substantial and immediate (generation-to-generation) evolution in the development of consciousness.

(13) This is a rather significant understanding for educators to comprehend and employ as they nurture the young. It is perhaps more significant at the junior college level or when dealing with college freshmen (or high school kids). You have to bring a less authoritarian I-know-best attitude to a classroom filled with minds poised to surpass your own.

(14) On a bad day, I sink below 4. For example, during the election year of 2016, when Donald Trump and Hillary Clinton dueled for the presidency, I fell to level 3 as I took to social media to rant and rave. I got caught up in the excitement of electing the first woman president. Mr. Trump seemed bombastic, ill-tempered, and unqualified for the highest job of the land. Mrs. Clinton was exceptionally qualified. This was a choice that was so clear it was inconceivable that Donald Trump could be president. I flooded Facebook with posts from comedians, past-presidents, college professors, and experienced journalists who supported my viewpoint. My level 3 ego was stunned when the final votes showed that Donald Trump was America's 45th president. I had hoped the American population had reached a level of consciousness that would champion feminine energy: the rise of the allocentric mind, a decline of the patriarchal views of the past, and an insistence on compassion and cooperation. Unfortunately, half of the American population is on the lower rungs of the evolution of consciousness. This is a serious threat to democracy and to the planet. My own behaviors during the election were not stellar; obviously the bombastic, self-righteous approach has a few flaws. On the other hand, bombastic, hateful, irrational memes *actually did* get an unqualified candidate elected.

(15) **The Enneagram** is a complex tool for self-understanding and for personal evolution—multiple pathways can lead to higher levels of consciousness. The Enneagram uses a geometric pattern to give an image-based understanding of behavioral evolution and interconnectedness. This Enneagram figure is composed of three parts; a circle, an inner triangle, and an irregular hexagon. In esoteric studies, the circle symbolizes unity, the triangle symbolizes the *Law of Three*, and the hexagon represents the *Law of Seven*. I was introduced to the Enneagram by my friend (and editor) Karen Horwath. Karen also introduced me to the works of Franciscan Priest Richard Rohr, who is also an authority on the Enneagram. However, only after Karen directed my attention to Susan Rhodes' book *The Integral Enneagram* and then *The Spiritual Dimension of the Enneagram: Nine Faces of the Soul* by Sandra Maitri, did I begin to see how the Enneagram was connected to levels of consciousness.

(16) I am drawing some of this discussion about Gurdjieff from the website of Chandrakala, a Canadian-based spiritual leader with extensive experience teaching Gurdjieff's sacred movements. She is also a former teacher at the *Osho Center* in Poona, India. I also used a website called *The Gurdjieff Legacy Foundation*, which led me to books and archival material about Gurdjieff's life and philosophy.

(17) **Sacred movements and dances** are, of course, an integral part of the Muslim esoteric tradition, most notably within the Sufi Mevlevi tradition. The whirling dervishes came from this tradition. The Mevlevi tradition was started by the great spiritual leader and poet Jalâluddîn Rumi.

Three

Poets and Duality

An empty mirror and your worst destructive habits,
when they are held up to each other,
that is when the real making begins.
That is what art and crafting are.
- "Childhood Friends," *A Year with Rumi*, 2006.

Poetic Meals

"Got any poetic meals, Surge?"
"You're kidding, right?"
"Why would I be kidding?"
"Because poetry is the essence, the main point, the issue at hand. The Third-Eye-Watching Café is a holy poetic space. I am a poetic waiter. You are a poetic writer. The reader is a poet. Poetry is the joy of experiencing and expressing the mysteries. *Of course*, we have poetic meals. Haven't you been paying attention to the menu?"
"I don't care for poetry, actually; usually I don't understand it."
"Get me my heart medicine. Where have we failed, Dutch? We put eons of effort into your creation and then you walk in here with that mindless grin on your face and confess that you are allergic to poetry. A sea of poetry all around you, poetry in every molecule, and yet you don't perceive it. God have mercy on the dead poets who tried so hard."
"Okay, sometimes I like poetry. Most of the time, I like the poems of Rumi. On the other hand, I often don't often get what Rumi is saying, for example, about empty mirrors and destructive habits. That's why I break out in hives. I feel dumb when I don't understand what poets are trying to express."
"Face it, Dutch, you are deaf, dumb, and blind—that's just the way it is, the bond you share with your kind. That's why poets are needed—to challenge

171

minds like yours. Anyway, we have just the meal for poets who don't know they are poets. It's called Baked Blah-Blah for the Half Wit. It comes in a wooden soup bowl that has been through the dishwasher once too often. We bake the Blah-Blah too long, at a ridiculously high temperature, and then we plunge it into lukewarm salt water. It is then served with stale saltine crackers and raw, slimy seaweed. But here's the catch. It tastes delicious, even though it looks unappetizing. Patrons are stunned by the contrast between the preparation and the tasting. Our customers arrive prejudiced and foolish, yet they depart in a poetic trance."

"Will I like poetry if I taste it more often, Surge?"

"Why don't you give it a try, Dutch? Look in the mirror at your worst destructive habits."

"I don't have any bad habits."

"Yeah, sure you do. Just hold that face-in-denial up to the empty mirror—see how reality stares back through you."

"All I see is my face."

"Keep looking."

"Nothing yet."

"That's it, Dutch. The face that denies its bad-habit energy is the face expressing and defending bad habits. That face that says, "I only see my face looking back at me" is the face of denial—that is your primary bad habit, Dutch. The root of your bad habits is the denial of bad habits. Fill your mirror with kindness to your essential *self*. Fill your mirror with the face of resolve."

"Okay. I'll try."

"There is no try. Do or don't do. Eat well or don't eat well. There is no diet, Dutch. Get yourself an empty mirror and begin the poetic journey."

Colin Wilson

Why should I describe the moon
coming up over our sleeping quarters?
~ "This Recklessness," A Year with Rumi, 2006.

In the first two chapters, we explored levels of consciousness and the dual-nature of human cognition. Many people have explored these concepts, including poets, philosophers, and psychologists. Indeed, it is hard to find any area of study that does not include an awareness of duality and a sense that consciousness is a process with many levels of complexity. The terminology used in each discipline is different, but a close look reveals that the same concepts are being continually re-discovered and assigned profession-specific terminology.

The rest of this book looks at these various disciplines and how they each discovered and explored dual-processing. In other words, every field of study in some way differentiates egocentric processing from allocentric processing. Every profession is as concerned with intuition, love, peace, wisdom, and mindfulness—the ingredients of the self—as it is with logic, laws, and analysis, the ingredients of the ego.

In addition, each set of dualities has an inherent understanding that accompanies the insight of duality: there is an allocentric background entity—the older or original mind—out of which foregrounds (egocentric "objects") manifest and then return. The final insight is that in every case there is an oscillation between one polarity and the other—this is a necessary oscillation which results in mutual exclusiveness.

In summary of the above, all disciplines of study find two minds which oscillate; these minds are mutually exclusive—when one is on, the other is off—and the two minds can be framed in terms of background and foreground.

Those who have seen our inherent duality have spoken out repeatedly to make the point that logic and evidence support the allocentric mind as an equal partner to the egocentric mind. Each field of knowledge insists that we cannot lose sight of love, joy, and wisdom, in our rush to be logical and technologically advanced.

No field of study and expression has so clearly and emphatically insisted on the equality of the ego with the self, as the discipline of poetry. Therefore, I will begin with the poets and with a champion of dual-consciousness, the prolific British author and poet Colin Wilson. After Wilson, I will discuss the poetry of Theodore Roethke, Mevlana Jelaluddin Rumi, and Owen Barfield—all of whom explored humanity's inherent cognitive duality.

Colin Wilson (1931-2013) was a writer and a philosopher who wrote about mysticism, the paranormal, and consciousness. He wrote over a hundred books during his prolific career. Wilson coined the phrase *New Existentialism*, which was his attempt to portray Existentialism as optimistic and positive, rather than dreary and noir. In the chapter on philosophy, I will look in detail at Existentialism. For the moment, I will simply point out that you cannot explore Existentialism without finding yourself knee-deep in cognitive duality.

In his book, *Poetry and Mysticism* (1969), Colin Wilson muses about dual consciousness. Wilson begins the discussion by making the important observation that consciousness shrinks to fit whatever time and space it finds itself within. Most often that space is familiar and is filled with duty and routine; therefore, most of the time consciousness is a boring affair that takes place in a cubicle or a small rectangular room in a house or office building. Outside under the stars (in nature), consciousness expands and the soul sings.

Poets hate the feeling that life is boring. Every fiber of their being screams at the mundane—they hate cloistered routines. Poets attack the ennui-creatures that live in small spaces, in small minds, in minds made dull and sick by habit. For a poet, life is stale, thankless, and even worthless if it is only filled with mundane responsibilities and cognitive repetition. Poets do not like to be boxed in.

Poets also know from experience that deeply felt emotions can arise from nowhere and for no-reason. At unexpected moments—usually triggered by some trivial happening—comes what the poet G. K. Chesterton called *absurd good news*. For no obvious or earthshaking reason, there are rare fleeting moments when we suddenly feel inexplicably wonderful, full of life, buoyant no matter what space we are in. We are suddenly calm, happy for no reason, jubilant to be alive as we experience a gifted-moment of bliss. We discover what the word *joy* feels like.

This contrast between the rare moments of absurd good news and our mundane every day, grinding-it-out consciousness is the duality that Colin Wilson discusses in his books. Poets feel the contrast between joy and boredom. They know that life can be exhilarating, passionate, on fire with joyfulness. They also know that most people are asleep as they go through the motions of everyday life. However, poets have learned to walk on the edge of insanity, at the border between the known and the mystery. According to Colin Wilson, poets are called to use their craft to awaken sleepwalkers.

Once a person has felt absurd good news—what Wilson calls an *intensity experience*—then they know something, a secret. They know intuitively, because they have been in that absurdly wonderful space, what the Buddhists call enlightenment. They know that enlightenment is real because they have glimpsed this miracle. If you are a poet and you experience just one of these moments, you spend the rest of your life searching for the next high, and the next. Then you try to capture that high in poetic language; you feel that you must share the gift in your poems—you are driven to paint a word picture of absurd, sudden, fleeting goodness.

Or, on less hopeful days, poets vent their anger and frustration because life seems more about aging, illness, angst, and fear; the absurd good news is a tease laughing at pitiful human poets as they stand waist deep in mental mud. Poets are allocentric creatures struggling like moths to leave the cocoon; they *feel* too much, too often. Here is how Colin Wilson puts it:

> . . . the poet, whether he knows it or not, is the antithesis of the logical positivist or the scientific specialist, since poetry is *by nature* a personal statement that aims at becoming a generalization about human existence. In the poetic experience, the everyday façade melts; the sense of a world that you know all-too-well gives place to a feeling of wider significances that we are normally too brutish and self-occupied to grasp. Whether he likes it or not, the poet cannot take a "specialist" view of nature, in which one single aspect is chosen for study.
>
> ~ *POETRY AND MYSTICISM*, Colin Wilson, 1969.

In the quote above, Wilson is insisting that the poetic mind is an exact opposite of the rational mind—the two minds are mutually exclusive. You cannot be poetic the same instant that you are being rational. The poet is not a specialist studying phenomenon; the poet participates in existence.

According to Wilson, who has had his share of absurd good news, an intensity experience is a moment where judging and being critical has no meaning. Negative, obsessive, tense states of consciousness are replaced by an innocent passivity. We just suddenly feel good, we feel happy; that is all there is to it, period. Moments of absurd good news arise in the allocentric mind. In contrast, our egocentric mind dwells in anxious, boring, routine moments that are confined to all-too-familiar spaces. Absurd good news presents us with a contrast. We see the evidence that we have two minds, one of which, the allocentric, is where "true reality" exists. Wilson (like many others) says that when we see things *as they really are*, we suddenly rejoice. But where does absurd good news come from, and where might the poet search to discover its source? How do we enter into the allocentric mind?

Wilson knew that our duality of consciousness had to do with perspective. He did not have the evolutionary and biological information that we are able to draw on today—he didn't see the connection between navigation and consciousness, nor did he comprehend the allocentric and egocentric minds in scientific terms—but he tried in various ways to explain the perspective:

> . . . "mystical vision" occurs when we get a sudden "bird's eye" view of life; when, for a moment, we "pull back" from it and see more of it, instead of being confined to the narrow focus of our usual worm's eye view. This seeing-more-of-it always brings a feeling of exaltation, delight.
> ~ POETRY AND MYSTICISM, Colin Wilson, 1969.

In other words, to enter into the allocentric mind, we must broaden our spatial awareness. We are at our *least poetic* when we are so ego-bound that our awareness does not extend beyond our body. If we are obsessed with emotions, problem solving, and physical issues, we lose awareness of the room we are in, the building we are in, the neighborhood, the city, the state, the country, the planet, the solar system, the galaxy, and the universe—or, as Wilson says, in his succinct way:

> His power of close-up vision condemns man to the trivial.
> ~ POETRY AND MYSTICISM, Colin Wilson, 1969.

Wilson is saying above that our narrowly focused egocentric attention system actually causes depression if we stay in that state for too long. He says that *close-up depresses* while *wide-angle exhilarates*. Use the allocentric mind and you feel great. Use the egocentric mind for too long, you will feel bad. One of the main jobs of a good poet is to wrench the reader out of his or her small world, to shock, to dislodge, to transform the reader. To wake up a sleeping reader, the poet uses images that broaden perspective, that yank the reader's mind out of the small space where dull ego-consciousness has taken root.

Wilson also understood the difference between attention and awareness. He didn't have the science to understand why this duality existed, but he could see it everywhere in his study of the poets:

What I am speaking of are two modes of apprehension of the real world, two states of consciousness. If I am ill—let us say, in a fever—my consciousness is hardly more than a mirror reflecting my environment, seeing things blankly, without attaching meanings to them. I confuse illusion with reality; dreams seem real, and reality seems a dream. What does this mean? It means that my consciousness and my "awareness" are two different things.

~ *POETRY AND MYSTICISM*, Colin Wilson, 1969.

Consciousness, as used by Wilson (above) and many others, is equated with the egocentric mind; it is a form of perception. But Wilson can sense something else beyond egocentric perception, *something that is aware.* The egocentric mind *pays attention* to a tree, for example, and understands the tree as a feature within an environment. The ego can label the tree, categorize the tree, and describe the tree to others. However, the allocentric mind *is aware* of the tree and feels a relationship with the tree. The allocentric mind is aware of being *with the tree* in a moment of existence. Awareness opens a portal; being awake causes a cascade of poetic emotions and insights to arise.

Wilson makes a very important observation when he discusses the difference between attention and awareness. He says that there is an energy within us that can be called forth; we can direct this inner energy (inner light) using either attention *or* awareness. Most of the time, consciousness is a dual affair because we don't use our inner energy correctly (or at all). If we have not discovered our ability to control our internal energy, if we have not developed the skill of energy-management, we will creep through life, according to Wilson, like mindless zombies:

In short, the difference between "poetic perception" and "ordinary perception" is a difference of chosen impulse, so to speak. I reach out for a tree as I might reach out for a sandwich when I am hungry. The act is as intentional—purposeful—as a boxer hitting a punching bag. And just as a boxer strengthens his arm by hitting the punching bag often enough, so I may strengthen the muscle of intentionality by making it punch out at a tree when I have no particular reason to do so. If I now merely narrow my eyes and concentrate hard on the tree, it becomes for a second more real, and a spark of delight shoots up inside me.

Young people tend to be passive; they hope for something interesting to turn up. On a dull day, they yawn, and the eyes grow dull: another way of saying that the muscle of intentionality is not being used. Inevitably, it gets weaker.

A Martian would find it very odd that human beings are not aware of the simple mechanisms of perception; that they think of consciousness as a mirror reflecting reality. However, once he had this key to human behavior, he would understand [William] Blake and [W. B.] Yeats. If one fails to recognize the simple intentionality of perception, then how does one reconcile the two states of mind—boredom and intensity?

~ *POETRY AND MYSTICISM*, Colin Wilson, 1969.

Wilson is saying that we have control over consciousness. If we are suffocating within a small physical space, a cubicle, a factory, we can choose to go outside in nature where consciousness is only confined by the earth and the sky. We can also move our mental perspective to ever higher levels simply by willing it to be. We are stuck in our egos by habit. We are voluntary prisoners in our own egoic reality. Ironically, we are also the warden of the prison—we have the key to the gate. We can quit this boring job with every breath.

Wilson uses the terms "poetic perception" and "ordinary perception" because he doesn't have an understanding of the allocentric mind (the source of poetic perception) and the egocentric mind (the source of ordinary perception). But he clearly is both a poet and a professor of poetry; he gets the essence: it is *how* we pay attention, combined with *our ability to be aware*, that matters.

Wilson also studied the mystic G. I. Gurdjieff and wrote about Gurdjieff's extraordinary insights. Gurdjieff said that human beings are machines on autopilot; we are sleepwalkers and robots, going about the day with very little awareness. What is needed, according to Gurdjieff, is some kind of shock (reboot) to cause the robot to wake up.

High energy is necessary to keep the allocentric mind active. When our energy level slips below a threshold, we drop back into robotic-mode. That is why meditating monks drive their frequencies into the gamma range—this is a much higher state of being than what we experience in everyday beta-wave consciousness. It is also why humans go to extremes (jumping from airplanes, climbing the highest mountains)—they are trying to raise their energy levels so that they can feel alive and keenly awake.

After writing 100 books, many about consciousness, Colin Wilson concluded that peak experiences come about through both *paying attention* and *being aware*. Furthermore, according to Wilson, the perception that creates absurd good news is a normal part of every human being; we can self-generate the feeling of absurd good news. We can be alive within a self-created aura of joy. We can be mindful, awake every moment if we know how to be *aware* as well as *attentive*. We have two minds, two kinds of consciousness, and two ways to know reality: the way of the poet and saint, or the way of the philosopher and scientist. They are equal powers (the self and the ego), yet the poetic self has been historically subdued by the egocentric mind—a mind that refuses to acknowledge an equal twin.

Colin Wilson, from a philosophical perspective, also knew that there was something beyond the ego that was being missed. He did not use the word *proprioception* and he did not have a navigational perspective, but he was on the hunt—he had a mission—to describe and understand this strange non-egoic mind.

As a young man, full of sexual energy, Wilson came at the investigation through an exploration of man's sexual appetite. He could see that sex was not—or should not be—a cognitive, head-bound phenomenon. Sex, if it is experienced poetically (using the allocentric mind) is a whole-body, totally consuming *union*. The egocentric mind interferes with the sexual experience and dampens the mystery and adventure of union.

We recognize today that poetic sex is proprioceptive and allocentric; Wilson did not have the advantage of these perspectives. But he did see (using my terminology) that sex ignites the allocentric mind and is a form of shock that can cause transcendence beyond the egocentric. For Wilson, sex is an avenue for transcendence, as Hindu devotees of Tantric spirituality know quite deeply.

Theodore Roethke

Great poets are insurgents. They are in revolt against the limitations of reason and logic.
~ *THEODORE ROETHKE: THE JOURNEY FROM I TO OTHERWISE*, Neal Brown, 1982.

179

Theodore Roethke grew up about a mile from where I write these words, in Saginaw, Michigan—I drive by his former home just about every day as I go about my routines. His house is preserved, occasionally open to the public, and he is honored at triennial celebrations where the good citizens of Saginaw read all his poems—a multi-day marathon of appreciation for our native son.

Roethke's family owned a large greenhouse in Saginaw, a business that struggled to survive in the harsh Michigan climate. Winters are often severe here; they are long, overcast, and depressing. Keeping plants alive in this environment requires heroic energy; even so, good intentions are often not enough to sustain the delicate lives of plants. Roethke's whole life was affected by depression mixed at rare times with a manic energy (when the sun made infrequent visits, I suppose). It was as if the weather and the losing battle with the greenhouse business shaped the young boy into the Pulitzer Prize winning poet he was to become.

When you grow up under overcast skies—unable to go for a stroll in the rare moments of sunlight because artic winds howl outside the greenhouse walls—you can become fascinated with light and warmth. Below is one of my favorite Roethke lines; he must have been searching for hope, for evidence that the light had not abandoned us:

Deep in their roots, all flowers keep the light.
- Theodore Roethke's journal

In the dark underground where roots travel unseen, there is a hidden illumination, a carefully guarded life-force. The flowering plants save this colorful light for us. They keep the light safe, hidden underground throughout the long gray days of winter. However, when the short overcast days of winter give way to spring, there is enlightenment: the cold gray earth explodes with colors and intense light. We feel more alive when the light comes out of hiding: we feel free, we love life, and we glory in the beauty of the light. Something inexplicably wonderful—a gift hidden in the light—makes us happy.

Meanwhile, in the cold winter darkness, sensitive poets like Roethke become manic depressives at the whim of the clouds and the arctic winds. Roethke tells us that he "can hear, underground, [the] sucking and sobbing" of the plant souls. "In my veins," he says, "in my bones I feel it." Roethke stares at the plants in the greenhouse as they struggle to grow or fight against decay. He sits down in the greenhouse, stops his analytical mind, and he simply experiences the plants. His technique is "to look at a thing so long that you are a part of it and it is a part of you." (*Theodore Roethke: The Journey From I to Otherwise*, 1982)

In other words, the poetic soul knows the difference between the two minds: you can analyze plant life, or you can smell the flowers. Furthermore, the source of poetry—of all creative energy—comes from the allocentric mind and from experience. The egocentric mind inhibits the joy of experiencing. Here is Neal Brown, author of *Theodore Roethke; The Journey From I to Otherwise*, commenting on this kind of poetry:

> Roethke felt that in this kind of poem, the poet should not "comment," or use many judgement-words; instead he should "render the experience, however condensed or elliptical that experience might be." The result is a line of varied length, sometimes cryptic, filled with aphorisms and enigmatic questions.
> ~ THEODORE ROETHKE: THE JOURNEY FROM I TO OTHERWISE, Neal Brown, 1982.

Here we find plain language, saying it like it is: render the experience and dwell in the enigmatic mystery of being alive. Don't dissect the innocent flowers or render judgment about their complexation or stature. Let them be and be with them in the moment. If you must write about the moment, do so as a participant, not as a surgeon with a scalpel.

Furthermore, both the poet and the mystic know that the background and the foreground oscillate. Neal Brown says that Roethke was part of a tradition that could see this oscillation between "absorbing or being absorbed:"

> . . . the contemplative experience varies little from person to person and even from century to century, being viewed most often as a process of absorbing and being absorbed, simultaneously—"driving and drawing," "sucking and sobbing."
> ~ THEODORE ROETHKE: THE JOURNEY FROM I TO OTHERWISE, Neal Brown, 1982.

We will encounter this oscillation again and again as we explore the Wisdom Tradition, psychology, and philosophy. This universal oscillation is a form of breathing in and breathing out. Everyone who encounters this awareness explains it differently. I much prefer Roethke's *sucking and sobbing* image because it brings needed emotion into the oscillation.

Brown says that Roethke was a student of mysticism; that his poems reached for a sense beyond the mundane. Brown offers this nice distinction between the occult and the mystical, which Roethke understood:

> The difference is obvious: the occultist exalts the will and attempts to gain control, whereas the mystic surrenders his will and allows himself to be absorbed into the truth he perceives. The dichotomy is as old as man himself, and to mistake the difference is to confuse two human impulses. Roethke's ultimate impulse, for example, is to discover his identity, his place in ultimate reality so that he can *be*, not so that he can control. In his own words, "Being, not doing, is my first joy." Mysticism and occultism, then, proceed from the same kind of awareness, an intuition that there is something more than the reality that is normally perceived. But mysticism seeks knowledge leading to being, whereas occultism seeks knowledge leading to power.
> ~ *THEODORE ROETHKE: THE JOURNEY FROM I TO OTHERWISE*, Neal Brown, 1982.

The paragraph above is a great summary of the distinction between egocentric perception and allocentric perception. According to Neal Brown, occult seekers are egocentric and they are probably not aware of the hidden mind. This is ironic since "occult" means "hidden." In their quest for power and control, the occultists miss the point. Brown identifies two human impulses in Roethke's poetry: one is *to be*; the other is *to do*. Mystics, saints, and religious teachers have been saying this for thousands of years as part of the Wisdom Tradition. A mystic is on a path towards *being*. The occultist is on a path towards *doing and accomplishing*.

As Roethke studies mysticism and reflects on his own mind, he comes to see that most of his essence is hidden. Like roots underground and out of sight, the greater part of him does not manifest as he goes about his daily chores. Roethke's poetry is often about the underworld of the plants, the roots, and the elaborate plant-brain that resides hidden below the surface. As his biographer, Neal Brown observes:

> . . . [Roethke] finds that he has more in common with the underground parts of the plant than with the rising, flowering stems.
> – *Theodore Roethke: The Journey From I to Otherwise*, Neal Brown, 1982.

Of course, the poet has more in common with the soul, with the underground mind, with the allocentric source. The flower cannot manifest without the support and nutriments stored in the root-mind of the plant. The ego, the flower, manifests from the self, the roots. Roethke knew this and wrote wonderful poetry—he gave us poetic flowers that arose from within his underground mind.

Roethke is like the phenomenologist Edmund Husserl who used the word *empathy* as if it was *a way of knowing*, as if there was a separate mental process that could *know* in an experiential way but was not rational. Poets like Roethke understand this hidden mind intuitively; *they just know*. But others *do not know*, and so the poets teach their readers through experiential stories and images. Poets have a gift, an unusual eloquence. They have an ability to shock their readers into a more mindful, less rational, allocentric consciousness.

Poets are also translators. They can hear the language of the non-verbal allocentric mind, but this background voice is oblique, opaque; it whispers silent gibberish that the ego cannot entertain without help. The egocentric mind uses written and spoken language to communicate. However, the ego is not an expert in music-like rhythms, images, and word-song. Roethke understood the "songs that words could sing." Many of his poems make no sense unless they are spoken out loud in a singsong voice. Speaking of Roethke's poem collection called the *Lost Son Sequence*, Neal Brown says we might miss the essence of Roethke's magic if we don't pay attention to the sound of his sentences:

> The critic who sits silently at his desk, defining words and digging for symbols without listening to the composition of sounds, will miss perhaps the most significant aspect of the poems—their auditory sug-

gestiveness. The word has power that goes beyond its connotations and denotations, and Roethke knew this, knew that the mind has an intuitive as well as a rational side to which those sounds could appeal. By manipulating the sounds in his poems and at the same time confronting the reader with insolvable verbal dilemmas, Roethke hoped to subvert the rational mind and elevate the intuitive faculties.

~ *THEODORE ROETHKE: THE JOURNEY FROM I TO OTHERWISE*, Neal Brown, 1982.

Yes, Roethke's quest was to subvert the rational mind and elevate his readers into higher states of consciousness. Roethke spoke the intuitive language of nature. Here, for example, is a line from one of Roethke's poems that makes no sense unless it is spoken (almost sung) out loud:

> *"God, give me a near."*
> ~ Theodore Roethke's journal

Like all the great poets, Roethke is trying to experience life in the raw. He then wants to share his pure experience, to craft a poetic language that reveals a deep truth: there exists a sane background in contrast to our crazed foreground. Poets don't want control over nature, over the lives of other sentient creatures; they want to appreciate, to feel, to be immersed in each of their moments—moments that go by too quickly. They also want to awaken their reader; to show them that there is a poetic, empathetic mind that is not the ego. Mystical poets like Roethke take their readers on a journey from *I to Otherwise*.

Roethke even used the techniques of the Zen masters to awaken his readers; he used koans, unsolvable riddles that can drive the problem-solving egocentric mind insane. Koans are not meant to be solved, their purpose is to show the egocentric mind that reason and logic cannot always arrive at solutions. Here are some Roethke koans:

> *Has the dark a door?*
> *Who is the bishop of breathing?*
> *What's a thick?*
> *Who untied the tree?*

Roethke is trying in his poems and through the use of koans and singsong language to dissolve the ego so that the poetic soul can be experienced. When you dissolve your ego and become one with nature, what you then experience is the *empathetic mind*, the allocentric mind. This mind participates and is absorbed—it moves from *I to Otherwise*, which is a phrase that Roethke invented. Here is Neal Brown commenting on a fundamental duality found throughout Roethke's poetry. This quote refers to one of the Roethke's earliest collections of Poems called *Open House*:

> *Open House* is constructed on two related dichotomies: (1) analysis and intuition, (2) flesh and spirit. These divisions are the very center of the mystical experience. The mystic believes that he can arrive at ultimate truth or reality only by rejecting logic and analytical reasoning, because they are the basis for the reality normally perceived and are therefore of no use to anyone who wishes to perceive *true* reality. He realizes that he must somehow transcend them if he is to apprehend things as they truly are, and he does this through the only means available to him, his intuition. Likewise, the mystic invariably perceives a tension between flesh and spirit. A close relationship is seen between the analytical mind and the flesh, which are undesirable because restrictive, and between intuition and the spirit, which can lead to new perceptions.
>
> - THEODORE ROETHKE: THE JOURNEY FROM I TO OTHERWISE, Neal Brown, 1982.

Roethke knows his allocentric mind is flesh, a whole-body animation. He doesn't use the term egocentric; he uses *analytical mind*. Likewise, he doesn't refer to an allocentric mind; he speaks of flesh and intuition. The words are different, but the understanding is the same.

In *Open House*, Roethke reveals that he understands that beyond *ego* is another entity called *self*. But he goes further by saying that this *self* is the soul that saints and seers have been telling us about for eons. There is this distinction: the self is the allocentric mind confined to a human body, but the soul is a universal entity that can be seen as the ego fades. The silent soul is the cosmic mind, the universal wisdom of which the self is a small manifestation. The departing ego leaves a deep sink hole in the mind, an energy sink that draws empathy from the cosmos. A soulful person is filled with compassion. A compassionate person can feel the sucking and sobbing that goes on below egoic cognition. To have an *open house* is to throw open all the doors and windows of perception so that the stale ego can dissipate as the fresh air of compassion flows in.

We are on a journey from *I to otherwise*, that is Roethke's conclusion. We are born with egocentricity, but every breath taken after birth is a journey away from self-centeredness into ever greater realms of something *other*.

In his last set of poems, Roethke talks about *The Far Field*. Perhaps this is the very same far away field where the poet Rumi suggested we meet:

> *Out beyond ideas of wrong-doing and right-doing,*
> *there is a field. I will meet you there.*
> *When the soul lies down in that grass,*
> *the world is too full to talk about.*
> *Ideas, language, even the phrase "each other" makes no sense.*
> ~ "OUT BEYOND," *A YEAR WITH RUMI*, 2006.

Judgments of wrong-doing and right-doing come from the egocentric mind. Fortunately, there is a field where we can meet that is free of judgment—a place where the ego cannot go. When the self (not the ego) lies down in the soft, safe grass of this far field, we are so overwhelmed with peace we do not wish the bliss to be contaminated by words and thoughts. Even the thought that we are separate creatures is too disturbing, and makes no sense, as we bathe in the glow of this joyful moment.

Mevlana Jelaluddin Rumi

> *Whatever put eloquence into language*
> *is happening here.*
> ~ "WHAT WAS SAID TO THE ROSE," *A YEAR WITH RUMI*, 2006.

The mind of right-doing and wrong-doing is the egocentric mind: the ego solves problems, analyzes, creates duality, and catalogs existence. Rumi has no desire to hang out with the egocentric crowd. He is a poet. He wants to feel life, to fall in love, to be overcome with joy and peace. So he whispers to you: Leave the rational mind here and meet me in the field that has no logic, no thinking, and no subject to dissect—let's just experience being alive together. The self is not the ego. The self is the hidden mind. Leave the ego in the busy city of the egocentric mind and come, meet me where souls (many selves) lie down in silence in the deep green grass of peaceful-ness. The world is overflowing with chatter, noise, with languages, ideas, and is filled with too many bodies and personalities. None of that busy world makes any sense to the soul—reason is irrelevant to experience.

According to Rumi Scholar Coleman Barks, Rumi sometimes called this hidden field where souls gather *The Friend*:

> The growing field between these two—the formless source and the form—Rumi calls the Friend.
> ~ *THE ILLUMINATED RUMI*, Jalal Al-Din Rumi, Michael Green, illus-trator, Coleman Barks, translator, 1997.

Friendship is about relationship. A friend can sit beside you in silence; no words are needed. The presence of a friend is also quietly healing—energy is mutually restored. Friendship is a shared participation in the act of living. Friendship arises from the formless allocentric mind, not from the anxious competition of the egocentric mind.

Rumi was more aware, 700 years ago, than most human beings are today. We are the ancient, primitive men who have forgotten our spiritual core. We are in power now—this is our moment to live upon the earth—we rule the planet, and yet we have lost our poetry. In our time, the ego has imprisoned the self; we can no longer sense the distinction between *being* and *doing*. And yet, the field still awaits anyone who wishes to connect with love, peace, joy, wisdom, and mindfulness. We can join together in Rumi's field of the soul any time we wish—regardless of the era we live in and the circumstances of our lives.

It is the poets who remind us about the field of the soul. It is the poets who meet together in the field of friendship and urge us to join them. It is the poets who are the soul-warriors, the insurgents, the voices in revolt against the ego. Rumi says:

Work in the invisible world
at least as hard

as you do in the visible.
[He also says]
Rise up Nimbly
and go on your strange journey
to the ocean of meanings.
~ "INVISIBLE WORLD," *A YEAR WITH RUMI*, 2006.

Your strange journey is your life on earth. On this strange journey, there is no arriving at a destination while you are alive. There is no moment when process stops—life is a flow through oceans of meaning.

But what does Rumi mean when he refers to the *ocean of meanings*? Perhaps his *ocean of meanings* refers to the flow of experiences that slowly define who we are. Experiences are events that elevate our consciousness on this strange journey. There is meaning in everything; therefore, *to experience* is to bathe in meaning.

You are not a teardrop in the ocean, as the ego would have you believe. You are the entire ocean in a drop. This is the essence of Sufi wisdom: the entire universe exists to instruct you—it is a book for you to learn from. You are the whole thing and everything else is part of you.

Rumi was born in Balkh, Afghanistan, but lived most of his life in Konya, Turkey. He was a poet, artist, scholar, and a religious leader with his own school. He became a Sufi master, an enlightened being. His poetry contains the message that the Sufi mystics give to mankind: that our inner beauty, the awe of our inner awareness, is the source that gives meaning to life. The outer world is put here so that we might comprehend our essence, our dual consciousness:

Sufis are loved the world over for reminding us that the glory is our inner reality, the outer being a kind of language that explains THAT. *Love is the religion. The universe is the book.*
~ *THE ILLUMINATED RUMI*, Jalal Al-Din Rumi, Michael Green, illustrator, Coleman Barks, translator, 1997.

In other words, "I am THAT" means: The universe is contained within my awareness, just as I am contained within the awareness of the universe. The world *out there* is a book put here so that my inner essence can evolve. I am that which exists.

The Sufis, Rumi especially, understood human duality. Here is a Rumi poem that his translator Coleman Barks has put into modern language in this poem called *Two Kinds of Intelligence*:

> *There are two kinds of intelligence: one acquired,*
> *as a child in school memorizes facts and concepts*
> *from books and from what the teacher says,*
> *collecting information from the traditional sciences*
> *as well as from the new sciences.*
> *With such intelligence you rise in the world.*
> *You get ranked ahead or behind others*
> *in regard to your competence in retaining*
> *information. You stroll with this intelligence*
> *in and out of fields of knowledge, getting always*
> *more marks on your preserving tablets.*
> *But there is another kind of tablet, one*
> *already completed and preserved inside you.*
> *A spring overflowing its springbox.*
> *A freshness in the center of the chest.*
> *The other intelligence does not turn yellow or stagnate,*
> *It is fluid, and it does not move from outside to inside*
> *through the conduits of plumping-learning.*
> *This second knowing is a fountainhead*
> *from within you, moving out.*
> ~ "Two Kinds of Intelligence," *A Year with Rumi*, 2006.

We can't state the case much clear than this quote: there are two minds. The two minds have different roles to play. We think we are the mind that memorizes facts and concepts; but there is another kind of tablet, one already completed and preserved inside you. The egoic mind pulls the knowledge from the outside world into the body. The allocentric mind, the self, is a fountainhead from within you, moving out.

Owen Barfield

> I wish our clever young poets would remember my homely defini-
> tions of prose and poetry; that is, prose equals words in their best
> order; poetry equals the best words in the best order.
> ~ Samuel Taylor Coleridge Best, Poetry, Words

Owen Barfield was a philologist as well as a poet and philosopher. In other
words, he was an expert on the words used in historic and literary texts. He
was also a Coleridge scholar—which is why I started with the quote above.
The *precise words* of the great poets were of special interest to Barfield.

As he studied the evolution of words used in ancient and literary documents,
Barfied found that poets were often the members of a society who had levels of
consciousness greater than the average citizen. They were *masters of words*, and
they could translate between the egocentric mind and the allocentric mind—to
use my terminology. Poets could play with words like a musician plays with notes.
Therefore, it became the job of the poet to help lift the level of an individual's consciousness and thereby, eventually, lift the level of consciousness of whole cultures.

Barfield saw his culture, the time in which he lived, as spiritually barren.
He blamed this barrenness on humanity's loss of contact with nature and on
the rise of an egocentric mind that turned mankind inward toward mental
abstractions. Barfield asked himself some questions:

- What is it about poetry that cannot be reduced?
- What is it in a metaphor that causes a thrilling feeling in the reader?
- Why do we suddenly "slam on the cognitive brakes" when we come upon
  a beautiful poetic metaphor?
- Why do we feel that some kind of magic has happened to us when we
  read a particular poetic phrase?

Barfield believed that the answer to each of these questions was the same:
The best metaphors cause an immediate transformation in the reader. A low
level of consciousness is confronted directly by good poetry; the reader is
forced (jolted) out of old, stuck habits and must face a new and strange perspective that seems to be coming from a place of higher wisdom, greater intelligence, and more compassion.

Usually it takes a severe experience, a shock—like a death, divorce, or the sudden loss of a job—to shut down the egocentric mind long enough to expose the allocentric mind. Such severe experiences can result in a shift to a higher level of consciousness. Often this shock pushes an individual upward on the spiritual journey. Barfield believed that poetry and powerful prose could also nudge a person upward on the spiritual path—poetry and prose cause lesser shocks, but they are nonetheless effective, especially for minds ready for transformation.

The primary tool of the poet, according to Barfield is the metaphor. Used effectively, the metaphor can cause the egocentric mind to pause or go silent, giving the allocentric mind room to expand and communicate.

The Importance of Metaphors

Metaphor involves a tension between two ostensibly incompatible meanings; but it also involves a tension between that part of us which experiences the incompatibles as a mysterious unity and that part which remains well able to appreciate their duality and their incompatibility. Without the former, metaphor is nonsense language, but without the latter it is not even language.

~ THE DISCOVERY OF MEANING AND OTHER ESSAYS, Owen Barfield, 1977.

In the quote above, Barfield (1898-1997) is saying that metaphors are a unique literary tool, commanding attention from both of our minds. A good metaphor allows the egocentric mind to dissect the words, to analyze, and to provide explanations about meaning and intent—so a good metaphor is satisfying to the egocentric mind. Yet that is only half the power of a metaphor. A good metaphor also—and equally—impacts the allocentric mind. The metaphor is *an experience in the moment* that leaves the allocentric mind with a sense of pleasure (or dread). Therefore, a good metaphor is also satisfying to the allocentric mind.

One mind—the allocentric or background mind—is about unity, about relationships and harmony, about images and story-telling, while the other mind—the egocentric mind—is about analyzing patterns arising from the background. There is a tug-of-war, a tension between these dualities: the background versus the foreground. Barfield is saying that our two minds can be poetically woven together through the skillful use of metaphor. Therefore, poetic metaphor is a unique tool for using analytical language—arising from the dualistic egocentric mind—in combination with images of raw experience, which arise from within the allocentric mind.

If it is true, as I would suggest, that *only experience can cause lasting change*—that rational thinking cannot cause long-lasting change—then metaphors and imaginative phrasing take the egocentric mind out of its comfort zone and force a kind of mental experience that enables change. As Barfield strongly believed, poetry can be a force that enables cognitive evolution—more so than prose. The magic of the metaphor is that both our minds are challenged to evolve. Metaphorical language becomes a method to raise individual and collective consciousness.

Metaphor, in my terminology, is a way for the allocentric mind to communicate. Spatial imagination comes from the allocentric mind. This *imagination* speaks a non-verbal, archetypal, dreamlike language that is symbolic and experiential; imagination is not literal—there is no linear logic to it. Contrary to this, spoken language, conceptualization, and abstract thinking are modern artifacts that evolved from the sophisticated egocentric mind. Poetry uses precise words from the egocentric mind to express the intuitive images that arise in the allocentric mind.

The poet Coleridge thought of *imagination* as the vessel by which divinity passes down into humanity. Imagination is, therefore, the energy-well for allocentric awareness. When we become one with nature, one with all that is, we open a portal into worlds beyond what our ego can understand. That world beyond the ego—if we accept what our spiritual brethren relate—is where we find the domain of spirituality. To understand how imagination works, to grasp the *spiritual science* (as Rudolf Steiner would say), to actually use *active imagination* (a term Jung used), is, therefore, a vehicle for attaining higher allocentric levels of knowing. I will discuss this in more detail in the chapters about philosophy and psychology, when I explore the contributions of Coleridge, Steiner, and Jung.

Barfield says that *metaphors are messages from a different and superior level of consciousness*—they force us to perceive and confront our lower level of being. Therefore, metaphors force us to reconsider what it means to be conscious (aware). Metaphors give us a glimpse into more sophisticated levels of meaning.

At the same time, some metaphors reawaken ancient memories of a time when we were part of nature. We retain a memory, Barfield tells us, of the time in our cultural and genetic heritage when we were *in* nature—when we were more allocentric and less egocentric. This allocentric inheritance provides us with a powerful and wonderful memory that refuses to be hidden or denied. This memory of being part of nature forces us to rebel against the meaninglessness of our modern material world. We have a memory of *another way of seeing*. This memory is embedded in our best poetry. Metaphors are a record of a new consciousness trying to inform (to communicate with) and transform an older consciousness. Therefore, metaphors are archeological finds; they are Rosetta-Stone inscriptions that tell us the history of our minds at the same time as they urge us to transform ourselves.

In his book *Poetic Diction* (1928), Barfield begins in the introduction by stating that he is struggling with a duality. This struggle has resulted in his writing the book:

> An early perception that poetry reacts on the meaning of the words it employs was followed by a dim . . . conviction that there are "two sorts of poetry;" and a series of unsuccessful attempts to rationalize these and other aesthetic experiences in terms of the various theories of language, literature, and life . . . resulted in the present volume.
> - POETIC DICTION, A STUDY IN MEANING, Owen Barfield, 1928.

We have two minds, an egocentric mind and an allocentric mind—it should come as no surprise that Owen Barfield would discover this duality. Good poetry affects the evolution of both of our minds. Barfield wrote an entire book trying to explain why there were two kinds of poetry.

Barfield believed that metaphor was the principle instrument used by poets and prose writers to affect consciousness. Words (communication) can get stuck at a low level of consciousness, but new words embedded within a metaphor, can free the mind and allow it to evolve. Take the word *God*, for example. Deepak Chopra wrote an entire book called *How to Know God* wherein he discussed seven stages of *knowing God*. These stages are essentially seven definitions of God—seven ways to express meaning about God—each definition representing a different level of consciousness. Therefore, seven people could be sitting around a table arguing about what God is and all seven of them would be defining God differently. What's the chance of that group coming to any kind of harmony?

Barfield would say that the right kind of poetry, using the right kind of metaphors, has the power to pull lower levels of consciousness into higher realms. This assumes that higher spiritual realms embrace a more loving, tolerant, peaceful God, while lower levels of consciousness perceive God to be less tolerant and less peace-loving. Spiritually evolved poets—those who are more tolerant and loving—can help others perceive the levels of meaning and the varieties of definitions that surround the concept of God. Indeed, that is one of the major jobs of the poet, to craft metaphors that cause cognitive transformation.

Our materialistic sophistication—the rapid evolution of the egocentric mind—has cost us an eye, says Barfield. We used to be binocular, with both allocentric vision and egocentric vision. We have now lost half our capacity to perceive. However, we can restore our cognitive binocularity, according to Barfield, through the skillful use of metaphor:

> Our sophistication, like Odin's, has cost us an eye; and now it is the language of poets, in so far as they create true metaphors, which must restore this unity conceptually, after it has been lost from perception.
> ~ POETIC DICTION, A STUDY IN MEANING, Owen Barfield, 1928.

There was a reason why we *lost an eye*. There was a purpose to nature's direction. The ego evolved because natural selection brought it about. The ego is a powerful survival strategy. However, in the process of becoming egos, we lost our participatory souls—we forgot that we were nature. We lost an eye, a way of seeing, a way of being aware—we became blind in one eye. Perhaps, Barfield suggests, this is the meaning of "falling from the Grace of God," falling from the kingdom of heaven. We tumbled into an evolved form. We became creatures with a sense of objective existence, but we forgot from whence we came. We forgot our allocentric inheritance.

The end result of this fall from Grace is the modern world we live in, which, according to Barfield, is lacking spiritual meaning, poetry, purpose, and hope—this is true because the egocentric mind is dominant in our technological world. We live in a spiritual barrenness now, without sufficient poetry, and it is going to get worse if we don't find a way out of the falling. We need a *final participation* (Barfield's term) that gets us back in connection

194

with nature. We have to do this without losing the miracle of ego and personality. Can we preserve both our minds as we evolve? That is the goal.

Barfield was a Christian; he felt that Christ was an example of the final participation, an example of how to be separate from nature at the same time we are part of nature. You are one with God even though you are a manifestation of God because the form never completely separates from the background. Barfield's faith and his Catholic beliefs prevented him from becoming a complete cynic—the world around him might have seemed barren and hopeless, yet he lived a life that was neither.

For Barfield, poetry is an actual and deliberate means for understanding and affecting cognition. It can be deliberately used to alter consciousness. Furthermore, we can use poetry to probe how consciousness has evolved throughout recorded history. It is beyond the scope of this book to explain Barfield's vast body of work. His thinking is complex and it evolved over a fascinating lifetime. I invite you to explore the metaphors of many poets from Barfield's evolutionary perspective.

The Work of the Poets

Now silence.
Unless these words fill
with nourishment from the unseen,
they will stay empty.
~ "How Minds Most Want to Be," *A Year with Rumi*, 2006.

I use poets in this section to represent all creative people, including dancers, writers, musicians, and painters. Innovative people use images and movement to speak a non-verbal language that talks directly through the allocentric mind. Artists are at home in this intuitive realm. When I use the word *poet*, I am also referring to an unusual kind of artist who has found a way to use language to reach the allocentric mind of others. This is not the direct language of the egocentric mind, nor is it the pure image and movement-based language of other artists. However, artists know about the allocentric mind—each in their unique way—and they communicate through the medium that inspires and directs their creativity.

Poets are usually sensitive and articulate souls; most of them overflow with intense and often conflicting emotions. Many poets are also bewildered—they often write to express a personal pain that they feel the need to share, as if crafting and sharing was cathartic. Poets can see bewilderment in the faces and actions of the people around them, and so they set about the task of mirroring in poetry the contrary emotions that their fellow human beings experience. Poets do not analyze bewilderment; instead, they simply express emotions and metaphors on the written page. The reader of poetry is often comforted to know that from time-to-time other sophisticated people feel lost, or unloved, or in awe. To the poets, it is clear that humanity is confused by something fundamental in the very nature of their humanity—poets see this paradox everywhere they look, and they feel the conundrum to their very core.

Many poets know intuitively that human beings are dual creatures. These poets also know that this mutually exclusive dual-cognition cannot be fully explained or felt through rational discourse alone. It falls to the poets to find the metaphors that speak to the emotions and instincts of the allocentric mind—bypassing, yet paradoxically including, the abstractions and logic of the egocentric mind.

Poets also know that the allocentric mind is hidden from the ego. Conflict and confusion arise from this fundamental blindness. There is also awareness that the earth is peopled by individuals who possess different levels of consciousness. Each individual has a different perceptual system; each has a different understanding of reality and meaning—each has a different worldview. Poetry can be tailored to address specific levels of consciousness; different metaphors can be crafted to expose and transcend cultural blindness.

Poets also know that a life out-of-touch with the allocentric mind is in trouble. The allocentric mind is the wellspring for love, peace, joy, wisdom, and mindfulness. To be devoid of these blessings is a psychological crime against humanity. Poets feel a desperate need to save the souls of mankind. The hidden mind—the true self, the allocentric mind—is the source of creativity, innovation, spirituality, soulfulness, compassion, and wisdom—all the best of human values reside hidden within. This essence of humanity cannot be allowed to be denigrated and ignored.

Poets also know that we help each other—poet-to-poet we hand down the mysteries. Poets and writers are driven to share what they have pilfered from the experiences of others. I love this personal sentiment by Poet Laureate Billy Collins about his own poetry:

But mostly poetry fills me
with the urge to write poetry,
to sit in the dark and wait for a little flame
to appear at the tip of my pencil.
And along with that, the longing to steal,
to break into the poems of others
with a flashlight and a ski mask.
~AIMLESS LOVE, 2013

Ah, yes, we *borrow* from each other. I have my ski mask and flashlight—I recraft the words of better minds than my own and place them here as if God chose me as the vessel. The truth is that the baton is passed back and forth; we just need to remember to be thankful to, and to acknowledge, those who came before.

Poets have a mission to awaken the less poetic among us to a wiser, kinder, more intelligent reality. Poets are the *more awake people* who walk beside us as we journey through life.

If you have a poetic gift, if there is an artist's voice that speaks through you, then you must join in the effort to raise the consciousness of humanity. Everyday life need not be sad and barren, devoid of mystery, humorless, joyless, the occupation of sleepwalkers. We need not be creatures who simply struggle to survive so that we can leave more offspring to struggle and survive. There is awe and magic in existence—at a certain level of consciousness, you can feel it and participate within the awe and the mystery. There is a sense of divinity. We must become ever more compassionate, creative, tolerant, forgiving, and nurturing creatures.

However, there is a force that works to keep minds and hearts at low levels of consciousness. This is an ugly, intolerant, racist, sexist, mean-spirited, soulless beast that is devoid of poetry. It devours all that is kind and good. This force cannot be allowed to win. The poet must be a warrior for the emergence of a better breed of humanity.

Therefore, as a poet, an artist, *you have an obligation*:
Do as much good as you can in the time you have left
Get up.
This is a work day.

197

Four

DUALITY AND RELIGION

Silence is the language of God,
everything written down is just a bad translation.
~ Fr. Thomas Keating, unknown source and date

Why Would the Pope Bless Adolf Hitler?

"I am troubled, Surge."

"Maybe you're just hungry."

"Why would a Pope bless Adolf Hitler?"

"You should try our Political Pie."

"Why do so-called Holy Men condone war and poverty?"

"Our political pie is coated with a surface of sickie-sweet syrup. It is made with charcoal and sand; the sweetness masks the raw truth."

"I prefer martyrs to morons."

"Okay. So try our Priest-less Soup—that might help your mood. Priest-less Soup is made with human blood and with the tears of mothers who lost sons in battle, or had their babies ripped from their hands by ruthless soldiers who then gang-raped the grieving mothers. Pungent croutons explode in your mouth and tears of grief flow from your eyes as you sip the acidic broth. You will never again be able to look upon religious icons without puking."

"I'm not sure this dialogue is helping, Surge."

"We have a hard time swallowing what the cruel past has done to our souls. By the way, we brew a famous tea here called Tea of Eternal Angst. You should try some. It causes collective weeping—all the vegans on the planet weep in unison for the loss of decency and kindness. There is a certain hopeless bitterness available for all to share. A cup of tea with friends is what you need."

"Okay. Bring me some Tea of Eternal Angst. Will it help me accept my fate: to live amongst the mental patients; to accept my spiritual limitations; to love those who vote against their own best interests and insist on being deaf-blind?"

"No. It doesn't perform miracles. It just tastes collectively good. Look Dutch, develop your own spirit. You can't solve the whole thing. Why don't you give acceptance and hard work a try?"

"I am still thinking about the soldiers who murdered and raped women and children. Hard work and acceptance seem not only lame but cowardly, Surge. I am not so sure that your meek advice is sufficient, given the gravity of the situation."

"Tread carefully, Dutch. Immoral soldiers and bad priests—and the powerful masculine minds that direct them—are agents of the Devil. They will not hesitate to take out a mouthy poet or two. There is a thin moment between martyrdom and the grave."

In this chapter, we will explore the crossroads where heaven meets hell, light meets dark, and goodness meets evil. As the chapter unfolds, we will cross paths with Albert Einstein, Adolf Hitler, Pope Pius XI, Helen Keller, Mary Magdalene, Thich Nhat Hanh, Rudolf Steiner, Richard Rohr, Cynthia Bourgeault, Jesus, Jacob Boehme, Deepak Chopra, G.I. Gurdjieff, and Hermes Trismegistus. Everywhere we look at this party, we will see the same duality that the poets embraced. Although the names for our two minds will vary with the era and with authors, the concept will remain exactly the same. Let us begin with Einstein, Hitler, and the Holy Ghost.

Adolf Hitler is the True Holy Ghost

Hans Kerrl, Reich minister for Church Affairs declared: "Adolf Hitler is the true Holy Ghost."
- *EINSTEIN AND THE POET, IN SEARCH OF THE COSMIC MAN*, William Hermanns, 1983.

No one is surprised that a henchman for Hitler would say something so stupid and mindboggling as the above quote, but that was not the main problem in Germany as war loomed on the horizon. The problem was the population of Germans who could believe drivel like this, even murder in the name of such outrageous lies. However, for me, the great sadness is that we have a population on earth today that is still susceptible to lies like this—they still vote. They still hate.

> I'll never forget how a teacher of mine took a long rusty nail from his pocket, held it up, and said, 'With such a nail the Jews crucified Christ.' Although I was very young at the time, I was already feeling the tragedy of being a Jew. That was in a Catholic school; how much worse the anti-Semitism must be in other Prussian schools, one can only imagine.
> - EINSTEIN AND THE POET, IN SEARCH OF THE COSMIC MAN, William Hermanns, 1983.

Why would seeds of hate and murder find a home inside churches and religious schools? It seems like the Devil is the headmaster at these so-called religious institutions, not a gentle, loving deity. What better place for the Devil to hide than behind the altar, speaking through the mouths of the priests. Soon their parishioners mouth the Devil's memes and take their sanctioned hatred to the streets. The great sadness is that even today, on the pulpits of fundamental Christianity, the same puppet-priests are being manipulated by the Devil, and the same parishioners express their hatefulness on social media—and they vote. It doesn't appear that we should look to established religions for our salvation:

> 'But, Dr. Einstein,' I protested, 'did you not experience their nationalistic fervor during the last war [WW1]? You could not change their minds, as hard as you tried. Didn't you feel isolated, if not betrayed? It was said that in the court of the Kaiser you were called a *moral leper'*. . . At these words, Einstein seemed to bolt ahead of me and I found it difficult to keep up with his long steps. Suddenly he stopped and said, "If you want to help the Germans forget and save the world from a new ward, don't count on established religions. We must

found a cosmic religion, one of unconditional love, not one which sells itself to those in power."
~ *Einstein and the Poet, In Search of the Cosmic Man*, William Hermanns, 1983.

The poet William Hermanns—quoted above—interviewed Albert Einstein on a day when Nazi youth marched ominously outside Einstein's apartment in Berlin, Germany—this was shortly before the Second World War began. Einstein felt that religious men had sold out to baser instincts; for him, this made their religions a lie.

"I predict," said Einstein, "that the Vatican will support Hitler if he comes to power. The Church since Constantine has always favored the authoritarian State, as long as the State allows the Church to baptize and instruct the masses." About a third of all Germans were Catholics and Hitler had been raised a Catholic. Einstein proved to be correct about the Catholic Church. William Hermanns writes:

> . . . I was shocked when I learned of the Concordant which Hitler signed with the Roman Catholic Church on July 20, 1933. Pope Pius XI then asked God to bless the Reich, and this after Hitler instituted the boycott of Jewish shops with the declaration, 'I believe that I act today in unison with the Almighty Creator's intention. By fighting the Jews, I battle for the Lord.'
>
> Hitler was relentlessly drawing the whole world into war to fulfill the proclamation of a group of German *Christians* in April 1937: Hitler's word is God's law, the decrees and laws which represent and possess divine authority.
> ~ *Einstein and the Poet, In Search of the Cosmic Man*, William Hermanns, 1983.

When I first read the paragraph above, I was stunned. William Hermanns is saying flat out that Hitler saw himself as a Catholic, a religious man, a Christian. Hitler felt his "faith" so strongly that he actually approached the Pope to get blessings for the Third Reich. Hitler needed to get God to sign a document that sanctioned the creation of concentration camps and genocide. In the name of God, and in the name of Christianity, Pope Pius XI signed on the dotted line.

The world's great religions have so far failed humanity—or, perhaps, the mission of religion to bring love, faith, and tolerance to the masses has only begun. The level of consciousness of many church leaders, and the level of consciousness of whole populations, has been insufficient so far in history to allow love to dominate hate and for peace to dominate violence. The institution we call religion has gone through cognitive evolutions just as cultures and individuals have evolved ever more complex levels of consciousness. However, religions seem to hold back the populous that is straining toward greater sophistication, tolerance, and kindness. I would contend that patriarchal religions still, to this very day, have Adolf Hitler as their true Holy Ghost. The egocentric mind has seized religion and inhibited the values of the allocentric mind where love, peace, joy, wisdom, and mindfulness reside. It is the egocentric mind, the ego, which can be convinced to validate the likes of Adolf Hitler, Stalin, and Mussolini.

A bumper sticker on a friend's car reads: "The Christian Right is neither." What passes for Christianity in the Western world, in its weakest form, is neither correct in its pontifications, nor does it follow the teachings of a compassionate, loving, tolerant, peaceful, and forgiving God. The modern political force in the United States called Christian Fundamentalism is no different than the Christians who supported Adolf Hitler. The same low level of consciousness is at work in undeveloped minds.

At the moment, in the world we currently live in (I write this in 2018), we have male minds that are out of control; they are ego-driven, violent, and cruel—at various levels of nastiness. Even Buddhist males are not immune to this cruelty. The Rohingya are a Muslim ethnic group living in the Buddhist nation of Myanmar—for centuries they have lived among the Buddhists. The Rohingya are currently being tortured and massacred by Buddhist authorities in Myanmar. Unfortunately, as this tragedy demonstrates, masculinity often trumps religious principles.

In my culture—the United States, 2018—in a political landscape currently controlled by fundamentalist "Christians," there is no avenue for moving through levels of consciousness, no insistence on addressing human suffering. Instead, there is denial and resistance to suffering, and there is no pathway to greater sophistication, intelligence, and compassion. Such a vacuous and failed "religion," must be rejected. I am not a Buddhist, but I find Buddhism to be a better philosophical and moral choice than the ashes of modern Christianity. However, *there is* a strong mystical and sane core to Christianity in the United States, exemplified by religious leaders like Richard Rohr and Cynthia Bourgeault, discussed below. The core of Christianity is still alive and is embraced by sophisticated loving people with highly evolved levels of consciousness.

Of course, I am not an isolated voice coming to this awareness that Buddhism is a sane spiritual voice in a world still manipulated by patriarchal institutions. Even hard-nosed scientists exploring spirituality have gravitated toward Buddhist practices. For example, neuroscientists and Buddhist scholars have compassionately gone forward to share their viewpoints. This joint effort has been going on through meetings, conferences, and shared research efforts for decades—since the 1980s. This collaboration has fostered a gentle exchange of worldviews and has been very fruitful for both groups. However, there is a necessary transition on the horizon as inevitable conflicts arise. I have written in detail about this collaboration in my earlier book *Consciousness: A New Slant on an Old Conundrum* (2017).

At some point, scientists and Buddhist practitioners will enter a debate that is more acute—to the betterment of both, I think. The patriarchal problem, for example, in both science and religion, cannot be turned aside. Gender and religion should be part of the debate, perhaps centrally so. Buddhism, Islam, and Christianity are patriarchal religions—defined and ruled by males for centuries. These male-dominated practices have a lot of work ahead to bring feminine energy into their essence.

Deaf-blind Helen Keller, who described herself as a militant suffragette, was disgusted with patriarchal religions, including the practices of Buddhists, Hindus, and Christians, even though she had a strong Christian *faith*. From Keller's perspective, women were doing all the work of nurturing, while the men hung out in the monasteries, mosques, and pubs searching for personal transcendence. Women, especially when Helen Keller lived, changed all the diapers and bandages; their opinions and passions were not included or valued.

I agree with Helen Keller. The patriarchal hierarchy needs to be dragged into the modern world where the allocentric (poetic, feminine) mind is an equal partner with the egocentric (logical, masculine) mind. Respectfully, I would assert that the *all-knowing males*—especially Jesus, Muhammed, and Buddha—are losing respect in the modern world. Dominant hierarchies are an ego-construction—the medium has unfortunately become the real message.

When males sit for two thousand years contemplating the mind, they end up contemplating and understanding the male mind, egocentricity. The egocentric masculine mind is a whole different animal compared to the feminine, allocentric mind. The solution to the problems of the male include: finding their feminine side and their intuitive wisdom; replacing aggression with compassion; and waking up enough to know what females already intuitively know—*stop analyzing life and live it.*

Despite the patriarchal administration of Buddhism, this spiritual system understands the need to balance the *ego* with the *self.* Buddhism is aware that egocentric males (especially) need help, and Buddhist practices offer that help. Ego-bound males cause most of the world's suffering. These unevolved males need to be more loving, peaceful, appreciative, wise, and awake. Perhaps the fate of humanity hinges on whether the discoveries of Buddhism (about how the mind works), and the contributions of mystics at the core of all religions, reach, nurture, and transform patriarchal males.

Patriarchal societies did a pretty good job of erasing all femininity (the allocentric mind) from the major world religions. Even the mother of Jesus was pushed out of the way in the eagerness of the male ego to portray the Divine as masculine. In her wonderful book *The Wisdom Jesus*, Episcopal Priest Cynthia Bourgeault gives a different version of the role of women in the teachings of Jesus. Here she looks at the role of Mary Magdalene:

Traditionally [Mary Magdalene] has been portrayed in the West as a repentant prostitute from whom Jesus cast out seven demons. But scholars have conclusively demonstrated that this picture is a medieval fabrication, largely attributable to sixth-century theologian and Pope St. Gregory the Great. What we do know now is that Mary Magdalene was almost certainly a full-fledged disciple of Jesus and quite likely his most advanced student. This statement in itself is challenging to the status quo, because many people will still insist that Jesus only had male disciples. (As a matter of fact, there's a whole line of theological argument that tries to bar women from the priesthood by saying that since Jesus's disciples were male, only males can be priests.) But the evidence emerging both from . . . new texts and from the canonical gospels themselves . . . is that this simply wasn't so. Jesus's band of close followers included both men and women, and the women were not merely camp followers or support staff. He taught them and enjoyed their fellowship; they were his intimates, Mary Magdalene most of all.

~ *THE WISDOM JESUS*, Cynthia Bourgeault, 2008

In Chapter Eight, I explore duality and gender. There is much more to be said about the repression of women and about the monumental effort that went into hiding our allocentric inheritance. The quote above would fit very well both in this chapter and in the chapter on gender. I left the quote here because Cynthia Bourgeault belongs primarily in this discussion of the Wisdom Tradition. Her work with Richard Rohr and James Finley at the Albuquerque, New Mexico *Center for Action and Contemplation* has global impact. The wisdom and compassion radiating from the *Center for Action and Contemplation* has helped form my worldview, for which I am very grateful.

History is being re-considered and re-written in places like the *Center for Action and Contemplation*, especially as we enter an age that is balancing the allocentric mind with the egocentric mind. The texts of all religions were written and interpreted by males, with little input from the feminine voice. This is in the process of being challenged—the contribution of the allocentric mind is finally being allowed.

When I was growing up in the Christian culture of Midwestern America (in Michigan), I slowly became disgusted with Christianity. This came about as I observed the behavior of so-called Christians. Jesus seemed to be supporting greedy, consumer-driven, narrow-minded materialists with low levels of consciousness—people quite willing to vote for war, people who looked the other way as the bombs fell from the sky. The American Christian seemed like a soulless, simple-minded, humorless creature to my naïve mind. These American Christians wanted young folks like me to love Jesus. The few true Christians who were loving role models, who worked to relieve suffering, who had a joyful spirit that you could feel, were in the minority in my small world—in the 1950s.

That harsh judgment being said let me quickly add that there are people like Franciscan friar Richard Rohr and Episcopal priest Cynthia Bourgeault who have faith in a more sophisticated Jesus, in a saint who is part of the Wisdom Tradition. The difference between *consumer Christians* (the spiritual capitalists) and sophisticated, loving human beings like Rohr and Bourgeault depends on their level of consciousness. Rohr and Bourgeault are up there on the high rungs of conscious-sophistication, while the "voting for war" Christians are down in the dark cellars where the light of mental complexity and compassion rarely penetrate.

Minds evolve, we go through cognitive and spiritual stages of development, and each of us is—like it or not—at some primitive level of consciousness. So how does this affect how we view traditional religion? Here are two ego-heavy questions for those who follow the patriarchal religious paths, Eastern and Western:

1. How do *Eastern religions* define reincarnation and karma when levels of consciousness are considered? Would we not *expect to have* bad Karma from our past lives since we are an evolving species with less and less conscious awareness, less tolerance, and less empathy as we stare back at our cognitively shallow history? Wouldn't we expect that the males, especially, would have bad karma, since they were predators (not the prey), warriors with a savage killing-consciousness? Would we not expect bad karma to be inherent, since compassionate, empathetic creatures, with broad worldviews and integral levels of consciousness didn't evolve until the last couple of nanoseconds on the cosmic clock? Everyone reincarnates with bad karma—and it's not their fault (they don't deserve to be punished).

2. How do *Western Religions* define the afterlife, considering levels of consciousness? When we get to the Christian Heaven, will the guys get to eat first? Will the female saints wash the feet of Holy males and leave the less holy female toes unwashed? Will the girl Christians make the beds every morning for the male Christians (forever)? Is the role of the female in Islamic heaven to be a prostitute for the sexual lust of male warriors? What level of consciousness will we take to Heaven, and will we keep our two minds and our complex and extremely limited perceptual structure? Is Heaven a static place where process has stopped?

There is something fundamentally bankrupt about established religions, Eastern and Western. The logic that supports Eastern and Western assumptions doesn't hold together and current practices fail to instill ongoing confidence that these religious structures should be blindly followed.

Current religious traditions are losing their followers at alarming rates. In part, this is because modern religions have failed to adjust to changes in human cognitive development; they are not keeping pace with advances in the cultural evolution of consciousness. But we *do need* spiritual pathways; we cannot let our religions—our spiritual, allocentric minds—dry up and die out. If we look back at our history, we will see that a reassuring philosophy and discipline, called the Wisdom Tradition, has existed for centuries. In other words, there is hope and there is reason to remain within our faith-based essence.

The Wisdom Tradition

> It is a bit embarrassing to have been concerned with the human problem all one's life and find at the end that one has no more to offer by way of advice than "try to be a little kinder."
> ~ ALDUS HUXLEY, WHAT ABOUT THE BIG STUFF? FINDING STRENGTH AND MOVING FORWARD WHEN THE STAKES ARE HIGH (2002) by Richard Carlson, p. 293.

Try to be a little kinder is a message arising from our allocentric mind. It is an answer to the questions raised in the discussion above. What are we to do about wars and cruelty and the dominance of the egocentric mind? The mystic core of our major religions has had the answer for centuries: change the biochemical and quantum nature of the human mind so that our essence is driven by compassion.

When we look back at the history of the various religions, when we explore the writings of the saints and mystics, and when we observe what was taught in the esoteric schools, it becomes clear that there is a repeating pattern—the same *curriculum* is experienced by various generations in diverse regions of the planet. In other words, there is a universal Wisdom Tradition that has been passed from the ancients, to the moderns, to you.

The term *Wisdom Tradition* is a synonym for *perennialism*, the idea that human beings everywhere and throughout history feel their essence and are driven to figure out who they are. This essence, this realization that we are complex and yet alike, that we are all immersed in this mystery, is found in all spiritual traditions regardless of the trappings, doctrines, and power structures associated with institutionalized religions.

The Wisdom Tradition teaches that human beings have two minds, one of which, the allocentric (to use my terminology), is hidden from the ego. To make the world a gentler, kinder, and more peaceful abode, the hidden allocentric mind needs to be at least an equal partner with egocentricity. In more forceful traditions, the view is that the ego should be monitored and managed by the allocentric mind.

The Wisdom Tradition often warns of the ascendancy and abuses of the egocentric mind. This evolutionary force (the egocentric mind)—as wonderful as it might eventually become if we learn how to ride the wild beast—is dangerous for ultimate survival *if left on its own*. The ego has no intrinsic compassion, no loving-kindness, no empathy, and no respect for experience. To be fair, these are not its biological mandates; by design, the ego is deaf-blind to experience. Ego is only about knowledge, patterns, and power. It has no wisdom—and it cares not one tiny bit about values, ethics, or nurturing. The mystics tell us, over and over, that the primitive, unevolved ego, as it is manifesting in our era especially, is a dangerous animal living in the glory of selfishness.

However, it is important to remember that we are paradoxical creatures. Within our egocentric minds *we are islands*. Isolated, alone, unable to communicate what we truly perceive, we are utterly lost. Like everything else it has analyzed, the ego has reduced itself to a separate, isolated object. However, within our allocentric minds *no man is an island*. We are one. There is no manifestation out of any background; we are the background as well as the manifestations from the background. If God is the background, then we are one with God.

The perennial philosophy, the spiritual, mystical, and compassionate teachings handed down to us through multiple avenues, is everywhere the same. If you are a Catholic, you are also a Jew. If you are a Jew, you are also a Muslim. If you are a Muslim, you are also a Buddhist. At their core, all these traditions tell us to locate and cultivate our allocentric mind. *Kindness is the esoteric heartbeat of every one of our great religions.*

> *Not Christian or Jew or Muslim,*
> *Not Hindu, Buddhist, Sufi, or Zen.*
> *I am not from the East or the West,*
> *not out of ocean or up from the earth.*
> *Not natural or ethereal, not composed*
> *of elements at all. I do not exist.*
> ~ "Only Breath," *A Year with Rumi*, 2006.

In the quote above, Rumi is trying to capture the essence of the Wisdom Tradition. He is defining the allocentric mind. *Identity*—who you think you are—is an egocentric construct. You are not really a Jew, Christian, Hindu, or Muslim; you are undefinable and part of a universal whole.

In his book *Perennial Philosophy*, Professor Arthur Versluis, Department Chair of Religious Studies at Michigan State University, adds this clarification [Italics mine]:

> . . . If one discovers what is true, then it is true not only for oneself, but also for others. Perennial philosophy does not claim that all religious or spiritual traditions are the same, but rather that the human search for and realization of truth is perennial, that is, it can be experienced by different people in diverse circumstance. Again, what is perennial in perennial philosophy is truth, or to put it another way, *what is perennial is experience of the ground of being*, which, if it is true, is true and verifiable by others.
> ~ PERENNIAL PHILOSOPHY, Arthur Versluis, 2015.

The *ground of being* is an experience of the allocentric mind; the sensation of transcending the ego—it is a feeling of oneness. Dr. Versluis also defines the words *perennial* and *philosophy* to make sure that we are getting his major point: that a transformational process is available to any human being. There is radiance, a change for the better that we might all share if we put in the effort:

> The word "perennial" . . . means that human beings can go through transformative and illuminative processes that are intrinsically open to us as human beings; that people have gone through, are going through, or may go through such a process in the past, present, and future (hence it perennially recurs). The word "philosophy" does not have the meaning of "an abstract theoretical system constructed by discursive reason," but rather that of "a virtuous life leading to the realization of love (philo) and wisdom (sophia)."
> ~ *Perennial Philosophy*, Arthur Versluis, 2015.

In other words, to be a philosopher means to love self and others, and to seek wisdom. Wisdom is how the allocentric mind *knows through experience*—because it has *been there and done that*.

The wisdom tradition has tried numerous ways to shock the ego into recognizing that it is not the only show in town (that there is a twin mind). The "wounds" we receive during our journey through life are lessons, messages for the initiate according to Franciscan priest Richard Rohr (in a blog post, May 23, 2016)—this is my edited version of Richard Rohr's post:

1. *Suffering is part of our life journey.* No one escapes. Life is hard. Accept suffering and learn from it. As Leonard Cohen put it in his song, *Anthem*: "There is a crack in everything. That's how the light gets in." Our wounds (the cracks in our armor) are the only things humbling enough to break our attachment to our false self (the ego) and make us yearn for our True Self.

2. *You are not that important.* In other words, the ego has to be taken down a peg or two. In the allocentric mind [my terminology], the ego is dissolved so there is no one there to be seen as important.

3. *Your life is not about you.* Life is not about egocentricity. There *is no you* within the hidden mind.

4. *You are not in control.* There is a flow to events, a flow in nature; a powerful river flows that pulls you along. You cannot control this flow. It is exhausting and futile to fight against it. Go with the flow. Become the river (because *you are* the river).

5. *You are going to die.* The ego came from *nothing*. Eventually, the ego returns to this *nothing*. Every moment you die to who the ego thought it was. Every moment you are a new being. Moment-by-moment you die and die and die. One day, the whole body dies, and there is no more egocentric attention. If anything survives death, it is allocentric, pervasive, and all encompassing.

The lessons above do not sit well with the ego, so it redirects, denies, and fiercely resists. During the times in life when we suffer, the ego goes through the cycles of death, denial, anger, bargaining, depression, and finally, if all works out, the ego finds acceptance and release. The ego must find and accept its twin before it can evolve. That is the core message of the Wisdom Tradition. Let's now look next at an Eastern religious perspective, starting with the famous Buddhist Heart Sutra.

Listen, Shariputra

As I followed my interest in the Wisdom Tradition, I saw clearly that Eastern philosophers knew centuries ago the difference between the rational mind (the ego) and our second mind, the self or soul. This understanding of our mental duality is illustrated in perhaps the most famous of Buddha's sermons, the *Heart Sutra*. This poetic wisdom, dating back over 2000 years, is chanted daily throughout the world. The most enigmatic stanza of the *Heart Sutra* for a Western thinker is this set of lines:

> *Listen, Shariputra,*
> *Form is emptiness,*
> *emptiness is form.*
> *Form is not other than emptiness,*
> *emptiness is not other than form.*
> *The same is true with feelings, perceptions, mental formations,*
> *and consciousness.*

Here is what I think this famous sutra means: Form is nothing but a manifestation out of a background field, a *figure* arising—but never separating—from the background. Form is egocentricity. Emptiness is allocentricity. The *ego* and the *self* oscillate—they are interconnected and yet they are mutually exclusive. Here is Vietnamese Buddhist monk Thich Nhat Hanh's explanation from his book about the *Heart Sutra*:

Form is the wave and emptiness is the water . . . So "form is emptiness, and emptiness is form" is like [saying] wave is water, water is wave.
~ THE HEART OF UNDERSTANDING, Thich Nhat Hanh, 1987.

Thich Nhat Hanh repeats a story below to illustrate the difference between the ego, a grain of salt, and the self, the ocean:

There is an Indian story about a grain of salt that wanted to know just how salty the ocean was, so it jumped in and became one with the water of the ocean. In this way, the grain of salt gained perfect understanding.

~ THE HEART OF UNDERSTANDING, Thich Nhat Hanh, 1987.

In other words, the ego is a grain of salt that knows only that it has an external form and an internal composition; it perceives that it is separate from all that is around its form. One day, the ego hears talk of oneness, of being part of a greater reality. But talk, analysis, systems thinking, and egocentricity fail to bring a feeling of understanding; nothing helps the ego get any closer to comprehending oneness. Then one special day the ego has a transformative experience of the ocean and only then becomes convinced that there is more to life and reality. If we fail to grasp this experience during our lifetime, we get to experience it at the point of death—no ego escapes the salt water.

If the grain of salt, the ego, is allowed to dissolve, the experience of *being* the ocean is expansive and peaceful. The letting go, the accepting that we are part of a greater wholeness—a vaster reality—results in a deep feeling of calm. This peacefulness is followed by a welling up of joy (all the stress of past and future is gone), and from this ocean of peace and joy comes a profound appreciation (love and compassion), as well as a kind of knowing called *wisdom* that can only be attained through experience. To be fully dissolved in the ocean is to be fully awake, to be Buddha, to be enlightened.

A famous new age saying is *Be Here Now*. In other words, we should live in the moment and not dwell in the past or future. *Be Here Now* means we should experience life rather than analyze existence. However, sitting in meditation, alongside others in a spiritual setting, we soon realize how difficult it is to be awake in every moment. As we sit in stillness and silence, usually with eyes partially closed, we get glimpses of the hidden mind; we realize just how hidden it is and why the ego might reject the idea of a twin—because the twin (the background, the ocean) is so extremely hard to egocentrically conceive. Perhaps the most significant reason that the ego rejects or is blind to the hidden self is because *ego* and *self* are mutually exclusive—to experience the self, the ego must be momentarily dissolved. The egocentric mind is not present—doesn't exist—when the self, the allocentric mind, is active.

When the Buddhists speak of the *absolute* and the *relative*, they often mean that there is an unchanging background (the absolute) that is the canvas for an ever-changing (relative) world. *The absolute* is free from concepts, beliefs, and dualistic ideas of *ego* and *other*. *The relative* is by nature dual since it arises from a groundless background—it becomes an object manifested from the unchanging background. The ego must have a foreground *and* a background to exist.

Christianity: The Wisdom of Jesus

> We have to accept that human culture is in a mass hypnotic trance. Plato already said this, as most religions do at the higher levels. We are sleep-walkers, "seeing through a glass darkly" (1 Corinthians 13:12). Wisdom teachers from many traditions have recognized that we human beings do not naturally see; we have to be taught how to see.
>
> That's what religion is for, to help us let go of illusions and pretenses so we can be more and more present to what actually is. That's why the Buddha and Jesus both say with one voice, "Be awake." Jesus talks about "staying watchful" (Matthew 25:13, Luke 12:37, Mark 13:33-37), and "Buddha" literally means "I am awake" in Sanskrit. Jesus says further, "If your eye is healthy, your whole body will be full of light" (Luke 11:34).
>
> ~ Richard Rohr's Daily Meditations, August 30, 2016.

Richard Rohr, the author of the above quote, is one of our national treasures. He reaches thousands of people daily through his blog—he is a prolific author—and he has helped to organize many conferences at the Center for Action and Contemplation in Albuquerque, New Mexico, which he founded in 1987. Rohr is a Catholic priest but his philosophy encompasses all religions. In the quote above, he is reminding us that the Wisdom Traditions of the world's greatest religions all strive to wake up the human being—there are higher realms of empathy and love to be discovered. To raise the collective level of consciousness of mankind requires individual transformations—one human being at a time.

In her book *Wisdom Jesus* (2008), Dr. Cynthia Bourgeault, an Episcopal priest who teaches at the Center for Action and Contemplation, discusses Jesus from a fresh perspective, as a great teacher of wisdom. She sees Jesus:

... first and foremost as a wisdom teacher, a person who (for the moment setting aside the whole issue of his divine parentage) clearly emerges out of and works within an ancient tradition called "wisdom," sometimes known as *sophia perennis*, which is in fact at the headwaters of all the great religions of the world today.
~ *THE WISDOM JESUS*, Cynthia Bourgeault, 2008.

For Dr. Bourgeault, Jesus is *foremost a teacher in the Wisdom Tradition*. In other words, Jesus is part of a long line of spiritual teachers whose wisdom-practices predate the dawn of egocentric time.

In the paragraph below, Dr. Bourgeault provides a clear connection between allocentric consciousness and Christianity, especially the wisdom teachings of Jesus. In this quote, she defines the esoteric concepts of *gnosis* and *sophia*:

... the word "gnosis" ... means knowledge. Sophia and gnosis are more or less synonyms. They both imply an integral, participational knowledge carried not in one's head but in one's entire being [allocentricity] ... "Gnosis" is a perfectly acceptable New Testament word; the apostle Paul uses it repeatedly in his attempts to describe the intimate experience of knowing and being known in Christ.
~ *THE WISDOM JESUS*, Cynthia Bourgeault, 2008.

In other words, to become the background, to let go of judgement, is to gain a new kind of knowledge; not the knowledge of the ego, but *gnosis*, a participatory, experiential knowing, which is wisdom.

As a teacher in the Wisdom Tradition, Jesus was trying to raise the level of consciousness of the population:

... he stayed close to the perennial ground of wisdom: the transformation of human consciousness. He asked those timeless and deeply personal questions: What does it mean to die before you die? How do you go about losing your little life to find the bigger one? Is it possible to live on this planet with a generosity, abundance, fearlessness, and beauty that mirror Divine Being itself? These are the wisdom questions, and they are the entire field of Jesus's concern.
~ *THE WISDOM JESUS*, Cynthia Bourgeault, 2008.

The irony here is that Jesus was so far beyond the tribal consciousness of his day that there was little hope many people would be able to grasp the full extent of his teachings. They would not have asked for answers to the questions that Dr. Bourgeault poses. It is no wonder that even to this day, people who say they are Christian still have levels of consciousness closer to tribal than integral.

Dr. Bourtgeault says that the word *repent* means to *go beyond the mind*; it is a commandment to move through states of consciousness, to reach for *Christ consciousness*, which is enlightenment in Eastern terminology. Dr. Bourgeault says that repentance:

> . . . doesn't mean feeling sorry for yourself for doing bad things. It doesn't mean to 'change the direction in which you're looking for happiness,' although it is often translated that way. The word literally means . . . to 'go beyond the mind.' The repentance that Jesus really is talking about means to go beyond your little egoic operating system that says, 'I think, therefore I am,' and try out the other way—the big one—'I am, therefore, I think.'
> - THE WISDOM JESUS, Cynthia Bourgeault, 2008.

In other words, repentance means to transform, to become more awake. Jesus is showing us that levels of consciousness exist and that it is possible, necessary, to move to ever more sophisticated levels of being. Become more loving, more tolerant, and more compassionate. Become ever more active as you seek to discovered and evolve yourself and your culture.

Looking at the religious movements from anywhere on the globe, we can see the evidence, again and again, for the same duality. Whatever the names given for the duality, and there are a set of names for each tradition, the core teaching is the same. My contention is that the core is always allocentric versus egocentric. Dr. Bourgeault, for example, discusses a duality held by early Christian church fathers that implies we are of two minds. Clearly, the Christian understanding of our two forms of perception, our dual consciousness, is part of the ancient Wisdom Tradition:

> The early church fathers used to speak of a pathway epinoia, which meant knowing through intuition and direct revelation [through

experience], not through the linear and didactic dianoia of logic, doctrine, and dogma.

~ *THE WISDOM JESUS*, Cynthia Bourgeault, 2008.

Here we find two more words for our duality: the allocentric *epinoia*, and the egocentric *dianoia*. The contrast between the foreground and the background, between the egocentric mind and the allocentric mind has many names. Perception of the gestalt has been called the allocentric mind, epinoia, floodlight attention, experiential knowing, and the spiritual path. Perception of the figures that arise from the background has been called the egocentric mind, dianoia, spotlight attention, categorical knowledge, and the rational path. It is the same two beasts—*twins* wearing different clothing.

In the gnostic Gospel of Thomas (logion 22), we find these words [edited, see the original text in Dr. Bourgeault's book *The Wisdom Jesus*]:

When you are able to make two become one,
The inside like the outside and the outside like the inside,
The higher like the lower,
So that a man is no longer male,
And a woman no longer female,
But male and female becomes a single whole . . .
then you will enter.

Saint Thomas writes solidly from within the Wisdom Tradition. His gospel is evidence that Jesus was familiar with ancient esoteric teachings. "Making two into one" is another rendition of background perception absorbing the objective world—including our egos. The commandment is to become not your "location in space at a set time" (where the ego hangs out), but become the whole environment, become everything and timeless, become both feminine energy and masculine energy.

From another Western perspective, another look at duality, the Christian mystic Jacob Boehme (1575-1624) created a cosmology that came about because of a revelation he had as a young man. Boehme's work is complex, spread over several ancient books, so the following short summary is obviously incomplete.

For Boehme, God was a creation, a figure/form that came out of a dark background; in other words, for Boehme, God had an origin. The Divine was a figure that evolved from a universal ground. God emerged as a bodily form from an unknowable field. To the contrary, Lucifer, the god of luminous beings, also came from a field, but it was a background world of eternal light. God did not know about eternal light, and Lucifer did not know about eternal dark, so they both hungered for what they did not have. According to Boehme, God has been busy evolving into light and manifesting ever more love and compassion. Meanwhile Lucifer, seeking to know the dark side, has been evolving toward entropy, evil, the pitch dark night. Two evolutions, according to Boehme, are going in opposite directions, playing out on planet earth—and inside of us.

Boehme also said that there was a force of contraction that was sucking everything inward. This was balanced by a force of expansion that was driving everything outward. This is a figure-ground metaphor: the background role is to flatten everything into a gestalt, while the egocentric role is to manifest three-dimensional figures out of the flatness. Jacob Boehme is another great thinker who discovered he had two minds inside himself that were battling it out for recognition and ascendancy. His allocentric mind created an expansive, floodlight world of ever more light and compassion—a syntropy, a well of creativity, a life force, an eternity of expansion, a God worthy of adoration and awe. His egocentric mind, to the contrary, was sucking the universe into a point; spotlight energy was freezing the life force. Egocentric energy is a force of contraction, reductionism, and the "work of a devil" that would reduce human beings and all life to meaningless, dispensable, entirely material stuff.

In the trilogy of Father, Son, and the Holy Spirit, we find a symbolism for the background (The Father), the Figure (The Son) that arises from the background, and the mysterious force (energy) that extrudes figures from the background or pulls figures back into the background (The Holy Spirit). However, this simple image does not do justice to the complex concept behind the trilogy. That which is manifest from the background is still part of the background, just a bulging outward that gives the illusion of separateness. Also, the energy that holds the manifestation to the background is not a one-time phenomenon. The energy that connects the figure and the ground can be understood to oscillate and to spiral (to breathe), and is a constant communication or relationship between the background (God) and the foreground (Jesus, or *you*, for example). In a way, we become a manifestation of

God (the mysterious background), but we also manifest or create God—from our perspective. Richard Rohr explains:

> You become the God you worship. In other words, your image of God creates you. If you get the image of God wrong, everything else that builds on it is going to be rather inadequate. That might seem like an overstatement, but let's recognize how that's been true in Western Christianity in particular.
>
> The operative image of God for most Christians (except for the mystics) is a powerful monarch, usually an old white man sitting on a throne. It's no accident that the Latin word for God, Deus, came from the same root as Zeus. At the risk of shocking you, let me say that Christianity hasn't moved much beyond the mythological image of Zeus. Yet this is not the image of God revealed to us by Jesus—a vulnerable baby born in an occupied and oppressed land; a refugee; a humble carpenter whose friends were fishermen, prostitutes, and tax-collectors; a political criminal executed on a cross. In other words, Jesus shows a vulnerable God much more than the almighty one Christians often assume.
> - Richard Rohr's Daily Meditations, February 27, 2017.

I would put this eloquent explanation another way: the image that a person has of God depends on that person's level of cognitive and spiritual development. When the level of consciousness is low, at the red warrior stage or the ethnocentric amber level, God is an old white guy—a big daddy who loves you but also doesn't trust you. At ever higher stages of consciousness, God becomes ever more tolerant, forgiving, and loving. At the highest levels, you realize that you—as a manifestation—are also God. This realization comes with a great deal of responsibility. For example, you must decide what is good and what is evil, and you must take wise action. But how do you decide what is good and what is evil? A decision might be good for one sentient creature (a predator, for example) but not so good for another sentient creature (the prey).

The concept of good and evil might also be related to the oscillation between background and foreground. If we suppose the background to be Heaven or Nirvana, then manifestation is a movement away from all that is good and holy—it is a *fall from perfection* into the egocentric world of separateness from The Divine; it is a form of evil. Turning this on its head, manifestation of the human being is a plan that the background perfection is bringing forth—an evolution of perfection. Therefore, manifestation is a gift of The Divine. This is why the form, the manifestation, can also be seen as good.

The key to Western Wisdom Traditions is the assumption that the background has to initially contain sentience for it to manifest sentient beings. Something alive, which the Western Traditions call *God*, must have existed before it could extrude figures that contain sentience. In esoteric terms, *As above so below* is at work here. The *unmoved mover* can manifest movement; therefore, movement must have been potentially contained within the background to begin with.

In the early 1960s, Jesuit priest Karl Rahner (1904-1984) stated that if Western Christianity did not rediscover its mystical foundations, we might as well close the doors of the churches because we would have lost the primary reason for our existence. The *mystical foundations* teach us to move from *belief* systems—or *belonging* systems—to actual inner experience, to *being* systems. All spiritual traditions, at their mature core, agree that such a movement is possible, desirable, and available to everyone.

I will give Richard Rohr the final word in this section about Christianity. Rohr's perspective is primarily Christian, and he and his fellow teachers at the *Center for Action and Contemplation* in New Mexico, clearly link Christianity with all religions across the globe:

> Until someone has had some level of inner religious experience, there is no point in asking them to follow the ethical ideals of Jesus or to really understand Christian doctrines beyond the formulaic level. In fact, moral mandates and doctrinal affirmations only become the source of deeper anxiety and more contentiousness! And then that very anxiety will usually take the form of denial, pretension, and projection of our evil elsewhere.
>
> You quite simply don't have the power to obey the law or follow any ideal—such as loving others, forgiving enemies, nonviolence, or humble use of power—except in and through union with God. Nor do doctrines like the Trinity, the Real Presence, salvation, or the mystery of Incarnation have any meaning that actually changes your life.

They are merely books on shelves. Without some inner experience of the Divine, what Bill Wilson of Alcoholics Anonymous called "a vital spiritual experience," nothing authentically new or life-giving happens.
~ Richard Rohr's Daily Meditations, January 24, 2017.

In other words, only through *union with God* will we discover the wisdom of the Wisdom Tradition. Only in the most evolved forms of spiritual development, at the highest rungs of consciousness, can we expect to see the teachings of the Wisdom Tradition adopted and understood.

Rudolf Steiner and the Wisdom Tradition

In my terminology, Rudolf Steiner spent his adult lifetime trying to find an approach that would allow for serious exploration of the missing allocentric twin. Steiner called himself a spiritual scientist, and he spent his career looking at the evolution of consciousness. He tried to convince the materialistic, scientific culture of his time (and our time) that the allocentric mind was a real entity that could be studied. For example, Steiner made a distinction in his presentations between "Jesus" and "Christ," to demonstrate the difference between our two minds.

Jesus, he said—using my terms here—is the egocentric mind's depiction of The Divine—words and events are chronicled in the Bible, words that paint an academic, time-based, literary picture. Jesus, according to Steiner, was a manifestation from God, an egocentric sentience that emerged from a Divine background. However, according to Steiner, there is something else, something hidden from the ego that is being overlooked and not understood, something words cannot convey, something that is different than the egocentric Jesus, something that feels like a union with The Divine. This, Steiner says, is *Christ*, the allocentric mind's depiction of union with Divinity—Christians call this awareness *the body of Christ*. Steiner insists that this union with The Divine is relatively lost in modern man, but that it was the main consciousness of an earlier man. This makes sense, since the modern egocentric mind became dominant as language, social complexity, and sophisticated technology became dominant.

Steiner is saying that *Jesus* is egocentric and *Christ* is allocentric. Here, then, we find another language distinction between the two minds. Just as we have egocentric belief, ego, and the concept of Jesus, we also have faith, soul, and the concept of Christ.

Steiner felt that Christianity never lived up to the mission that Jesus brought down to the common man. Steiner saw his own mission on earth as a time to teach how this failure of Christianity impacted the Western World. Steiner remains fascinating and mysterious even to modern readers who do not understand our inherent duality. His teachings made an impact, but his effort to bring compassion, wisdom, and peace to the world—which early allocentric Christianity preached—has so far failed.

Steiner was ahead of time—he was not understood or appreciated by his contemporaries. I will discuss Steiner in more detail in the chapter about philosophy and duality.

Samsara and Lila

Just like in Christianity and Buddhism, the Hindus perceive a duality to the universe. We find the now familiar division between being and doing:

> Brahman, though undifferentiated, is dynamic and creative. From its ultimate "being" comes the temporary "becoming" of things in the manifest world, with their various attributes, functions, and relationships. The samara of being-to-becoming, and then of becoming-to-being, is the Lila of Brahman: its play of creation and dissolution.
> - *THE SELF-ACTUALIZING COSMOS*, Ervin László, 2014.

The Hindu philosophy that Ervin László describes above refers to a fundamental background out of which the universe arises—there exists an oscillation between emergence and dissolution, breathing in and breathing out, a sine wave going up and down. This background is called the Akashic Field in Hindu literature. Here is how László describes this background field:

> The Akasha is the fundamental element. It holds the other elements in itself, but is also outside of them, for it is beyond space and time. According to Paramahansa Yogananda, the Akasha is the *subtle background against which everything in the material universe becomes perceptible.* [my italics]

. . . The same concept is present in the Upanishads. "All beings arise from space, and into space they return: space is indeed their beginning, and space is their final end." (Chandogya Upanishad 1.9.1)

David Bohn (Nichol, 2003) enunciated an identical concept: What we experience through the senses as empty space is the ground for the existence of everything, including ourselves. The things that appear to our senses are derivative forms and their true meaning can be seen only when we consider the plenum, in which they are generated and sustained, and into which they must ultimately vanish.

~ *THE SELF-ACTUALIZING COSMOS*, Ervin László, 2014.

This quote is important for three reasons. First, it recognizes the limitations of human perception. These limitations color and influence our ability to truly understand the reality we are in—we are facing something that is beyond space and time. Second, the quote understands that empty space is a key to our ability to navigate through the world. There is an emptiness without which nothing could emerge. Third, the quote recognizes the philosophy of quantum physicist David Bohm who defined the fundamental duality in quantum language as an implicate order (the background) and an explicate order (the manifest).

There is a chicken and egg problem that confronts the Wisdom Tradition. Either our perceptual system is made from the same algorithm that repeats throughout the cosmos, or our perceptual system is making up all this background-foreground business based on its own biological mandates—we see backgrounds and figures everywhere because that is the only thing our biology (especially vision) knows how to do. Perhaps we copy and mirror that which came before and that which will come after. Here is how Colin Wilson describes the Buddhist perspective that understands human beings as enormous mirrors reflecting reality:

According to Gautama, the feeling of delight and freedom is only the beginning. In such a state, your mind can focus on wider horizons, broaden its beam of perception from its usual narrow compass to a sense of the richness and complexity of the world. But there is no reason why a man who has made this discovery should not continue to broaden his beam of perception until it embraces the whole universe, until it contemplates the notion of infinite space and time, and recognizes the total unimportance of human existence against this vast canvas. Man ceases to be man: that is, a creature who pursues his own aims and desires, and becomes an enormous mirror that reflects reality. And if ordinary "contemplation" brings a sense of delight and relief, then this universal contemplation would bring a detached ecstasy beyond anything we can imagine, the ultimate peace and delight of Nirvana.

~ *POETRY AND MYSTICISM*, Colin Wilson, 1969.

Wilson is showing us once again the universality of figure and ground. The Wisdom Tradition is found in Buddhism, Hinduism, and Christianity—different words are used, but the concept is the same. The practices of the great religions encourage greater spatial awareness, and they teach the following of a spiritual path that leads to ever more expansive allocentric awareness—toward enlightenment.

East Meets West: The Fourth Way of G.I. Gurdjieff

I am like the Japanese poet
who longed to be in Kyoto
even though he was already in Kyoto.
~ "BASHO IN IRELAND," *THE RAIN IN PORTUGAL*, Billy Collins, 2016.

That which you seek you already have. Everywhere you go your essence goes with you. But your ego doesn't know this, it doesn't know about its twin, so it must go on many journeys seeking for the "lost" sacred knowledge. That is what happened to G.I Gurdjieff when he was a young man.

The spiritual teacher George Ivanovitch Gurdjieff traveled through Asia as a young seeker; he was a man on a quest. He went from wisdom school to wisdom school studying ancient philosophy, trying to understand what the early religions wanted to preserve and pass on to future generations. What Gurdjieff found were the practices, meditations, and strategies used to transform the human mind from an egocentric level of consciousness to ever more allocentric stages of awareness. These practices were designed to expose our hidden mind, to move people out of their heads and into their hearts. Gurdjieff synthesized all the perspectives, invented his own esoteric terminology, and then set up his own schools in the Western world.

Gurdjieff, in his own books and words, is hard to understand because he runs abstract sentences together that fill half pages. In his defense, he didn't know English well and so his complex and esoteric thoughts had to be translated more than once—he insisted on editing and re-editing. We can thank the persistence and devotion of Olga De Hartmann, his friend and informal secretary for the hard work of translating. Olga was Thomas De Hartmann's wife—the two were avid followers of Gurdjieff's school. On the other hand, Gurdjieff deliberately crafted his books to be arcane and mysterious. To tease out the essence of his writing takes hard work, and Gurdjieff was all about hard work. He knew that only experience changed human essence. His books are not an easy read, they are a difficult experience.

One of Gurdjieff's most devoted students took up the quest to transform human consciousness after Gurdjieff (1949). Unlike Gurdjieff, Jeanne De Salzmann writes with great clarity, and thus she reveals Gurdjieff in a way that Western readers can readily comprehend. In the paragraph below, De Salzmann seems to capture the distinction between what I call the *egocentric attention system* and the *allocentric awareness system*. Italics are mine:

> Gurdjieff taught the necessity of self-observation, but this practice has been mostly misunderstood. Usually when I try to observe, there is a point from which the observation is made, and my mind projects the idea of observing, of an observer separate from the object observed. But the *idea* of observing is not the observing. Seeing is not an idea. It is an act, the act of seeing. Here the object is me, a living being that is not that of a fixed observer looking at an object. It is one complete act, *an experience* that can take place only if *there is no separation between what sees and what is seen*, no point from which the observation is made.
>
> ~ THE REALITY OF BEING; THE FOURTH WAY OF GURDJIEFF, Jeanne De Salzmann, 2011.

De Salzmann is making a distinction—as did Gurdjieff—between egocentric attention that involves a seer and a seen (a subjective and objective worldview)—and allocentric awareness which is pure experience, being in the moment. To self-observe means that you don't lose track of the self, as opposed to the ego.

Gurdjieff discovered and taught that perception from an egocentric perspective was all that most humans understood about themselves. Human beings do not know how to consciously perceive allocentrically even though the ability is within each person; indeed, it is part of our very anatomical structure. However, the ego obscures this twin mind, and most people go through a lifetime without realizing their true potential—they miss half their essence.

Gurdjieff knew that talking and thinking—using the rational mind, the ego-perspective—would never cause a transformation to a higher level of consciousness. Only experience can cause change. That's why he made people dance—just as Rudolf Steiner did—to get them into a frame of mind that would bring about an alteration in a person's level of consciousness. Gurdjieff made people participate (in music, dance, in all forms of creative expression, and through hard physical labor) rather than rationalize or debate.

Here is another quote from *The Reality of Being; The Fourth Way of Gurdjieff* that is about as clear as you can state the issue: We have a double nature: twins reside within one body, oscillating between manifestation as an ego and absorption (involution and evolution) back into the self:

> I have to know that I have a double nature, that there are two forces in me: the descending force of manifestation and an ascending force returning to the source. I have to experience them here at the same time in order to know myself as a whole. There must be some reason why I am here, something that is needed for a relation between the two. This is the meaning of my Presence.
>
> In each event in life—whether family, professional or inner life—there is a double movement of involution and evolution. The action is directed toward an aim, toward manifestation, but behind it is something that has no aim, that does not project itself but returns to the source. These two currents are indispensable to each other . . .
>
> I cannot be aware at the same time of two movements going in opposite directions. I am taken by one movement and ignore or oppose the other. Nevertheless, I have to accept that the two currents

determine my life, and that I have two natures in myself. I must learn to see the lower nature and at the same time remember the higher. The struggle is in living the two together.
~ THE REALITY OF BEING; THE FOURTH WAY OF GURDJIEFF, Jeanne De Salzmann, 2011.

I am amazed how clearly the above paragraph explains the conundrum of human existence. There is a background that Gurdjieff calls the source (self or soul). If the source is seen as a spiritual field, then it pulls forms, including human beings, back into its undifferentiated self. The forms ascend into the background, as if the background was above, like heaven. However, when forms manifest from the source, it is as if they are descending downward from above. Evolution is the manifestation, the outward breath of the cosmos. Involution ("backward evolution," Rudolf Steiner called it) is a return to the source, a movement away from form into formlessness, what the Buddhists call *emptiness*.

Deepak Chopra and the Wisdom Tradition

Deepak Chopra (born 1947) fits everywhere in this dialogue; putting him in this chapter on the Wisdom Tradition is as good as anywhere else. Dr. Chopra is a spokesman for consciousness, a medical doctor, a teacher, and an eloquent and sincere modern guru.

It's unnerving to realize that Deepak Chopra explored many of the same insights that I present in this book, and he did it decades before I began my musings. In 1989, for example, he wrote a book called *Quantum Healing* (revised in 2015). In that book, Dr. Chopra used a simple logic to begin his thinking: physicists have deduced a fundamental field out of which quantum forms manifest; the whole world is quantum. This means that human beings are also quantum—we are made from the same universal soup. Western minds may go on living in the "biochemical universe," but logic doesn't support this. Our microscope-based perspective cannot be sustained given the evidence from quantum physics.

Dr. Chopra routinely goes after the egocentric mind, finding fault with reductionism and materialism. In Western culture, egocentricity is seen as the only "home base" from which to explore spirituality. If we stay in this barren landscape of the egocentric mind, however, we miss the miracle that is life; we fail *to experience* our lives as full, mysterious, and holy. Essentially, Chopra is saying that we have ignored the allocentric mind—we refuse to even entertain the prospect that such a hidden mind might exist.

In a review of the Wisdom Tradition (available on YouTube), Dr. Chopra outlines seven stages of consciousness. This looks very much like the levels of consciousness that came out of integral psychology, the work of Ken Wilber and his colleagues. In the following categorization, Dr. Chopra is associating brain waves with *states* of wakefulness (rather than *levels* of consciousness). Dr. Chopra is exploring degrees of wakefulness:

1. The first conscious state is deep sleep. In this realm of reality we are almost totally egocentric. This level of consciousness is dominated by delta brain waves.

2. The second state of consciousness is the dream state. Here we can envision the ego in relationship with an illusory environment peopled with other sentient creatures; ego-and-other manifest in this magic state of reality. This level of reality is confined to the body; our dream movements probably come from proprioceptive sensory-motor memory.

   States of deep sleep and dreaming can be subdivided and yield amazing realms—like the theta state that is a zone between sleep and waking where psychics and mystics get many of their powers and insights. This twilight zone of consciousness is dominated by theta waves.

3. State three is waking consciousness, what we call everyday reality. In this state of consciousness we are still pretty much ego-creatures who identify with the body. We are a solid form in this reality, and this form is separate from others and from nature. Beta waves dominate this level of consciousness. Between theta and beta waves is a level of consciousness dominated by alpha waves—here we get a feeling of deep relaxation, which is primarily induced when the eyes are closed.

4. The first three states of consciousness are familiar to everyone. However, the fourth level goes beyond our ordinary understanding of reality—it is a big leap. State four is called the transcendental state where we glimpse the soul as something beyond ordinary wakefulness. We become aware (primarily through meditation, but also through various kinds of shock—like near-death experiences or drug-induced highs) that we can *project our self* beyond our body. At first we get just a glimpse of this abil-

ity, a tease. However, with practice, this different level of reality becomes more easily achieved at will.

When you drive a car, your brain makes a body map that includes the car. That's why you don't crash into everything as you drive on the highway. The brain knows the boundaries of the car, just as it knows the boundary of the body. In other words, the mind has expanded to include the car. That is your glimpse of the soul moving beyond the body. However, Dr. Chopra says that what is transcended in state four is space, time, and causality. The soul is not in space, not in time; it is acausal. It has no spatial location and no cause. It is not in this material world. "When" and "where" are not relevant for the soul. Everything is perceived as related and synchronized.

This spiritual perspective is different from a proprioceptive perspective. A scientist might say that it takes brain maps and proprioception to give a person a sensation of out-of-body awareness, of going into realities beyond the material body. Dr. Chopra instructs us to have patience because what is happening at state four is just a glimpse, a hint. It is just the first step on a spiritual journey.

5. During the fifth state of consciousness, called Cosmic Consciousness or Christ Consciousness, the soul expands beyond your car and beyond the four walls of any room; it is on an ever-expanding journey outward—it has left the familiar hometown and headed on an adventure. Our minds are now much more sophisticated. We can remain in state four wakeful consciousness at the very same time as we experience being in state five consciousness. The watcher, the witness, shows up at this time; we can watch ourselves going about our daily activities. We can observe our lower soul-states and yet remain in the higher state. Our human role-playing is not who we really are because there is "someone" who can watch us as we play the roles. The witness has awakened.

Our goal at this level of consciousness is to keep the witness active. It is very easy to "fall asleep" and then lose contact with the witness—we tumble back into level four consciousness. We have to cultivate the witness so that we eventually reach a state of constant witnessing. We cannot move beyond state five consciousness unless the witness is ever-present. We eventually learn to be engaged and detached at the same time. At state five consciousness, coincidences start to happen, synchronicities arise. We ask for help and we get the help. Insights arrive just in time. Imagination takes us where we need to go. We learn to communicate with the flower, the bee, with nature; nature can then communicate with us. This is true because we now understand that "mind" includes the environment—we *are* the environment.

6. Level six is called Divine or God Consciousness. The soul is fully awake now, able to witness at all times, and it has expanded (at state five) to include the whole universe. But now we see that witnessing is not just what we are capable of doing; it is not just a human ability. We now observe that the "objects" of the world also witness. There is intelligence in the universe. Nature is awake, alive, and it can communicate its intelligence to you (which is why your questions get answered). Nothing is dead in the universe, everything is a process. *As above so below*—the universe can think, have emotions, mirror what we are, as we mirror what it is.

   When we take a picture, we freeze reality—which soon is history; everything in the picture has passed away the moment after the picture is taken. The picture freezes a time and space that no longer exists. Likewise, when we perceive, we are taking a biochemical picture of a world that is also dead and gone—from our view. But the spirit is preserved. There is nowhere to go that doesn't contain the spirit of things past.

7. Level 7 is called Unity Consciousness, the merging of the subjective witness with the objective witness; subjective and objective are the same "entity." There is just one witness and that is you. You are the whole thing. You are the miracle that creates the whole universe; or any universe.

   For Dr. Chopra, individual human beings are somewhere on this scale of consciousness—most of us sleepwalk through existence, some of us have discovered that there is a spiritual pathway to be followed, and a few of us are actually well along on the journey. There are levels of blindness inherent in this worldview: we are blind to levels above our own and there are those who are blind to our spiritual attainment. In other words, the ego perceives from a level of consciousness, but is blind to hidden dimensions. The Wisdom Tradition is a set of teachings (practices, experiences, knowledge) that reveal that which is hidden.

Richard Rohr

When I was a young man, I was very skeptical about religion and spirituality. I was not raised in a home that followed any religious doctrine—I absorbed the Christian values of my mid-Western culture, but my family never went to church. There were no mornings or evenings filled with exploring the teachings of Buddha, or Jesus, or Muhammad. It wasn't until I retired that I discovered the writings of Richard Rohr, a priest who expressed what I felt:

> I will go so far as to say that the more you can connect, the more of a saint you are. The less you can connect, the less transformed you are. If you can't connect with people of other religions, classes, or races, with your "enemies" or with those who are suffering, with people who are disabled, with LGBTQ folks, or with anyone who is not like you—well, to put it very bluntly, you're not very converted. You're still in the kindergarten of faith. We have a lot of Christians who are still in kindergarten, walking around the world with their old politics and economics. They have not allowed the Risen Christ to fully transform their lives. Truly transformed individuals are capable of a universal recognition. They see that everything is one.
> ~ Richard Rohr's Daily Meditations, December 14, 2016.

When I looked around at the values and wisdom of the "Christians" in my community, I was not impressed. The "holy ones" were almost all males—especially the ones who held power. The preaching was often simple-minded, and the followers quite often ignored what they professed to hold so dear. Hypocrisy ruled over piety. Don't covet your neighbor's wife—right. Coveting was, and still is, a favorite pastime of masculinity—no matter what the religious uniform. The male sex drive has made a mockery of all the creeds. Priests, evidently for hundreds of years, have had sex with altar boys and vulnerable women. Indeed, sex dominated in the shadows no matter what the sermons. And lying and cheating were just part of business—the way things were done in the real world. And there were wars, lots of wars, incessant wars—all justified from beginning to end by the "holy men." *Thou shall not kill*, they said from the pulpit, except when killing was sanctioned by the state or one's religion. So capital punishment was okay, and being a soldier was okay, an eye-for-an-eye was okay, and if your wife cheated on you, you could kill her according to the Christianity that I observed.

The religion I saw all around me as I grew up was a contrary mess of ignorance and cruelty. It was stupid, and anyone with a brain and a heart would naturally reject the whole sorry edifice. That was my youthful stance on religion and on spirituality.

My youth might have been much more tolerant and forgiving had I found the teachings of Franciscan priest Richard Rohr when I was younger. When I did encounter Rohr—I was already in my 60s—Rohr made Christianity sensible to me, finally. He discussed a God that was big enough and wise enough for my scientific, secular mind to consider. Rohr made Christianity sane and understandable; he was able to fit Christianity into a historical perspective. It wasn't that Christianity was ignorant and harmful, as I had perceived in my youth, it was rather that Christianity had never honestly been tried. The Christians only took the parts of the teachings of Jesus that fit their levels of consciousness—most of which was at a very low level.

So why is Richard Rohr such a modern treasure? For many reasons, of course, but I will mention the four that are most pleasing and relevant to me:

1. Richard Rohr can see that masculinity is a major problem on the planet. He has held many retreats at his center in New Mexico designed to help men find their empathy, their inner spirit, and their need for relationships. He has helped men explore what love is. He has explored personality—mostly through his understanding of the Enneagram—and he has a solid comprehension of levels of consciousness and Integral Philosophy. He could see the need for a Men's Liberation Movement, and he has quietly gone about the process of bringing it about.

2. Richard Rohr understands the Wisdom Tradition. His God and his religion exclude no one. His God is all of creation, all the sentient creatures, all the planets, the void—eternally forward and forever backward. His God is big enough to hold all of us.

3. Richard Rohr understands human duality. He understands our fundamental dilemma. He perceives the dual mind—egocentric consciousness—and he understands that there is a second mind, an allocentric consciousness that knows no duality. He sees our conundrum:

The work of spirituality is to look with a different pair of eyes (non-dual eyes) beyond what [Thomas] Merton calls "the shadow and the disguise" of things until you can see them in their connectedness and wholeness. In a very real sense, the word God is just a synonym for everything. So if you do not want to get involved with every-thing, stay away from God. But then you need the nondual or mysti-cal mind to love and fully experience limited ordinary things and to peek through the cloud to glimpse infinite and seemingly invis-ible things. This is the contemplative mind that can "know spiritual things in a spiritual way" (1 Corinthians 2:13), as Paul says.
~ Richard Rohr's Daily Meditations, August 17, 2016.

4. Richard Rohr understands the importance of being in the moment. He grasps the significance of the allocentric mind:

Of all the things I have learned and taught over the years, I can think of nothing that could be of more help to you than living in the now. It is truly time-tested wisdom.

So many leaders in so many traditions have taught the same thing: Hindu masters, Zen and Tibetan Buddhists, Sufi poets, Jew-ish rabbis, and Christian mystics to name a few. In the Christian tradition, we have heard it from Augustine, the Franciscan Francisco de Osuna, the Carmelite Brother Lawrence, and more recently, Paul Tillich and Alan Watts. Contemporary teachers Thich Nhat Hanh and Eckhart Tolle have done much to help us understand the impor-tance of living in the now.

Jesuit priest Jean Pierre de Caussade (1675-1751) called it the "sac-rament of the present moment." His book, *Abandonment to Divine Providence*, was the book most recommended by spiritual directors for many decades. His key theme is: "If we have abandoned ourselves to God, there is only one rule for us: the duty of the present moment." To live in the present is finally what we mean by presence itself!
~ Richard Rohr's Daily Meditations, November 21, 2017.

Rohr is not saying in the quote above that we should shun our egocentric minds. He can see that we need the egocentric dualistic mind to function in practical life to do our work as teachers, nurses, scientists, or engineers. However, the egocentric mind can be (and is) very dangerous when it inhibits basic human attributes like love and charity:

> It [the egocentric mind] is helpful and fully necessary as far as it goes, but it just doesn't go far enough. The dualistic mind cannot process things like infinity, mystery, God, grace, suffering, sexuality, death, or love; this is exactly why most people stumble over these very issues. The dualistic mind pulls everything down into some kind of tit-for-tat system of false choices and too-simple contraries, which is largely what "fast food religion" teaches, usually without even knowing it. Without the contemplative and converted mind—honest and humble perception—much religion is frankly dangerous.
> ~ Richard Rohr's Daily Meditations, January 29, 2017.

It is quite refreshing to read Richard Rohr say that much of religion is frankly dangerous. Not his religion, which is an honest, compassionate, hopeful lighthouse, but so much of what passes for spirituality in the United States (and in most of the world) today.

Nondual consciousness is about receiving and being present to the moment without judgment, analysis, or critique, without the ego deciding whether something should be labeled good or bad. Reality does not need our opinion. Here are a couple more Richard Rohr insights that appeared in his blog as I was writing this chapter:

> The nondual, contemplative mind is a whole new mind for most people! With it, you can stand back and compassionately observe the self or any event from an appropriately detached viewing platform. This is the most immediate and practical meaning of "dying to self" I can think of. As a general rule, if you cannot detach from something, you are far too attached to it! Eventually, you can laugh or weep over your little self-created dramas without being overly identified with them or needing to hate them. Frankly, few people fully enjoy this emotional freedom.
> ~ Richard Rohr's Daily Meditations, January 30, 2017.

Whenever you move to a higher level of consciousness, you would be wise to include the previous stages; do not waste much time reacting against the previous idea or generation, which had some level of truth to it. Rejecting the lower level was the dualistic mistake of almost all reforms and revolutions until very recently. This is why most reformers tended to repeat the same ills but just in a new way. Because nothing really "new" happened, most reforms quickly needed reforming themselves. The ability to "transcend and include" is the sign of a higher (or deeper) level of consciousness.
~ Richard Rohr's Daily Meditations, January 31, 2017.

This is an important insight: when you transcend, don't look back where you were with distain, especially when you survey the people still stuck where you just were. Use the new compassion you gained to be tolerant and helpful, and hope that those more sophisticated than you will do you the same favor. Hope is necessary for our survival. This is Rohr talking about hope as it relates to Noah's ark and what the allegory really means:

The story of Noah and the flood is filled with insight. (Although I do not really believe God killed all the people on the earth and saved only one family!) God tells Noah to bring into the ark all the opposites: the wild and the domestic, the crawling and the flying, the clean and the unclean, the male and the female of each animal (Genesis 7:2-15).

Then God does a most amazing thing. God locks them together inside the ark (Genesis 7:16). Check it out.

Most people never note that God actually closed them in! God puts all the natural animosities, all the opposites together, and holds them in one place. I used to think it was about balancing all the opposites within me, but slowly I have learned that it is actually "holding" things in their seemingly unreconciled state that widens and deepens the soul. We must allow things to be only partly resolved, without perfect closure or explanation. Christians have not been taught how to live in hope. The ego always wants to settle the dust quickly and have answers right now. But Paul rightly says, "In hope we are saved, yet hope is not hope if its object is seen"

(Romans 8:24). The virtue of hope widens and deepens our foundation.
- Richard Rohr's Daily Meditations, August 30, 2017: Forgiving Reality for Being What It Is;

We are all floating in Noah's ark. We are stuck in this space, this vehicle, with all these inevitable, irreversible contradictions—dangers, confusions, and possibilities. This is the human condition. This is the given we are born into. We have no choice except to create a holding space for this dilemma where we can practice compassion and communion.

I look forward each day to the short blog posts of Richard Rohr. If only Christianity could become the clear, sane, loving force that he so genuinely expresses. I also sincerely thank Richard Rohr for opening my heart—his wisdom and articulate words have impacted my life in a positive, hopeful, and joyful way. Check him out.

Hermetic Philosophy

Hermes Trismegistus is a legendary (perhaps mythological) *Master of Masters* who is purported to be the author of a series of sacred texts that are the basis of Hermeticism (secret doctrines). It could be that because Hermes Trismegistus was a Greek god of wisdom, anything written about wisdom (or with wisdom) was attributed to Hermes Trismegistus during the early centuries of recorded history.

Hermetic philosophy is a perspective thought to be so powerful it could only be understood by very sophisticated minds, by people with high levels of consciousness. Consequently, the hermetic teachings became secret, passed through oral traditions or, when the teachings were written down, presented through obscure symbolism that only the initiates could comprehend. In modern terms, only those who reach a certain level of consciousness—a threshold of comprehension—begin the search to move beyond knowledge into wisdom. Above this threshold of wisdom, rites of passage allow the seeker to transcend to ever higher levels of consciousness.

Consequently, in the beginning, hermetic philosophy was not a religion for the masses. However, the collective consciousness of mankind is accelerating, so perhaps the masses are about to be let in on the secrets. It certainly feels as if more and more people are meditating, doing yoga, dialoguing about spirituality, and disciplining their minds—eating right, exercising, and working in the community to cultivate compassion.

Below, I summarize the seven hermetic principles. These foundational ideas underlying hermetic philosophy are amazingly modern. They are even a foreshadowing of the quantum world. Certainly, they represent the fractals that are repeating in this world. I will review these seven principles below because they strongly support dual-process theory. I drew heavily from the book *The Kybalion: A Study of The Hermetic Philosophy of Ancient Egypt and Greece* (2012) for this information and I invite you to refer to that publication for greater elaboration.

Principle One

The universe is intelligent. Beyond everything is a Mind. Whatever this Mind "thinks" becomes manifest, just as thoughts are manifest in human minds. From a dual-process perspective, what this principle says is that there is a fundamental (original) background out of which features (like universes) manifest. Therefore, a fundamental law of existence is this notion of backgrounds and foregrounds (figure and ground). The ancients figured this out centuries ago.

The question that always arises for my egocentric, systems-oriented mind is whether our notion of a universal principle of figure-ground was created because our own minds are constructed this way (did we create Hermeticism in our own dual-image?), or rather that our minds were the creation of a fundamental first principle. It is a chicken or egg paradox—which came first? The answer, I think, is that we are smack in the middle of a process. From an egocentric perspective, we chickens came first and then we created all our conceptualizations, our religions and our sciences. However, from an allocentric perspective, we are part of a process where only the whole (background) is real—all figures are illusions that arise from the whole. In the beginning was the egg, the seed, the frequency code for creation. The paradox is true.

The logic of this becomes clearer when we understand Principle Two. If there is a mind below, then there is a mind above (a more universal intelligence). Likewise, if there is a human mind above, there are other minds being created below.

Principle Two

This is the principle of correspondence: *as above, so below.* There are planes (levels) of existence. What happens in each plane is mirrored in all others. If there is intelligence above, there is intelligence below. If our minds can manifest thoughts, then greater minds on a higher plane can also manifest thoughts. If a fundamental Mental Background exists out of which everything manifests, then on a physical plane there should also exist a "Higgs Background Field" out of which all material existence manifests. If there are two kinds of consciousness below, then there are two kinds of consciousness above. If quantum physics reveals a fundamental duality, then that duality corresponds (carries over) to other planes of existence and was itself caused by a more fundamental duality. *As above, so below.*

Laws discovered above or below inform us about all levels of existence, all scales, and all planes. The principle of correspondence is a portal into understanding the mental, material, and spiritual planes; it is a portal that contains *insight-generating pathways.* We search for the unknown above us and below us as we explore the fractals that build our world.

For example, if we observe that life seems to arise from seeds or cells, and if we find that these seeds contain informational codes that are a plan for creating organisms, then we can look everywhere for seeds and codes—above and below. DNA is a code that creates lifeforms. Therefore, we would expect that there are seeds and codes that came before DNA; these codes create molecules. Below the molecular level are seeds and codes that make atoms. We would also expect to find seeds and codes for making universes—as if a seed, filled with instructions for making a universe, was the source for the big bang.

Furthermore, if there is a fundamental on/off fractal building our universe, we would expect to see the same process operating above, as well as below. We would expect to find movement alternating with no-movement at all scales. Indeed, in our own biology, we find stacks of examples of inhibition alternating with excitation, suppression alternating with expression, genes switching on and genes switching off, and so on. Of course, Sir Isaac Newton figured this out long ago when his mathematics led him to conclude that *for every action there was an equal and opposite reaction.* It should come as no surprise that this universal oscillation—this law of our universe—is found in the construction of the human mind.

There is also a fundamental blindness build into our plane of existence. When I was reading *The Blind Watchmaker* (1986) by Richard Dawkins, I came across this sentence:

> Natural selection is the blind watchmaker, blind because it does not see ahead, does not plan consequences, has no purpose in view.
> *THE BLIND WATCHMAKER*, Richard Dawkins, 1986.

This struck me as a perfectly good definition of the allocentric mind. One half of what we understand to be Nature contains a process that goes about its business blindly. Something that was blind created the fractal that made natural selection. That fractal is busy creating whatever comes after this incarnation of natural selection—perhaps artificial (manmade) selection. Blindness, the inability to perceive one-half of existence, is a correspondence: as above, so below. Attention and observation are always accompanied by levels of ignorance.

Scientists—led by the mathematicians—discovered that chaos and pattern-building were inherently linked in our physical universe. Out of seeming chaos emerge spontaneous and never-quite-the-same patterns. In other words, Nature can self-organize. Patterns emerge from a background of randomness. This is the familiar primal pattern: figures emerging from a patternless ground. Notice that these emerging patterns are always moving—this is not a static law; it is a process. Forms emerge and then forms dissipate.

Flow (process and movement) is another fractal that is being used above and below in creation. Everything is moving relative to everything else. There is a cosmic allocentric mind at work. As this mind flows, it is not able to comprehend. If it pauses to comprehend—to be an egocentric mind—it can no longer flow. Vision cannot see itself in the act of seeing. The universe cannot observe itself observing. We cannot observe another mind at the same time we are observing our own mind.

If there is something (a fractal) that creates allocentric minds, then what is the fractal that makes egocentric minds? What is the logic of the universe? Where is the other half of the fractal equation that balances the blind watchmaker? What is it that is transcending blindness? What is building minds capable of knowing? What is the fractal that is building "knowing that we know?" Where is self-awareness coming from? It must be coming from something above that is aware and attentive. Something below is also building awareness and attentiveness.

Using the principle of correspondence doesn't provide answers to *why* things operate in this fashion. But it does point out fractal building blocks. For example, it does suggest that self-consciousness, arising from the egocentric mind, came before human minds, and that self- consciousness is being created beyond the level of the human mind. Perhaps, as the internet combines with cloud computing, with robotics, and with brain-creation engineering, human beings will be the agents for creating the next generations of entities that are self-conscious. These evolved creatures will then create the next series of self-conscious organisms, and so on.

Principle Three

> Every action, either physical or mental, every movement occurring either on the plane of gross matter or on the plane of the mind, causes an emission of energy.
> ~ *The Secret Oral Teachings in Tibetan Buddhist Sects*, Alexandra David-Néel, 1967.

Nothing manifests or de-manifests *unless energy is applied*. Since life is a process—always in a state of change—this energy requirement is a constant. The word *energy* is just another term we have for movement: potential movement or manifest movement. *Process* is another word for movement or energy. Constant, relative *flow* is another term for movement. *Frequency* is yet another term for relative movement. Everywhere we look—above or below—we see this relative movement, this balance between movement and no-movement, between the non-manifest and the manifest.

Everything is moving relative to everything else. This relationship of energies is fundamental to our reality. Nothing is ever at rest. Existence is created out of these various relative vibrational states. Furthermore, high states of vibration seem to correlate with high planes of consciousness.

The material plane is composed of relatively slow vibrations. Higher planes of vibration, beyond the material, are in the mental plane. The highest vibrational planes, beyond material and mental, are spiritual.

Principle Four

Everything is dual. If there is "on," there is also "off." Inhibition is balanced by excitation. The egocentric mind has an allocentric twin. The reason that human beings are bilateral and cognitively dual is because we are constructed from a universal fractal that creates duality. "Nothing" is a background. "Something" is the illusion that arises from the background. Nothing can manifest that doesn't have a background.

Everywhere we look, in our books, our movies, our history, our biology, our psychology, in quantum physics, all of it is dual, all of it follows the fourth principle of hermetic philosophy.

Principle Five

There is a universal rhythm. Everything breathes in and breathes out. Everything manifests and de-manifests. The tides come in and the tides go out. Genes are activated and genes are shut down. For every action there is a reaction. What rises like a sine wave must fall like a sine wave. The pendulum swings left and then the pendulum swings right. Synchronicity is a universal law. Coherency is law. Harmony is law.

Principle Six

Every cause has an effect, and every effect has a cause. There is no such thing as chance. There is a law for everything. Nothing ever just happens.

Of course, this is an egocentric perspective. To the ego there is a past and a future; causation is a universal truth. However, the allocentric mind lives in an acausal, potential, unmanifest state of probability. What seems to arise from the allocentric mind are acausal events. Reverse causation pulls forms back into the hell of nothingness—inactivity that lacks purpose. There is no past, present, or future in the background mind—it is pure potential. Therefore, this law is egocentric—and for me, the sixth law is incomplete because it omits allocentric non-causality.

Perhaps this law simply points out that every manifestation has an origin. That which became a manifestation is predestined to return to the unmanifest state.

Principle Seven

Gender is everywhere and in everything. The masculine is balanced by the feminine. The yin cannot be separated from its opposite, the yang. The feminine is the background womb from which "organisms" (the masculine) are given birth. The allocentric mind—that is one half of our cognition—is modeled after a universal background mind. The egocentric mind, our other cognitive half, is modeled after universal manifestation.

A Quick Summary

As I mused about the seven principles of hermetic philosophy, the principles themselves seemed to be an example of background and foreground, unmanifest and manifest, not-moving versus moving; as above, so below. Here are some cosmic invariants (in no particular order), a summary that flowed into my mind as I mused about the hermetic principles. I realize that they are all a single principle (the operation of a rule-bound universe) that the egocentric mind is struggling to categorize and conceptualize:

- everything is moving relative to everything else
- everything seeks a balance, an equilibrium
- everything arises from seeds (cells, embryos)
- everything contains everything else; *as within, so without* (Holographic)
- everything is made out of everything else
- everything "eats" (consumes) everything else
- everything has a life cycle; everything is "born," everything "dies"
- everything changes form; life is building up, while the universe is tearing down
- everything is a composite of that which has switched on (like genes)
- everything spirals from a point (an eye) and widens outward
- everything gets more complex as time passes
- everything in existence is exponentially generating more variables with which to create
- everything is created and everything is destroyed
- everything resides within a domain or a dimension
- everything is a fractal
- everything is a mirror-image; duality is everywhere

The Work of Spiritual Teachers

Why would I serve my friends bowls
with no food in them?
~ "How Minds Most Want to Be," *A Year with Rumi*, 2006.

Saints, mystics, and spiritual people see pain and confusion in all the world's cultures—wherever they look they see suffering. The Buddha had *suffering* as the heartbeat of his teaching because he could perceive and feel the pain that was inherent in the human condition. All humanity, all sentient creatures, by the very fact of living and breathing, are destined to age, get sick, and die without understanding what existence is about. Life is full of hardships, unexpected tragedy, and long years of toil and woe. Good times were few and far between for most of history and for most of humanity.

Why was there so much bewilderment and confusion in the past? Why do we *still* live lives filled with bewilderment, suffering, and dissonance? Perhaps, duality itself is the seed of ultimate bewilderment and hardship—especially for human beings. By design, we have bodies that oscillate, that are polarized, and that easily swing out of balance. We are destined to never gain a firm footing, not in this dimension. This means that we are always tenuous, never sure, never emotionally settled. That is just the way it is. However, this innate insecurity means that joy must oscillate with sorrow, knowing must oscillate with never knowing, perception must forever alternate with blindness.

If our very nature, our essence, is rooted in suffering and the confusion of duality, then we need our saints, mystics, and spiritual healers; these are the people who know how to maintain relative balance. There are those among us who have spiraled upward through the stages of consciousness and are now able to guide and nurture those below who are stuck in the hells and mood swings of lesser awareness. The spiritually blind need the seers of their age to bring comfort, hope, and a spiritual plan. Those who have attained greater wakefulness do not need to be encouraged or convinced to do their magic. There is something in the attainment of spiritual vision that contains the energy needed to bring guidance and comfort. There is a light that illuminates the way toward ever greater degrees of consciousness—saints can perceive using this light.

So what is it that spiritual teachers down through the centuries have been trying to communicate to mankind? That there is a Wisdom Tradition, a pathway to follow that leads toward joy, love, and peace—away from despair, discomfort, and turmoil. The pathway begins with an understanding that the path swings round and round, that repetition and duality are encountered at every turn. We are dual creatures. We have two minds, two kinds of consciousness. God, through infinite wisdom, chose for all sentient creatures a double life—heaven and hell together within the same cognitive planet.

We only seem to know one of our minds; the other mind is hidden. Religion has its source in the hidden allocentric mind. Understand the allocentric mind and you will comprehend the popularity and necessity of spiritual practice. The hidden mind is the source of creation, innovation, spirituality, soulfulness, compassion, and wisdom—all the best of humankind's attributes reside hidden within.

There is wisdom—a pathway, a discipline—and love at the heart of every traditional religion. Waves of kindness (love) sustained the saints. It is this saintly force called love that spiritual teachers maintain, model, and share.

Spiritual teachers have a responsibility to awaken the less spiritual to a kinder, wiser, and more accepting reality. Religious leaders should have been—down through the ages—humanity's main ally in the battle against hatred and violence. But this has not played out. Instead, so-called religious groups have created actual armies that fostered violence and hatred.

The irony drips from the ceiling. An ocean of cruelty and ignorance has spread over the landscape of the planet, with the help and active participation of the very organizations that use the language of love as their foundation. Fortunately, there are still compassionate seekers like Thich Nhat Hahn, Richard Rohr, Deepak Chopra, Jeanne De Salzmann, Cynthia Bourgeault, *and you*, to carry on the Wisdom Tradition.

Therefore, as a spiritual teacher, *you have an obligation*:
Do as much good as you can in the time you have left
Get up.
This is a work day.

Five

WESTERN PHILOSOPHY AND DUALITY

You that give new life to this planet,
you that transcend logic,
come.
~ "A TRACE," *A YEAR WITH RUMI*, 2006.

Philosopher's Toast

"Surge, I've been thinking about philosophers."
"Perfect. Our special this morning is Vegan Corn Dogs served on Philosopher's Toast with Word Jam. Does that sound good?"
"I mean, is thinking a good thing, or is silence golden?"
"Do you want the answer in words or silence?"
"Let's try words."
"There you go! Now you are a philosopher."
"You give me a headache, Surge."
"Yes, of course. That's what happens when you try to read Hegel. Or Kant. Or Sartre. Or, well, any of the systems-oriented word-guys."
"It's the abstractions that make me crazy. The philosophers keep redefining their terms, and they have a fondness for making up hard to pronounce words. I don't feel any different after I finish reading one of their treatises. All I feel is numb and dumb."
"Yes, that's a very normal philosophical reaction, Dutch. However, the philosophers aren't concerned about your personal transformation. Their job is to think."
"What does it mean *to think*, Surge? Because I am not so sure that thinking is something I am good at or fond of."

"Join the chorus. At least you hang out at the Third-Eye-Watching Vegan House of Subtleness. If you surround yourself with non-philosophers, the mysterious types, you will enjoy your moments more and have fewer migraine headaches."

"Well, okay, sure, except I like the philosophers. I get that they are trying to figure themselves out using language."

"What do you want to eat, Dutch? That is the question. What will you feed your mind? If you eat too many slices of Philosopher's Toast, you might come down with brain crust and hardening of the neurons."

"Okay. I am feeling courageous. Give me two slices of Philosopher's Toast and a half dozen of those vegan corn dogs."

"Do you want vegan coffee with abstract cream?"

"What can be abstract about cream, Surge?"

"When you taste the brew, you detect a complexity that is so obscure you aren't sure you tasted anything. Abstract cream is not experiential, but it is heavy cream, and it will cause indigestion if you try to comprehend its obscure flavor."

"Skip the cream, then."

"We can't skip the cream. If we leave out the cream, it won't be philosopher's coffee. It will just be your normal dishwasher brew, simpleminded and weak."

"Please don't make me eat my abstractions, Surge. I hate abstractions."

"Life is hard, Dutch. You know that feeling of dread when you stare at a long sentence full of hard to pronounce words? That is the egocentric mind struggling to digest breakfast."

"You are not making philosophy sound very enjoyable, Surge."

"I am just being honest. Consume your abstractions slowly—with mindfulness—and keep a dictionary of vagueness on the end table. Get some toothpicks for your eyes and take drugs. You can do this, Dutch."

"Okay, here we go."

"That's the spirit. Damn the Existential Portobellos, fuck the Phenomenological Teacakes—full speed ahead into the Metaphysical Bouillabaisse. Give me Abstract Liberty, or give me Cream-filled Death! Get back in there, soldier. Die for the cause."

"I'm going to skip the dying part."

"No, you are not."

New Words Arise for the Same Two Concepts

Philosophy is a vast complex ocean. No one person can grasp the entirety of this impressive body of knowledge. Therefore, in this chapter I look to those who have studied various disciplines of philosophy, and have identified perspectives supporting my thesis that human beings evolved to be cognitively dual creatures. I drew the following discussion from a few favorite books whose authors I found especially readable and thorough. Sarah Bakewell, for example, was especially helpful when I was researching existentialism and phenomenology. Her book *At the Existentialist Café* (2016) is delightful. John Armstrong's wonderful biography of Johann Wolfgang von Goethe, *Love, Life, Goethe* (2007) was also inspirational and was the main source for the section on Goethe. Gary Lachman is one of my favorite authors especially because his passions parallel my own. I drew from Lachman's research on the Romantics and also from his books on consciousness, esotericism, and mysticism. I will acknowledge other authors as the discourse unfolds.

The conflict between the egocentric mind and the allocentric mind is universal for human beings. Therefore, it comes as no surprise that philosophers—when they use logic to understand human behavior—also find humanity's cognitive duality. Philosophers, by the very nature of their discipline, have faced a unique dilemma throughout history. The philosophical domain is primarily verbal and logical—it is not a philosopher's job to be a champion of the intuitive arts. Philosophers are primarily champions of the egocentric perspective—they use language to understand and explain the world. It has been difficult for men of words to accept a second mind where words and sequential logic hold no meaning.

The allocentric mind is purely experiential, a silent flow, and it has no analytical capacity. For a philosopher, the spiritual mind seems like a ghost or a flaw in logic. Nevertheless, the ghost has refused to go away, and every serious philosopher has had to turn and face this strange cognitive apparition.

When the egocentric mind of a philosopher takes on the task of dissecting itself, it immediately gets into a quandary—the ego finds duality wherever it looks:

- The ego only functions through attachments—a *seer and a seen* (a perceiver and an object to be perceived) are required by the egoic mind. This kind of duality is not hard to understand, but the second duality—from the perspective of the egocentric mind—is an absurdity. The ego cannot grasp that it has a mutually exclusive deaf-mute twin and, therefore, is itself a partner in a fundamental duality. The ego's job is to dismantle puzzles and to study the pieces. The ego is blind to its twin mind, the allocentric mind, which is busy taking the puzzle pieces and putting the picture back together again.

- The allocentric mind cannot conceive of duality—for this intuitive mind everything is a unity. So philosophers are faced with a mysterious twin sister (the self) that cannot know duality and is blind to the ego. In other words, philosophers repeatedly uncover a paradox: the egoic mind *cannot perceive* its twin, although it *can conceive of the twin*—and write about it, as I am doing here. However, the allocentric mind does not have the capacity to conceptualize. The allocentric mind *cannot perceive nor can it conceive* of a twin. We have two minds that are blind to each other; that enigma has bewildered generations of our best philosophers.

Therefore, from the onset of philosophy, the conundrum of duality has confused and alienated the debaters. Unfortunately, each great thinker has created a unique language to discuss the two poles of duality. Indeed, our entire philosophical literature is a reverberating redundancy. And that is exactly a major point in this book: Our inherent duality—our two minds—through a steady, unrelenting evolution, have dictated what and how we write about our world and about ourselves.

Thomas Hobbes and Rene Descartes

Let's begin this journey with two great Western philosophers, *Thomas Hobbes and Rene Descartes*. These two contemporaries appear to have been the first modern Western philosophers to explore how cognition might work. It is not possible to pin down exactly when the duality of *ego and other* first appeared in the discourse of Western philosophy, so I will start this discussion at an arbitrary time in the 1500s when the words "conscious" and "consciousness" are first found in written documents in the West. (*Revolutionaries of the Soul: Reflections on Magicians, Philosophers, and Occultists*, by Gary Lachman, 2014.)

Thomas Hobbes (1588-1679) wrote "Where two or more men know of one and the same fact, they are said to be "conscious of it one to another." Rene Descartes (1596-1650) is famous for saying "I think, therefore, I am." In modern Western philosophy, these two men, Hobbes and Descartes, are the philosophers who opened the discussion about consciousness.

From Hobbes' perspective, there is an objective reality that can be perceived and verified by others. He does not start the discussion by saying that everything we perceive is an illusion created by the mind (solipsism). He assumes objectivity as a given. Notice that *this is a strictly egocentric perspective*—egocentric cognition requires a seer and a seen (an objective reality).

Descartes also defined consciousness as an egocentric phenomenon. He doesn't say "I experience" therefore "I am." While the word *think*, as used by Descartes, could mean both "experience" and "internal dialogue" (introspection, reasoning, subjectivity), that is not the way it has come to be understood. Defining consciousness as "knowing that I know" or "knowing that others know," is purely egocentric. The allocentric mind *does not* "know that it knows," yet it is one half of who we are and how we operate in the world.

Therefore, hidden within the philosophies of both Hobbes and Descartes is the assumption of a seer and a seen. What both Hobbes and Descartes identify as consciousness is the egocentric mind. Therefore, the opening discussion about consciousness starts with the questionable assertion that there is only one mind and one kind of consciousness. The debate begins with little appreciation of allocentricity. However, as more philosophers entered the discussion, the debate quickly divided into two now familiar camps.

Given our inherent biological duality, this "dividing into two camps" comes as no surprise. In the 1500s and 1600s, for example, we had rationalists (Descartes, Leibniz, and Spinoza) in opposition to empiricists (Locke, Berkeley, and Hume). The rationalists observed their own egocentric minds in action and argued for the realities that the egocentric mind routinely uncovers—objectivity versus subjectivity, for example. Contrary to this perspective, empiricists saw the allocentric mind at work; they argued that experience was the key to understanding reality. Similarly, in the 1800s in Europe, two schools squared off with oppositional philosophies. The continental philosophers—as they were called—were in the allocentric camp, while the so-called analytical philosophers were in the egocentric camp. Essentially, the continental philosophers argued that experience and synthesis was the key to understanding consciousness, while the analytical philosophers held that rational thought and analysis was the key. What we see as we examine the history of philosophy is the repeated discovery of two minds in conflict.

In the 1600's we find another example of a brilliant thinker who conceived of cognitive duality. Blaise Pascal (1623-1662), a mathematician and philosopher, wrote in his unfinished notes about the contrast between the analytical mind of the mathematician and the intuitive mind of the artist. In Oliver Sack's autobiography *On the Move* (2015), he compares two of his early medical advisors, using Pascal's insights (italics mine):

> Kremer and Gilliatt made me think of Pascal's comparison between intuition and analysis in the opening of the Pensees. Kremer was preeminently intuitive; *he saw everything at a glance*, often more than he could put into words. Gilliatt was primarily analytical, *looking at phenomenon one at a time* but seeing in depth the physiological antecedents or consequences of each.
> ~ On the Move, A life, Oliver Sacks, 2015.

The allocentric perspective—*perceiving everything at a glance*—is the intuitive, religious, spiritual, and artistic side of the equation of life. It is the *experiencing* of sensations and emotions, instantly. On the other hand, the egocentric perspective—*looking at phenomenon one at a time*—examines the world in search of meaning. As such, the egocentric mind is analytical and reflective; it gave birth to the sciences. Pascal could perceive this split in humanity's cognitive makeup, even in the 1600s.

If we jump ahead to the 1900s, we see that the same confusion (of being your own twin) continues. Now it takes the form of Phenomenology and Existentialism—two hefty abstractions that were the latest attempt to understand and label mankind's inherent cognitive duality. Edmund Husserl, for example, was a German philosopher who had one foot in the rational or analytical camp of the mathematicians (Gottlob Frege, Alfred North Whitehead, and Bertrand Russell) and the other foot in the continental philosopher's camp (Kant, Fichte, Schelling, and Hegel). Phenomenology, as developed by Husserl, Martin Heidegger, and Maurice Merleau-Ponty, was an attempt to acknowledge both minds. Phenomenology stubbornly kept the ghostly allocentric mind at the heart of its discourse—the egocentric mind was not allowed to dominate.

If we now move ahead to our era, we see that science has entered as a key player in the debate about consciousness. It is acceptable now to use both biological principles and quantum theory to explore the nature of mind. In other words, the new philosophers are scientists. Again, it comes as no surprise that what these new scientific philosophers quickly discover is that cognition is inherently a duality. An entire field of scientific inquiry has evolved to address this cognitive conundrum. When I looked at this new field of study, I found that two men had been instrumental: professors Keith Frankish and Jonathan Evans. These two pioneers have collaborated during their professional careers and provided an overview of what has come to be called *dual-process theory*. "The Duality of Mind: An Historical Perspective," by Frankish and Evans is an excellent online overview of the history of thought as it relates to our cognitive duality. When I speak of dual–process theory, I am often using the research conclusions of Keith Frankish and Jonathan Evans. Each of these talented men has written about our dual minds. Jonathan Evans with Keith Frankish co-edited the book *In Two Minds: Dual Processes and Beyond* (2009). Evans has written two books *Thinking Twice: Two Minds in one Brain* (2010), and *Thinking and Reasoning: A Very Short Introduction* (2017). In 2007, Keith Frankish wrote *Mind and Supermind*.

Dual-process theory generated two new (rather boring) abstractions for our inherent cognitive duality: System One and System Two. This new terminology is just the latest set of labels that correspond to allocentricity and egocentricity: System One is composed of all the processing elements that result in the allocentric mind; System Two is composed of all the processing elements that result in the egocentric mind.

In their essay "The Duality of Mind: An Historical Perspective," Frankish and Evans show that the philosophy of dual-process theory has historical roots that go back thousands of years. In the paragraph below, they give an example drawn from the philosophy of Leibniz. Leibniz (1646-1716) compared what he called "pure, or bare perception" (allocentric processing) with "conscious appreciation" (egocentric processing).

Philosophers often refer to a *conscious mind* and an *unconscious mind*. Leibniz uses this terminology, but I suggest—from our modern perspective—

that he is grappling with the egocentric frame of reference (the conscious mind, or "conscious appreciation") in contrast to the allocentric frame of reference (bare perception, the unconscious mind). In the quote below, Frankish and Evans give us a list of ancient philosophers who struggled to explain cognitive duality:

> References to unconscious memories and perceptions crop up in various ancient and early modern philosophers; examples can be found in, among others, Plato, Plotinus, Augustine, Aquinas, Pascal, Spinoza, and Leibniz. Leibniz, in particular, made important observations about unconscious perception and memory. He distinguished between bare perception and conscious appreciation, and held that we continually experience a multitude of unattended petite perceptions, which are below the threshold of consciousness but which collectively shape our conscious experience. He also noted that we often retain information which we cannot consciously recall but which nonetheless influences our thoughts and behavior.
> - "The Duality of Mind: An Historical Perspective," Keith Frankish and Jonathan Evans

I selected this quote for two reasons. First, it is an example of a Western philosopher (Leibniz)—writing over 300 years ago—who had discerned and tried to explain dual cognition. His conclusions fit with what we know today, even though he created his own terminology. Second, in the above quote, Frankish and Evans assert that other philosophers throughout history—even well before Leibniz—created dual-process theories.

I made the observation in the first chapter that one of our questionable scientific memes is that there is an unconscious mind and a conscious mind. I suggest these are just early names—another labeling—for allocentric and egocentric. The so-called unconscious (or subconscious) mind is really allocentric processing. The conscious mind is really egocentric processing. Many of our most revered philosophers and writers discussed our inherent duality—using the terminology of *unconscious mind* versus *conscious mind*—long before Freud made that duality commonplace. In the quote below, Frankish and Evans trace this evolution to the 1700s and 1800s:

With the rise of German idealism and Romantic aesthetics in the late eighteenth and early nineteenth centuries, claims about the unconscious became common. There are frequent references to unconscious mental states in Herder, Schelling, Hegel, and Schopenhauer, among others, as well as in creative writers such as Goethe, Richter, and Wordsworth. These writers' conceptions of the unconscious had strong metaphysical and mystical overtones.

~ "The Duality of Mind: An Historical Perspective," Keith Frankish and Jonathan Evans

The above quote points out that philosophers two hundred years ago had identified a hidden mind—a kind of processing system different from, and more subtle than the egoic mind. In other words, these early philosophers had introspected allocentric processing—without having the biochemical and quantum evidence we have today. As Frankish and Evans observe in the above quote, the unconscious (allocentric) mind has been historically connected to mysticism and metaphysics. This is not surprising because the allocentric mind is non-verbal—it communicates through facial expressions and body language (animation). The allocentric mind also speaks through artistic renderings; we see the allocentric mind at work in dance, paintings, and metaphorical poetry. There is something powerful and yet frightening (for the ego) about non-verbal expression.

The above timeline is a brief overview, but I want to look now in greater detail at various movements in philosophy to show how dual-process thinking is pervasive in history. In particular, I will look at a few important philosophers who took on the conundrum of dual consciousness. Phenomenology and Existentialism are especially relevant to this discussion.

Romantics and Existentialists

I drew heavily from the books of British author Gary Lachman as I crafted my comments on the Romantic Movement. Gary Lachman is a prolific writer whom I have come to admire for his explorations into the nature of human consciousness. Lachman was a friend of Colin Wilson—another avid student of consciousness—who I wrote about in the chapter on poetry and duality. Lachman has written over 20 books about consciousness and mysticism, including books on Carl Jung, Colin Wilson, Rudolf Steiner, Hermes Trismegistus, Aleister Crowley, Madame Blavatsky, and P. D. Ospensky. According to Lachman, from roughly 1800 to 1850, a group of writers and poets produced works that had a common theme, an opposition to the materialism arising during the industrial revolution, with roots going back to the Enlightenment (1600-1700). In the quote below, from his book *A Secret History of Consciousness*, Lachman explains that the Romantics—like so many others before them—discovered a hidden (allocentric) mind:

> The Romantics discovered that there was more to human beings than the "daylight" rational picture of them that had emerged in the Enlightenment. There was a whole "other world" *inside their own mind*, a world that the shallow, rational Enlightenment picture ignored or suppressed. The Romantics had discovered an entirely new dimension of reality, and had taught their readers how to explore it.
> - *A SECRET HISTORY OF CONSCIOUSNESS*, Gary Lachman, 2003.

The new dimension of reality that the Romantics discovered was the allocentric mind—appearing in one of its many disguises. Even though the world's religions, especially within their esoteric branches, had asserted for thousands of years that we had a hidden mind—and the Greeks had rationally and extensively examined duality—it wasn't until the early 1800s that Romantics in Europe awakened popular Western thought to this revelation. As Lachman points out in the above quote, the Romantics were rebelling against the new science of the Enlightenment which was giddy with the power of the scientific method.

The Enlightenment was an intellectual movement that occurred between the late 1600s and the early 1700s. The movement emphasized reason and individualism. During the Enlightenment, scientists were busy banishing the allocentric twin sister from the discussion. This horrified the Romantics. As artists—poets, writers, and musicians—they knew intuitively that they had creative, innovative minds. They could feel the barrenness of materialism and they rebelled against it.

The Romantics were followed by the Phenomenologists and Existentialists who also recoiled against the rise of materialism. Philosophers and authors like Soren Kierkegaard, Jean-Jacques Rousseau, Fyodor Dostoyevsky, Fried-rich Nietzsche, Jean-Paul Sartre, Simone De Beauvoir, and Albert Camus argued that the rise of science (rationality, the egocentric perspective) was causing a spiritual bankruptcy in the world. Existentialists knew that human beings made decisions based on subjective meaning that arose from expe-rience, rather than from pure knowledge-based rationality. As Kierkegaard said, "Human reason has boundaries." Meanwhile, Rousseau insisted that the rise of modern society was murderous to something natural and primal in the human being. In essence, what Rousseau saw was the sudden disregard for our allocentric inheritance. From my perspective, the intellectual class of the Enlightenment embraced the power and glory of the evolving egocentric mind so completely that they tossed out the essence of their core humanity in their eagerness for knowledge and understanding.

Jean-Paul Sartre was a novelist and a philosopher. Existentialism, the phi-losophy that Sartre (and others) created using Phenomenology as a foundation, was a weapon of the artist to fight back against the angst and barrenness of the ego-world. "Be *in* the world," Sartre said, in various ways. Use your artist's eyes to peer deeply into the substance of Nature. Develop your own completely per-sonal description of phenomenon. Never mind what others have said, or how others have constructed their philosophies about the world. When you look at a tree, for example, drop all the knowledge you have about trees. Forget what the biologists have told you about tree parts and tree processes. This tree and you are alone at a specific time in all of existence. This tree and you are at a certain time in your biological development. This tree and you will not be the same when you visit each other tomorrow. Instead, look at everything fresh; experi-ence life in this fundamental way. Perceive with acute curiosity in the moment.

According to the Existentialists, a wonderful feeling envelops us when we perceive the world with ever-fresh perception. When we are able to blend with the environment in the moment, we feel elated, joyful, and free. An indi-vidual who experiences life in this way is liberated from the anxieties brought on by the guilt and grief of the past, and by the dread and fear of the future. This is what Sartre means when he writes about "description" and "libera-tion." Being in the moment, *describing* that moment as you are in it, opens a portal through which joy and *liberation* can flow. This insight comes from an essay that Sartre wrote in 1939 called "A Fundamental Idea of Husserl's Phenomenology: Intentionality."

Sartre also stated in that same essay that *existence preceded essence*. This is contrary to a religious view that God created us with an initial essence; that we had a purpose in life ordained at birth. Sartre did not buy that worldview. There is no initial, innate, God-derived essence, according to Sartre. We arrive here without a purpose and without an essence. Therefore, we are free to create our own meaning, our own self-evolved essence. We are free of any pre-conceived notions. We are free. And *we are responsible* for our words and deeds.

Clearly, as Sartre's essay shows, the Romantics and Existentialists were aware of human duality. They could see that one-half of their reality was being threatened by the rise of a purely egocentric perspective. Science and materialism were delivering great advances for humanity, but the cost was too high to pay if it meant the denial of the empathetic mind. Something dried up and died for the artist in the scientific world of measurement and categorization. What died was liberation and joy. The *experience* of feeling alive every moment didn't seem important to the proponents of the scientific enlightenment.

The "loss" of the allocentric twin caused a swelling of grief, angst, and despair that the Existentialists wrote about vividly. For example, in the writings of author Albert Camus he suggests that suicide might make sense in an absurd world devoid of compassion and soul. Camus was a regular at the "Existentialist Café," but he refused to be categorized within any philosophical camp.

Many of our most famous creative geniuses struggled to live in the "real" world as they pursued their art; many crashed and burned, drinking or drugging themselves into oblivion. The allocentric (creative) minds of the Existentialists fed their existence so splendidly and so fully that they recoiled at the ascendancy of the materialistic worldview.

Existentialism came out of Phenomenology. Indeed, *Existentialism* appears to be the French "translation" of *Phenomenology*—or rather, what French minds did with German discourse. The abstract, idea-heavy German version of Phenomenology was eagerly filtered through the French emotional-romantic mind and out came Existentialism. The two perspectives—Existentialism and Phenomenology—are actually the result of very specific minds laying down and defending their definitions. Later, philosophers took up the abstractions and turned them into sub-disciplines, into arcane areas of study.

As I stated at the beginning of this chapter, I have drawn heavily from a book written by Sarah Bakewell called *At the Existentialist Café* (2016). Bakewell is a British author of non-fiction. Her depiction of the individuals who crafted philosophy after World War Two—like Jean Paul Sartre and Simone De Beauvoir—vividly describes the lives of the young philosophers. Bakewell turns philosophy into a study of personalities and relationships. In the quote below, Bakewell asserts that the abstraction called *Existentialism* was crafted in the coffee houses of France by individuals with different emotions and beliefs:

> The key thinkers disagreed so much that, whatever you say, you are bound to misrepresent or exclude someone. Moreover, it is unclear who was an existentialist and who was not. [Jean-Paul] Sartre and [Simone de] Beauvoir were among the very few to accept the label, and even they were reluctant at first. Others refused it, often rightly. Some of the main thinkers . . . were phenomenologists but not existentialists at all (Husserl, Merleau-Ponty), or existentialists but not phenomenologists (Kierkegaard); some were neither (Camus), and some used to be one or both but changed their minds (Emmanuel Levinas).
> ~ *AT THE EXISTENTIALISM CAFÉ*, Sarah Bakewell, 2016.

The young French intellectuals that Bakewell wrote about were influenced by the German philosophers who created Phenomenology, especially Edmund Husserl and Martin Heidegger. Therefore, it comes as no surprise that the fundamental message of Existentialism is essentially same as Phenomenology: live in the moment. In our era, we have created phrases that echo "live in the moment." For example, we say "Be here now. Go with the flow. Experience your life. Take charge of your being. Decide to love your moments." The reasoning mind searches for answers, solutions, and relevancy. But the creative, poetic mind searches for joy, love, peace, wisdom, and mindfulness. This second mind, the allocentric mind—to use my terminology— strives to evolve toward ever more wakefulness and kindness.

Human beings, according to the Existentialists, are like no other animal because we have free will. We decide where we will be and what we will do moment-to-moment. This freedom brings on angst and doubt, and exposes our level of consciousness, which is never as sophisticated as it could be. Being loyal to experience, rather than being ruled by rationality is surprisingly revolutionary, according to Sarah Bakewell:

[Phenomenology] ought in theory to free us from ideologies, political and otherwise. In forcing us to be loyal to experience, and to sidestep authorities who try to influence how we interpret that experience, phenomenology has the capacity to neutralize all the "isms" around it, from scientism to religious fundamentalism to Marxism to fascism. All are to be set aside in the epoche—they have no business intruding on the things themselves. This gives phenomenology a surprisingly revolutionary edge, if done correctly.
~ AT THE EXISTENTIALISM CAFÉ, Sarah Bakewell, 2016.

In other words, if we stay focused in the moment—unencumbered by the past and future—we won't be swayed by belief systems, by the "isms" that have caused so much violence and hardship throughout history. Belief systems are tools of the egocentric mind. Living in the moment is a tool of the allocentric mind.

On the other hand, living in the moment can be dangerously irrational—a foolish way to address the problems facing individuals and cultures. For this reason, and because the philosophers themselves appeared to be gloomy souls, Existentialism got a bad reputation as a dreary, angst-ridden philosophy. This perspective was mostly coming from post-war American culture. The French saw the Existentialist crowd as free spirits, artists, and free-thinkers, but post-war Americans did not share this romantic perspective. Here is Sarah Bakewell's explanation for Existentialism's bad reputation:

For the French in the 1940s [Existentialism] tended to be seen as new, jazzy, sexy, and daring. For Americans, it evoked grimy Cafés and shadowy Parisian streets: it meant old Europe. Thus, while the French press portrayed Existentialists as rebellious youths with outrageous sex lives, Americans often saw them as pale, pessimistic souls, haunted by dread, despair and anxiety, a la Kierkegaard. This image stuck. Even now, especially in the English-speaking world, the word "Existentialist" brings to mind a noir figure staring into the bottom of an expresso cup.

~ AT THE EXISTENTIALISM CAFÉ, Sarah Bakewell, 2016.

I prefer the French version—that the post-war philosophers were free-thinkers and bohemian souls. They were young, idealistic, and brilliant. They had a lot of poetic energy. Furthermore, 50 million people had just died because of a second world war. Looking at the past was too painful and the future didn't seem all that hopeful, given the stupidity of the species—especially the males. Existentialism was a response to a stark and depressing era that had just ended. The new philosophies of Phenomenology and Existentialism were calls for the soul to be awakened. The hidden allocentric mind was a spiritual, cooperative, romantic, peace-loving mind. There was nothing gloomy in it, and it held the hope for a better future.

Perhaps, from my perspective, the chief flaw in Phenomenology and Existentialism is the failure to realize that we have two very real minds inside of us. Like so many others, the philosophers could sense it and talk around it, but no one ever nailed it to the wall. One of the first and enduring problems was the confusion over the term *intentionality*.

In Sartre's essay "A Fundamental Idea of Husserl's Phenomenology: Intentionality," the Phenomenological concept of *intentionality* is praised as the key to both Existentialism and Phenomenology. Sartre says that consciousness is not a thing, not a substance encased within the head. Consciousness is a flow that reaches out and touches the world. It is like a whirlwind thrown from our minds into the world. This whirlwind swirls around the phenomenon of the earth, isolating them for study, defining them, describing them. Sartre quotes Husserl as saying "All consciousness is consciousness *of* something."

According to Sarah Bakewell, the concept of intentionality as it is found in both Existentialism and Phenomenology can be traced back to the philosopher Franz Brentano:

Husserl had picked up this idea from his old teacher Franz Brentano ... In a fleeting paragraph of his book *Psychology from an Empirical Standpoint* [1874], Brentano proposed that we approach the mind in terms of "intentions"—a misleading word, which sounds like it means deliberate purposes. Instead it meant a general reaching or stretching, from the Latin root in-tend, meaning to stretch towards or into something. For Brentano, this reaching towards objects is what our minds do all the time.

- At the Existentialism Café, Sarah Bakewell, 2016.

In my opinion, Brentano was only half right when he expressed this very egocentric perspective. The ego does, indeed, need to point to an object of regard because the ego operates under the assumption of objectivity and subjectivity— *a seer* and *a seen* are necessary for egocentricity to manifest. However, it is clear to me that what Brentano, Husserl, and Sartre were discussing was not "*in*tention," but "*a*ttention." It is *attention* that reaches out from our bodies to touch the world. Intention is the egocentric will to do something. Intention comes prior to the act of attention. In the actual world of perception the two processes (intention and attention) are integrated, part of the same overall system for gathering information about our surroundings—so they are easily confused. Clearly, however, when these philosophers used the word *intention* they should have used the term *attention*. History would not have been confused for so long and the Phenomenologists would have been far less obscure had this distinction been understood.

Furthermore, like so many others, the Phenomenologists failed to understand that there were *two kinds of attention*. They identified only egocentric consciousness which, indeed, needs something to latch onto (an object-of-regard). In the quote below, Sarah Bakewell gives an example Sartre used to describe the egocentric perspective. Sartre shows how easily we become distracted simply because others also have egocentric perspectives:

Sartre asks us to imagine walking in a park. If I am alone, the park arranges itself comfortably around my point of view: everything I see presents itself *to me*. But then I notice a man crossing the lawn towards me. This causes a sudden cosmic shift. I become conscious that the man is also arranging *his* own universe around himself. As Sartre puts it, the green of the grass turns itself towards the other man as well as towards me, and some of my universe drains off in his direction. Some of me drains off too, for I am an object in his world

as he is in mine. I am no longer a pure perceiving nothingness; I have a visible outside, which I know he can see.
~ *At the Existentialism Café*, Sarah Bakewell, 2016.

We can make two important observations from the above quote. First, using an egocentric perspective, we manifest each other—our sense of being an entity grows as we come to see ourselves as an object in the eyes of others. Second, once we come to know that relationships define us, we get a glimpse of the allocentric mind which is based on the flow of relationships. Sartre can see that we are locked into our egocentric frame of reference—we are the center of the universe no matter where we go or how we move about. However, whenever we acknowledge the existence of other minds, especially *many* other minds, we are forced into a different frame of reference where relationships prevail. This is the beginning of the ego's conceptualization of its allocentric twin.

⌇⌇───────⌇⌇

The inability to actually live in the moment, as Phenomenology and Existentialism dictated, while having to practically exist in the everyday world of mundane work left many of the Existentialists and Romantics in despair. Below is a quote that I like from John Armstrong's biography of Johann Wolfgang von Goethe, *Love, Life, Goethe* (2007). John Armstrong (born in 1966) is an eloquent British author and philosopher. I used Armstrong's viewpoint and many of his quotes to explain the contributions of Johann Goethe.

The allocentric mind, wishing to dwell within the peace and bliss of *being*, struggles in the ordinary world of duties, in the time-obsessed egocentric world. Here is a quick list compiled by Armstrong of some great artists who couldn't cope:

Mozart: no idea about money, pauper's grave; *Beethoven:* his friends had to take his money away from him because he was so irresponsible; *Balzac:* dressed as a monk, drank forty cups of coffee a day, economic basket case; *Baudelaire:* drug addict, compulsive gambler, squandered his inheritance; *Wagner:* insanely egotistic, borrowed from all his friends, never paid his debts; *Tolstoy:* wanted to be a penniless serf; *Nietzsche:* didn't make a penny from writing, later royalties went to his horrible sister; *Proust:* didn't know how to open a window or boil a kettle, lost lots of money through extravagance and inept speculation; *Wittgenstein:* tormented ("If just one person could understand me I would be satisfied"), ate mainly bread and cheese, gave away all his money—to his rich siblings so as not to corrupt the poor; *Jackson Pollock:* everyone else made money out of him ("If I am so famous why ain't I rich?").

~ *LOVE, LIFE, GOETHE*, John Armstrong, 2007.

In the above quote, we see the damage done when we ignore our duality—especially our allocentric minds. In my opinion, the failures highlighted above are more a product of unevolved cultures than a defect in the artists. The battle between creative energy and practical duties, this schism between the creative and the practical, between the spiritual world of experience and the material world of knowledge, continues still (of course); it is *the* challenge of our inherent duality—how to live a balanced (joyful) life with a fundamental cognitive duality boiling inside our bodies. In other words, how do we survive the confusion caused by being our own twin?

Another *philosopher of duality*, Johann Wolfgang von Goethe existed before the Romantics but he had found a way to accept his dual fate. Goethe is not among the list of artists and philosophers who crashed and burned. His writing is a history of his steady and sane mental development. Goethe was practical, a full time administrator, a father and a husband, and yet he found a way to keep a creative fire burning all through his life.

A central theme of Goethe's mature mind was the need to transform *"knowledge by description"* (the egocentric perspective) into *"knowledge by acquaintance"* (the allocentric perspective). Goethe saw that the egocentric mind needed to be balanced or even dethroned by experience, by allocentric consciousness. His best friend, the poet and philosopher Friedrich Schiller, said that Goethe was remarkable because he wasn't fragmented by his intellect. Although Goethe was splendidly complex, a Renaissance man, he was natural and whole. Goethe, according to John Armstrong, shows us how we ought to behave:

The centrality of Goethe—Schiller is saying—arises because he reminds us, shows us, what we ought to be like. We ought not to be specialized fragments of competence—we ought to be whole, rounded people. Thus Goethe lives out—constitutes in himself—the goal of proper development.
~ LOVE, LIFE, GOETHE, John Armstrong, 2007.

In the above quote, Armstrong suggests that a developed egocentric mind results in specialization and fragmented competence. Goethe somehow escaped this narrow fate. Armstrong believes that Goethe's maturity slowly evolved as he grappled with his awareness of humanity's dual cognition.

Armstrong contends that Goethe's mental development was written out in his stories and plays for all of us to marvel at and examine. According to Armstrong, Goethe's early novel *Wilhelm Meister* is a long exploration of the internal battle between one mind that seeks material comfort and responsibility in the world—a mind steeped in tradition and habit, a mind that has inherited a cultural mandate to do that which was done before—and the artistic mind that seeks to be "its own person," a mind that is inventing a new culture and, therefore, must reject the past.

As the two minds struggle to state their positions, they both must win the battle before they surrender to each other and form an alliance that makes a whole person from these two twin universes:

This is a central pattern of the book *[Wilhelm Meister]*. We start with a black-and-white opposition. Poetry (or the theatre) is good; commerce is bad. Then there is a reversal: poetry is bad and commerce is good. Finally there is a development towards integration: perhaps theatre and commerce are not directly in competition.
~ LOVE, LIFE, GOETHE, John Armstrong, 2007.

Armstrong is suggesting above that the two minds battle it out as they mature, until a time is reached when a balance begins to settle in. Goethe was simply struggling to understand his own duality. Instead of using "allocentric consciousness," he speaks of art and comfort or *experience*. Instead of using "egocentric consciousness," Goethe speaks of being high minded or of sanity. Armstrong says that Goethe pondered a fundamental question:

What Goethe achieved in his life—even more than in his literary and intellectual work—is the closest we have yet seen to a solution to the problem of modern life. The question can be posed in various ways—one is "How do you combine art and sanity?" Another: "Can you be both comfortable and high-minded?" Or, to put it in its oldest form: "How can our spiritual and material needs be allies rather than enemies?"

~ *Love, Life, Goethe*, John Armstrong, 2007.

Or to put it in modern terms: how do we develop an astute, brilliant egocentric mind at the same time we evolve a poetic, spiritual, compassionate allocentric mind? Goethe could see his two minds at work and he struggled to express his bewilderment in his artistic creations.

Goethe's most famous work is *Faust* in which the egocentric mind finds a way to conquer the world, to gain immeasurable riches, to glory in all things egotistic. All the ego has to do to get these splendid rewards is to sell his twin sister, the soul, the allocentric mind, to the devil. It seemed like a good idea at the time the deal was cut.

When the devil comes to carry off the soul (the sister mind) Faust has second thoughts and lots of guilt about the dreadful deal he has so callously made. Below Armstrong reflects on the damage done when we sell our souls:

Such a story dramatizes one of the most basic and annoying features of the human condition. Anyone who has had a hangover knows only too well the feeling that the pleasures gained before seem shallow and silly when faced with the consequences of a burning head. "Selling your soul" merely projects this sequence on to the larger canvas of eternity. You get a lifetime of pleasures and then you get the hangover from hell—in Hell.

~ *Love, Life, Goethe*, John Armstrong, 2007.

In other words, Faust is holding a mirror up to the egocentric face, asking "Is this all there is to you?" "Are you just this busy, greedy, alien thing? Are you so immoral that you could sell your own sister for profit?" Goethe's answer is "Yes, that is, indeed, who you are. However, you also have a twin sister who can balance your narcissism with compassion—you have to embrace your twin, not sell her to the pirates." *Balance* is Goethe's answer—follow the middle way between cognitive extremes.

⁓————⁓

Immanuel Kant (1724-1804) was a contemporary of Goethe (1749-1832). In his philosophical tomes, Kant made it quite clear that we swim in a sea created by our senses, a fabricated world of phenomenon. The real world, the-thing-in-itself (noumenon), cannot be accurately perceived, and so we can never know for sure what is objectively real. Reason, the egocentric mind, cannot tell us anything about the world beyond our senses. Thus, we are stuck with a fundamental duality: the world inside of us which can be "known," and the world outside of us that must forever remain dubious and questionable. This is an egocentric conundrum since the allocentric mind is not ambiguous and not concerned with dissecting existence.

In the mid-1800s, British scientist Sir Francis Galton (1822-1911) concluded that most of what the mind did was automatic and unconscious. He even went so far as to suggest that consciousness (the egocentric mind) was a "helpless spectator" to the automatic operation of the (allocentric) mind. Galton's observations show that he also saw the mind as dual, mostly automatic (allocentric), but with a small amount of self-awareness coming from the egocentric mind.

After Goethe and Kant, in the 1800s, Eastern philosophy flowed heavily into the Western world influencing philosophical doctrines. Some philosophers, like Schopenhauer (1788-1860), were directly affected by Eastern thought, and drew comparisons between Western rationalism and Eastern non-duality. Drawing on Eastern philosophy, duality became a central concern for the Western philosophers and each in his own way tried to come to terms with this fundamental split in the fabric of mankind. It is probably the case that these philosophers were not enlightened or anywhere near that high level of awareness. Consequently, they boiled in the stew of their own intellect as they tried to explain the dual world. They also had no idea that there were two anatomically distinct minds inside of them, so they used their egocentric rationality to examine their predicament.

Schopenhauer gave more credence to the phenomenological side of the equation. He saw that the intellect was dividing the world into fragments and then crafting concepts using the pieces. This was a world totally different from the immediate awareness of the self. Schopenhauer, using Eastern language, asserted that there is a basic unity linking all things and this understanding arises only within the *will* (allocentric consciousness), not the intellect (not the egocentric consciousness).

Language gets in its own way often, and the debate over the nature of two minds is a mess of foggy, abstract terminology. Schopenhauer, for example, explains duality as a struggle between "the world out there, full of space, time, cause, effect, events, and objects, versus an inward world (our mental state, the location of the "will)" where there is no time, or space, or cause, or effect. German philosophers saw a fundamental duality arising simply because "mind" is such a foreign strangeness compared to everything else in the cosmos.

Schopenhauer, who gathered his mystical overtones from Hindu philosophy, sounds like Freud when he discusses mankind's animal nature, especially the will to live, and the power of sexual impulses. According to Schopenhauer, primal urges are embedded deep within the human body and are expressed through whole-body animation. Essentially, Schopenhauer described the development of the allocentric mind, a whole-body consciousness (awareness) of the present moment, compared to egocentric reflective consciousness. Schopenhauer described the egocentric mind, as having a "blind, insatiable, and malignant metaphysical will." (*On Human Nature*, Arthur Schopenhauer, 1897). In other words, Schopenhauer—like other philosophers of his era—wrote about human cognitive duality. He also tried—as I have done—to illustrate the characteristics of the two minds (the ego and the self).

Unfortunately, after examining this fundamental duality, Schopenhauer did several mental backflips and, using his egocentric mind, concluded that we were compelled by instinct to survive in a world of suffering and death, and furthermore that we human beings were locked in a battle for resources, and inevitably had to battle each other for survival. Given this dire consequence, according to Schopenhauer, life is incurably evil. Ouch. (Armstrong, 2007). Fortunately, the mysterious Schopenhauer was followed by the wonderful and tragic Nietzsche.

In his first popular book, *The Birth of Tragedy*, philosopher Friedrich Nietzsche (1844-1900) contrasts Dionysus with Apollo. The Greeks knew two thousand years ago that human beings held within them two minds— both minds were the sons of the Almighty god Zeus. One mind, Dionysus, was the playful artist in love with experience, and the other mind, Apollo, was a brilliant rational thinker. Greek tragedy is a repeating story about how the rational mind comes to know that another mind lives within the same body. This is a tragedy, from the rational mind's perspective. Having to live with a contrary, illogical, inexplicable, invisible double is unbearable. Only conflict, war, and tragedy can come from this disastrous plan. Nietzsche comes down hard on the side of the Dionysian spirit, especially since he felt that his culture was suffocating in rationality. His conclusion is that the two forces (two minds) must merge into a "superman," a higher level of consciousness; the twins must eventually unite as one into a supermind.

Nietzsche agreed with Eastern thought, and with Kant, that the objective world is an illusion. But he said if this was true, the internal world was also an illusion since our bodies and brains are part of objectivity. All our ideas, our dualities, paradoxes, conundrums, everything internal and external, are not as they seem. Therefore, there is no use talking about whether God exists because the "beyond" is just as much an illusion as "the world." Even experience is a product of the mind at work. We cannot know anything given our illusory paradox. Not only is "God dead," according to Nietzsche, but so is everything else. Actually, Nietzsche said that God was dead because we killed him. He believed humans had turned the world into such a rational, sterile place that there was no room left for a loving, vibrant God of creativity and joy.

Nietzsche was influenced by popular materialistic writers of his time, notably the monism theory of Ernst Haechel, which held the belief that there was only a single reality in the universe: matter playing out through Darwinian evolution. This is a very egocentric conclusion. However, Nietzsche also felt the power of the hidden mind; he was in conflict as he pondered our duality.

To elaborate further, Nietzsche's logic—that *all* is illusion—holds up in an egocentric world. It is a valid conclusion that can be derived from an egocentric, reductionist perspective. However, Nietzsche didn't know that there *actually were* two minds in his body, and it was the other mind, the allocentric mind, that "knew" he was only fifty percent correct when he was thinking egocentrically. On the other hand, Nietzsche *did* conclude there was a potential superhuman within us that didn't need to look to God; human beings were far more creative and capable then they realized, and holding out hope for a joyous afterlife sapped the strength from *this* life. From my viewpoint, Nietzsche went to war against the weak spirituality and complacent mediocrity of the era in which he lived. Nietzsche is saying that we need to balance our need for rational order and reason with our spiritual nature.

Nietzsche was a Greek scholar. He understood very clearly the divide between the rational mind and the hidden mind. He knew there was a part of us which could be subsumed by experiences so deeply that the cares and stresses of the everyday world evaporated. He knew *intuitively* (but he did not have the biological evidence) that there was a hidden mind that was every bit as important to our humanity as the wakeful ego.

For Nietzsche, there is a background melody, primarily found within music (a Cosmic Harmony), where ego can be dissolved:

> "The unending melody—you lose the shore and surrender to the waves."
>
> ". . . I know you and your secret, I know your type! You and I are of one and the same type!—you and I, we have one and the same secret!"
> - NIETZSCHE, A PHILOSOPHICAL BIOGRAPHY, Rüdiger Safranski, 2000.

This is the secret: We can become aware of a hidden mind wherein we are safe from the (potential) insanity of the ego. The background heartbeat, the primal music, can be heard when we get lost in experience, without judgment. However, if we live in egocentricity, we struggle with past regret and future angst and fail to find peace (and the music) in the moment:

> One of these secrets is the intimate relationship of wave, music, and the great game of life that Goethe called "expire and expand." It is a game of growth and fade, rule and be overruled. Music transports you into the heart of the world, but in such a way that you do not die in it. In *The Birth of Tragedy*, Nietzsche called this ecstatic life in music the "rapture of the Dionysian state, which eradicates the ordinary bounds and limits of existence." As long as the rapture persists, the everyday world is carried off, only to be regarded with disgust when it returns to one's consciousness. This sobered ecstatic succumbs to a "will-numbing frame of mind." At this moment, he

resembles Hamlet, who is similarly revolted by the world and can no longer brace himself to act.

~ *NIETZSCHE, A PHILOSOPHICAL BIOGRAPHY*, Rüdiger Safranski, 2000.

In the above quote, Rüdiger Safranski eloquently brings together the brilliance of Goethe with the brilliance of Nietzsche. There is a universal pulse—like breathing in and out—that defines our essence. We now know that the in-breath is egocentric, the out-breath is allocentric, and the two must oscillate to sustain life. As the quote suggests, Goethe saw this oscillation as an alternation between expansion and contraction (as have others). Music uses this universal pulse, these frequency waves, to create. Nietzsche saw that the Greeks knew this duality 2500 years ago. The allocentric mind is capable of rapture as the ego dies (as we breathe out). But when the ego returns with the in-breath—as it must—the come-down is depressing.

Having pure experiences, untarnished by rational thinking, is such a powerful high that the body hungers for more. Living the wonderful life can become an addiction. When the ego comes back into the skull, hauling its heavy armor, with that humorless scowl all over its grim face, the come-down from the high, from the joy of living, is painful, almost suicidal. This is the bane of the sensitive artist and the religious saint.

In the discussion above, I provided an introduction to the Romantic Movement, Phenomenology, and Existentialism, and I looked at several important individuals who identified, categorized, and labeled our inherent duality. However, the creation and evolution of Phenomenology is of particular interest since it was a serious attempt to perceive the allocentric mind as an equal partner to the egocentric mind (using my terminology). Therefore, more needs to be said about this significant philosophy and its relationship to cognitive duality.

Phenomenology

Phenomenology has been called the first-person science of consciousness. Edmund Husserl founded the discipline—he called it transcendental Phenomenology—and he spent a lifetime defining and clarifying his conceptualizations. His ideas stand on the shoulders of earlier philosophers like Franz Brentano, David Hume, and Bernard Bolzano. Brentano introduced Husserl to the medieval theory of *intentio, the mind aiming at objects in thought or perception* which eventually led to Husserl's notion of *intentionality.* On Husserl's shoulders stand other great thinkers who took his ideas about Phenomenology and created their own refinements to the theory, including Martin Heidegger, Edith Stein, Jean-Paul Sartre, Simone de Beauvoir and Maurice Merleau-Ponty.

Husserl grappled with what we now call *the hidden senses.* These are whole-body sensations, which, in a broad sense, we call proprioception (including the vestibular system, kinesthesia, and what I call photo-sensitive processing). However, Husserl did not have the biological and psychological research that would have enabled him to understand the allocentric mind. He could sense that the body as a whole, the processing that was all-at-once and immediate, was somehow very different from the rational mind. He was, it seems to me, struggling to comprehend the two ways we pay attention. Sometimes he tried to explain egocentric processing and at other times he spoke of the *Phenomenology of embodiment,* clearly referring to whole-body processing and the allocentric mind.

From a dual-process theory perspective, Phenomenology is a mixed blessing; it is supportive of dual-process theory from one perspective, and yet it is opposed from another perspective:

- *Paying attention to experience* seems at first glance to include an awareness that an allocentric mind exists. Experience is participatory; when we are experiencing we are not using our analytical mind. Yet Husserl defines consciousness as a process that is "always directed at something or someone." The important concept of *intentionality,* as it is used by Husserl, is a process of "aiming the mind at an object-of-regard." This is clearly a definition of egocentric consciousness and fails completely to acknowledge allocentricity, which does not aim the mind at objects of regard. To be fair, Husserl was not searching for a biological substrate for his theory, and the goal of Phenomenology is to have pure experiences, not to explain or develop hypotheses. Also, of course, dual-process theory is a modern proposition; I write this in 2018.

- Husserl turns our attention from the object-of-regard to the process of regarding (to our state of consciousness). However, what he misses, in my opinion, is an understanding that *attention* itself is the key to a greater understanding of how our minds work. The biological processes that give rise to two forms of "paying attention" determine what kind of meaning we are extracting.

- *Dependency and ground* are fundamental concepts in Phenomenology. Husserl seemed to grasp the Buddhist awareness of interconnectivity (simultaneous arising)—that everything is related. The figure is dependent on the background; features are perceptual manifestations from the background.

- *The theory of parts and wholes* is also fundamental in Phenomenology. Everything is part of something larger and at the same time everything large can be reduced to ever smaller parts. A background can be part of a greater background, and a feature (a figure) can itself be a background for smaller entities.

- In a 1935 lecture in Vienna, Husserl stated that *science had lost touch with reality*; it had lost sensitivity and empathy for the everyday world of human beings. There is an irony here that drips off the ugly walls of world history. Husserl had lost a son fighting for Germany in World War One; a second son had been wounded. Husserl was ethnically Jewish, but his religion was Protestantism. The Nazis, of course, did not care—for them, Husserl was just another ethnic Jew. They stripped him of his rights at the university and denied him the opportunity to lecture. He died before they could do more damage to his life. His widow was smuggled out of Germany as was the bulk of his written work. Martin Heidegger, the Phenomenologist who is often quoted as saying that humanity had lost contact with *being* (with sensitivity, loving-kindness, and empathy) is remembered by some as a Nazi Sympathizer:

In 1933 the Nazis came to power in Germany and issued new regulations prohibiting Jews and other non-Aryans from holding positions in government or in the universities. Husserl was thereby effectively locked out of the university. The *Rektor* of the University of Freiburg in the spring of 1933, the official who enforced the Nazi decree, was none other than Martin Heidegger, Husserl's former assistant and now his successor. Although Heidegger remained as *Rektor* only a few months, Husserl was shocked by the actions of both his former friend and his adopted country. Ethnically Jewish (though a Protestant), Husserl was loyal to Germany, noting that his son Wolfgang

was killed and his son Gerhart wounded while fighting for Germany in the First World War. Heidegger's magnum opus *Being and Time* was dedicated to Edmund Husserl "in admiration and friendship" when published in 1927; subsequent editions dropped the dedication. Heidegger's actions and political ambitions in Nazi Germany remain a subject of scrutiny, but his acquired enmity for his forerunner was expressed in no uncertain terms in letters years later.
~ *HUSSERL (THE ROUTLEDGE PHILOSOPHERS)*, David Woodruff Smith, 2013.

What leaves me cold and bewildered is the image of philosophers like Heidegger hunched over their desks penning their memorable abstract phrases about being awake, caring, and appreciating the moment, and yet failing completely to put any of their convictions into their daily lives. The masculine mind, especially, can grasp concepts and systems theory, can even define and debate empathy, and yet can have no connection to the actual experience of living, or of displaying empathy. Evidently, a man can uphold the supreme importance of compassion, empathy, hope, and charity, and yet perceive no conflict with a membership in the Nazi party. The irony is bewildering and disheartening.

There are many parallels in our own era. I am looking at the American political scene in the year 2018, feeling that egocentricity (narcissism) has temporally won out in the hearts and minds of a significant core of the American populous. Christian fundamentalists in America preach love and faith and yet vote as a block in a way that denigrates love and faith—the devil could not ask for better allies. The same brand of egocentric hatefulness arises repeatedly through history. In my opinion, this is because we fail as a culture to grasp our basic duality—we fail to balance the allocentric mind with the egocentric mind. The devil has purchased the souls of shallow people who wave their religious texts overhead as they pontificate—like warrior clans of old—about the glory of the regime.

I have to be careful when I make my sentiments known, as I have above. The world is complex, emotions are complex, and nothing is as simple as it seems on the surface. For example, I have intelligent, kind friends who are Christian fundamentalists. I have other secular friends who voted—for various complex reasons—for what I perceive to be a narcissistic political movement. If we look back at Martin Heidegger and his brief stint as a Nazi collaborator, we see a situation more complicated than we first surmised.

Heidegger was the successor to Husserl; he carried the torch of Phenomenology further along during his life. Perhaps we need another opinion, this one a tongue-in-cheek look at Martin Heidegger:

> [Heidegger] was (with the possible exception of Wittgenstein) the greatest philosopher of the twentieth century. He was (with the possible exception of Hegel) the greatest charlatan ever to claim the title of "philosopher," a master of hollow verbiage masquerading as profundity. He was an irredeemable German redneck, and, for a time, a gullible and self-important Nazi. He was a pungent, if inevitably covert, critic of Nazism, a discerning analyst of the ills of our age and our best hope of a cure for them. Each of these claims has been advanced, with greater or lesser plausibility, in Heidegger's behalf.
> ~ *HEIDEGGER, A VERY SHORT INTRODUCTION*, Michael Inwood, 2000.

To be fair to the complexity that is Heidegger, we must admit that the evidence and debate are inconclusive. Like all of us, we ebb and flow—emotions and rationality oscillate just as ego and self alternate. There is an inherent confusion caused by being our own twin.

- One reason that Phenomenology is sometimes supportive of dual-process theory and sometimes not is because each student of Phenomenology took the theory further than their teachers. Husserl, according to his biographers, went through four phases in his career, each time evolving his understanding of Phenomenology. Jean-Paul Sartre and Simone de Beauvoir turned Phenomenology into Existentialism, while Heidegger and Merleau-Ponty each redefined and elaborated on aspects of Phenomenology—it is not one animal, it is several.
- Husserl spoke of empathy as if it was a separate way of knowing from the rational mind. Therefore, as we already know, he was aware of humanity's cognitive duality. However, it was actually his talented student Edith Stein who took Husserl's ideas and applied them to empathy. Stein was writing at a time when empathy was just being defined and debated; her doctoral dissertation—published posthumously—*On the Problem of Empathy* (1964, 1989) comes closer to allocentricity than Husserl

272

expressed. The antithesis of empathy is the Nazi (egocentric) mind—the Nazis killed Edith Stein in a concentration camp.

⟞⟞⟞⟞⟞⟞⟞⟞⟞⟞⟞⟞⟞⟞

Let me turn now to Sarah Bakewell's book *At the Existentialism Café* for a more personal look at Phenomenology—as it plays out in everyday life. According to Bakewell, Phenomenology is *a method* for experiencing the world—it is more method than it is theory. It is more about *experiencing* moments (objects, events) than it is about analyzing them. The *method of Phenomenology* is to perceive without conceptualizing. In my terminology, Phenomenology is about using the allocentric mind to experience existence and not about the egocentric mind's mission to understand the world.

Phenomenology is also not concerned with history. Here is an example using Sarah Bakewell's analogy: If we consider our morning cup of coffee, we don't care how the coffee arrived on our table top. We don't care about the science behind caffeine, about the molecules that create aromas, or about workers getting slave wages on the coffee plantations. We don't care about the evolution of the coffee plant or about varieties of coffee beans created by genetic manipulation. We don't care about chemistry, botany or the international coffee trade. All we care about is *this cup* of coffee.

Taking Bakewell's analogy further, if I slow down enough to smell and taste my coffee, I will enjoy the moment. If I linger over my coffee, fully immerse myself in the experience, notice the birds singing and the tropical breeze, I might even get a sense of joy, gratitude, and awe about being alive in this particular moment.

Phenomenology, according to Bakewell, is also not concerned about the future. I don't want to know what my digestive system thinks about my decision to drink coffee or what my bowels do several hours later. I don't want to know how my purchase of the coffee has affected the local economy. I don't even want to know how the caffeine will affect my mood and decisions the rest of the day. I just want to enjoy the coffee right now and without interruptions.

From an egocentric perspective, if I decide to analyze my coffee experience, I will end up reducing the moment to quarks, emptiness, and depression. Emotion and wakefulness fade as we objectify, analyze, categorize, and make moral judgments. We are left with a cold, odorless cup of "meaning" rather than an experience that makes life worth living. The history and future of a cup of coffee are dead issues for Phenomenologists. All we have are moments. We either savor these moments, or we turn them into tasteless memories or senseless possibilities.

If we want to live fully in the moment, we must learn to suspend judgements, at least while in the act of experiencing. In other words, the egocentric mind needs to *butt out* of the moment. The Phenomenologists used abstractions and extra verbiage to define "butt out" (which is too bad, but that's my opinion). Husserl said we had to "bracket out" history and future; we had to suspend the ego's desire to dissect experience. Husserl used the word *epoche* which was coined by the Greek skeptics to mean "suspension of judgments." But it comes down to this: enjoy your coffee. Stop analyzing everything.

Bakewell further explains that Phenomenology doesn't care if something is objectively real or not. Phenomenology is about personal experience, about what's going on inside a person. It is a philosophy that gives validity to subjectivity. According to Phenomenology, the world of the mind is as real and important as the physical universe, more so actually from the perspective of an emotional, curious human being. When consciousness evolved it became another dimension. Besides space, time, and physical stuff, a new entity came into being when self-awareness appeared on the evolutionary stage. What this new dimension did moment-to-moment became the concern of Phenomenology.

Phenomenology does, of course, search for "meaning." However, Phenomenology uses an allocentric definition of meaning as well as an egocentric definition. Indeed, according to author and Philosophy Professor David Smith, Edmund Husserl felt that *meaning* was the heart of Phenomenology:

> Meaning is central to Phenomenology: meaning is the significant content of conscious experience, which we ascribe in saying "what" a person sees or thinks or wishes. It is meaning that distinguishes nearly all of our experiences, and it is meaning that renders experience a consciousness "of" anything at all. Only through meaning, Husserl held, does consciousness present us with a world, an organized structure of things around us, including ourselves.
> - HUSSERL (THE ROUTLEDGE PHILOSOPHERS), David Woodruff Smith, 2013.

In the above quote, there is a hidden conundrum that highlights the confusion caused by dual cognition. Essentially, in my opinion, Husserl is using *meaning* two ways. There is an egocentric definition of meaning that renders "consciousness of something," which depends on the duality of a seer and a seen. However, there is an immediate awareness and reaction to environmental changes that is allocentric. *Meaning*, in an allocentric context, requires an innate (unthinking) prior knowledge of how the world works and how the body should best respond to insure survival.

As we move forward in time beyond the philosophers who created Phenomenology and Existentialism, we arrive at 20th century thinkers who stand on the shoulders of Husserl, Heidegger, Sartre, and Simone De Beauvoir. One of these contemporary philosophers is British author Owen Barfield. For me, no discussion of the philosophy of duality would be complete without Barfield's insights. He was a personal friend of C. S. Lewis and J. R. R. Tolkien, and a member of the now famous literary group called *The Inklings*. The friends, The Inklings, met weekly in *The Eagle and Child* pub near Oxford University in England. Many people know the trilogies and mythologies that Lewis and Tolkien gifted to the world, but few know about their friend Owen Barfield and his contributions to the study of human consciousness. Barfield was also a Coleridge scholar and a follower of the philosophy and esoteric writings of Rudolf Steiner—I will elaborate on the contributions of Coleridge and Steiner in the discussion to follow.

Owen Barfield

Of all the authors I read for this book Owen Barfield was by far the most intriguing. Because he dissected human consciousness from several diverse perspectives, I spent many hours trying to summarize his complex thoughts.

Barfield's favorite book of all his works was *Saving the Appearances: A Study in Idolatry*. He felt that this book was a summary of his most important ideas. He was quoted as saying that the ideas in *Saving the Appearances* would not be completely understood for fifty years after his death—not until a time came when cultural levels of consciousness had risen sufficiently for most people to understand. He strongly felt—I find this true of most great thinkers—that his conclusions were central for the survival of humanity. His ambition was to set mankind free from "false habits of thought and common sense." He knew, as he wrote his books and his poetry, that he was writing for a future generation with an evolved consciousness.

As a philologist, Barfield studied the evolution of language in ancient texts. As he gained ever greater understanding, he realized that ancient languages contained a history of cognitive evolution. Author Gary Lachman explains Barfield's insights about the evolution of language:

> . . . the world the ancient Greek saw and the one we see are not the same. The kind of consciousness we enjoy—if that's the right word for it—is very different from that of an ancient Greek, a Greek of late antiquity, a person from the Middle Ages, or even one of the early modern age. Not only our ideas about things, Barfield tells us, but our consciousness itself has evolved over time. And if we are to take seriously the contention of philosophers like Immanuel Kant—that the world we perceive is a product of our perceptual apparatus— then a world produced by a different consciousness at a different time will be, well, different.
>
> ~ REVOLUTIONARIES OF THE SOUL, Gary Lachman, 2014.

This is a very significant observation. The Greeks, for example, were not modern humans with our level of consciousness. Thinkers from centuries before our own did not have the cognitive and perceptual apparatus that we enjoy. Unless we somehow comprehend how earlier minds worked—relative to our own—we will misinterpret their discourse and intent; we will falsely read into their words our modern viewpoints. Consciousness has evolved and continues to evolve—therefore, later generations may falsely attribute to us abilities and insights which we do not have.

If Barfield is correct, and I think he is, then we are faced with a remarkable insight: consciousness is evolving at a very fast—perhaps exponential—rate of speed. The consequences that face us, as we confront this new understanding are rather profound. In our era, unlike at any time in the past, we can now perceive levels of consciousness—the mechanism of consciousness is being studied and understood. Technology is also directly invading our nervous system and our genetic makeup is being altered with ever greater accuracy and ease—we are learning to engineer consciousness. Knowledge is exponentially expanding and being automatically cataloged and shared. But the scope of knowledge is beyond human capacity to fully comprehend—so we are creating cyborgs, part human and part technology, a blend of biochemical consciousness with artificial consciousness. We are becoming experts in ever more arcane and specialized arenas. It is getting harder and harder to accurately communicate—evolved minds exist side-by-side with less evolved minds.

As I said above, Barfield felt that his last book *Saving the Appearances* was his crowning achievement, and no doubt it is. However, having read several of Barfield's books, I can advise the reader *not* to start with *Saving the Appearances*, a book difficult to understand at times. *Saving the Appearances* assumes that the reader has a wealth of knowledge gathered from Barfield's earlier publications. Start with *History in English Words* (1926), or *Romanticism Comes of Age* (1944), or any of his earlier books. I would read *Saving the Appearances* (1957) last so that it can be appreciated for the literary treasure that it is.

Let me suggest, using dual-process theory and my terminology, what Barfield was getting at when he came up with the title *Saving the Appearances: A Study in Idolatry*. There are two ways to comprehend our existence, according to Barfield. Science has grasped one of these approaches (the egocentric) and rode the scientific method into the void. But there is another way to comprehend our essence and that is allocentrically. According to Barfield, we will have to allow these two contradictory hypotheses to share in the glory. What Barfield is trying to explain is the allocentric mind—although he did not have that term in his lexicon. Barfield wants us to "save the appearances," our moments, our sense of being alive. He is saving this allocentric perspective from the coldly rational decisions of the egocentric mind. My point is that Barfield, like so many others, discovered humanity's dual cognition and quickly grasped the paradox and challenge of having twin minds.

The study of idolatry, the second half of his title, refers to the scientific, egocentric habit of reducing everything into ever finer filaments of potential:

vacuums and voids nested inside each other. Then we idolize these filaments and fragments, forgetting the big picture and forgetting the relationships that hold everything together. Barfield's point is that we have come to worship science just as the Greeks worshiped the all-powerful Zeus. The study of idolatry is the study of how we set our theories on pedestals and then worship them as truth—forgetting they are tentative propositions. The wonderful and vital scientific method becomes scientism and materialism if we forget that the scientific method is based on eternal doubt. The spiritual mind remains hidden if we deny its existence and if we don't take spirituality (allocentricity) seriously. This is what we discover when we study our unfortunate and quite egocentric habit of idolatry.

Owen Barfield came to his conclusions about duality through his scholarly studies, especially his work with the insights of the poet Samuel Coleridge (1772-1834). We can almost discern a philosophical thread that weaves through history: Barfield found duality in Coleridge who found duality in the plays of Goethe and the idealism of Kant, who found duality in the philosophy of the Greeks, who probably got their notion of duality from contact with Eastern esotericism. The thread continues into the future: I am unraveling the thread in my discourse as I speak to you of an allocentric mind and an egocentric mind.

Owen Barfield and Samuel Taylor Coleridge

The following discussion about the life and ideas of Samuel Coleridge draws heavily from the works of philosopher Owen Barfield, especially from Barfield's books *What Coleridge Thought* (1971) and *Romanticism Comes of Age* (1944).

Samuel Taylor Coleridge (1772-1834) was an English poet and philosopher. He lived in the Lake District of England near one of his best friends, the poet William Wordsworth. Along with the poet Robert Southey, Wordsworth and Coleridge helped create the Romantic Movement in England—they were part of a literary group, a collection of romantic writers who became known as the *Lake Poets*. In 1798, Wordsworth and Coleridge published a joint collection of poems called *Lyrical Ballads*, which had a far reaching impact on poets and writers as far away as America—it was treasured by American transcendentalists like Ralph Waldo Emerson. Coleridge is remembered today for his poems *The Rime of the Ancient Mariner*, *Kubla Khan*, and *Christabel*, and for a two volume autobiography called *Biographia Literaria*. Coleridge also published a weekly journal entitled *The Friend*, which influenced many writers and philosophers, including John Stuart Mill and Emerson.

In 1798, Coleridge and Wordsworth traveled to Germany together where they encountered the theories of German idealists. When they returned to England, they incorporated the critical philosophy of Immanuel Kant into their writing and poetry. Coleridge studied German during this visit and later translated the poetry of German poet Friedrich Schiller. In 1814, Coleridge was asked to translate Goethe's *Faust* into English, which he initially took on but then abandoned. Whether or not Coleridge completed the translation is unclear (and controversial), but he no doubt spent many months on the project and was influence by Goethe's ideas.

Coleridge was a meticulous craftsman who careful reworked and honed his writing. He created a philosophy of poetry that is still valued today. Coleridge was critical of the literary competence of his fellow poets and novelists—he felt that quality literature was on the decline, while bad taste and imprecise writing were on the rise. This description of Coleridge sounds a lot like the character of Owen Barfield who also insisted on the meticulous use of language. It's no wonder that Barfield became a Coleridge scholar. Coleridge is one of the most important figures in English poetry—his poems influenced all the major writers of his era.

As a philologist, a word-smith, Owen Barfield was fascinated by Coleridge. Both men came to similar conclusions: that consciousness evolved and that there was a duality at the heart of the evolution. In an essay by Trevor Levere about Coleridge, Barfield asserts that Coleridge was looking in the opposite direction to the gaze of science:

> Coleridge's face, Barfield explains, was turned . . . in the opposite direction to the one which natural science was taking in his time . . . it was [Coleridge's] firm conviction that, if knowledge was to advance, there must be a science of qualities as well as quantities.
> ~ REVIEW OF POETRY REALIZED IN NATURE: SAMUEL TAYLOR COLERIDGE AND EARLY NINETEENTH-CENTURY SCIENCE, "Coleridge in the Twenty-First Century." Trevor H. Levere, 1982.

In modern terminology, the egocentric mind is concerned with quantities and measurement—as such, it gave birth to science. However, the allocentric mind is concerned with qualities, with experiences that come from participation—this mind gave birth to the arts and religion. Coleridge could see that these two ways to understand the world were different—they were mirror images, opposites, what he called *polarities*. What both Barfield and Coleridge saw, therefore, was that there was an active process occurring during the act of participating within the environment, during the moments when we were experiencing our lives. This non-verbal active process they called *imagination*. As simplistic as it might seem, we can best understand what Coleridge and Barfield are getting at if we substitute the word *mind* for *imagination*.

Coleridge said there were two kinds of imagination (mind), primary and a secondary. When I look at how he defines these distinctions, I find myself staring at the usual revelation: we have an allocentric mind (primary imagination) and an egocentric mind (secondary imagination). The primary imagination is concerned with quality, while the secondary imagination is concerned with quantity and measurement. Coleridge (and Barfield after him) felt that English culture had sold out to the secondary imagination (egocentric mind), and had devalued or ignored the primary imagination (the allocentric mind). This sentiment, this conclusion, is found in Wisdom Traditions everywhere in the world, and in the various postulates of philosophers throughout history—Eastern as well as Western. We will see this duality again as we later explore the ideas of psychologists, scientists, and educators.

Coleridge sees that the secondary imagination is a higher kind of processing that can evolve to create innovative works of art. The artist, the poet that both Barfield and Coleridge find so fascinating, is a person who is able to use a finely-honed intellect to create exquisite poetry—the right words in the right place. Add these exact words to an emotion, a quality like love or joy (add the allocentric component) and you get tasteful poetry that endures through the ages. Thus, in a way, Coleridge and Barfield are saying that a true artist is one who can balance two highly evolved minds to create the most beautiful creations. What the two men sought to understand was how the ability to craft great poetry cognitively evolved. They could see that the ability to use the human mind to forge creations was akin to the ability of a cosmic mind to similarly forge creations—as above, so below. Furthermore, the two men wanted to understand the stages of cognitive evolution that led to the ability to generate exquisite poetry (or any kind of great art). This is similar to the efforts of modern developmental philosophers to comprehend the evolution of the stages of consciousness.

Coleridge was a contemporary of Johann Goethe and was influenced by him. Goethe formulated a different kind of scientific method that sounds a great deal like what Coleridge had formulated when he differentiated primary from secondary imagination:

> Goethe brought his own scientific method, which is really none other than the method of imagination. Now Goethe with his method of "exact percipient fancy," as it is often translated, really transferred the esemplastic [to mold into one] imagination from literature and art to science. His method differs from the ordinary method of induction in that the observer, when he reaches a certain point (the "prime phenomenon"), stops there and endeavors rather to sink himself in contemplation *in* that phenomenon than to form further thoughts *about* it. . . . The blue of the sky, said Goethe, *is* the theory. To go further and weave a web of abstract ideas remote from anything we can perceive with our senses in order to "explain" this blue—that is to darken counsel.
> - *ROMANTICISM COMES OF AGE*, Owen Barfield, 1944.

I have a feeling that something was lost in translation of the above quote from the German to the English—especially since we are dealing with abstract concepts. To put it simply, Goethe suggested we study the living process—of a flower, for example—rather than dissect the flower and then study the dead parts. Goethe was not reducing the object of study into its constituent parts, as science would do, he was sinking himself into the animation of the object, trying to feel (intuit) how it worked. He was looking at process, trying to *become the flow* of that which he studied. Both Coleridge and Barfield understood this distinction and applied it to their own theories.

Coleridge seems very close to comprehending the physiological differences between our two minds. He seems to sense that—for some underlying physiological reason—life needs to be dual. In the quote below Coleridge defines a "universal law of polarity," which is an "essential dualism" found in nature:"

> Every power in nature and in spirit must evolve an opposite, as the sole means and condition of its manifestation. [Furthermore] all opposition has a tendency to re-union. This is the universal Law of Polarity or essential Dualism.
> ~ THE FRIEND, a periodical written by Coleridge in 1809 and 1810.

In the quote above and in his writing, Coleridge uses a very specific definition for *polarity*. He asserts that there is a connection that links two inseparable forces together—a symbiotic balance that keeps two forces together at the same time as it forbids them blending together. Two forces are so entwined that they cannot exist alone even though they appear to be opposites and separate. Therefore, in a paradoxical way the two polar opposites are part of a unity.

Another way to say this is that a figure—which has manifested out of a background—is still part of that background even though it appears to be a separate entity. Coleridge's writing indicates that he "discovered" humanity's two minds and then tried to explain what was happening through his observation of polarity. Here is a paragraph that seems to affirm my interpretation:

> O said I as I looked on the blue, yellow, green, & purple green Sea, with all its hollows & swells, & cut-glass surfaces—O what an Ocean of lovely forms!—and I was vexed, teased, that the sentence sounded like a play of Words. But it was not, the mind within me was struggling to express the marvelous distinctness & unconfounded personality of each of the million millions of forms, & yet the undivided Unity in which they subsisted.
> ~ THE NOTEBOOKS OF SAMUEL TAYLOR COLERIDGE, 1827-1834, edited by Kathleen Coburn & Anthony John Harding, 1962.

In the above quote, the ocean is the background, while the appearances (the colors, the currents, the reflections) are "forms" that seem separate from the ocean, but are not. The world is one unity even though there appears to be a polarity between the emerging figures and the ground.

To conclude this brief look at Coleridge through the eyes of Barfield, I will restate the obvious: great thinkers, great artists, and great religious figures all found a fundamental duality within their area of study. Whether we follow the path of intuition, or emotion, or intellect, the same image appears: a background manifests (displays) forms. Something perceivable emerges from something unperceivable. Coleridge saw it in Goethe and Barfield saw it in Rudolf Steiner and Coleridge. Great minds discover duality (the figures) at the heart of unity (the background).

Aquinas, Kant, and Barfield

Barfield and his father ran a law office together. After his father's death, responsibilities at the family law firm weighed heavily. Barfield didn't like what he was doing—he was happily a writer and philosopher, but not so much a lawyer. A domestic crisis at home added more pressure and Barfield, in his own words, was close to a nervous breakdown. He turned to cathartic writing and completed a semi-autobiographical novel called *This Ever-Diverse Pair* (1950), which is "a somewhat comical and philosophical exploration of the legal profession." (Quoted from Barfield's official website: Owen Barfield: philosopher, poet, author, thinker, sage). Barfield said that writing this book "stayed-off" the nervous breakdown. I especially like Barfield's title, since it seems to intuitively point to our two minds as an *Ever-Diverse Pair*—separate but connected.

After the pressure at work lessened, Barfield found time to do what he loved: writing, synthesizing ideas, and using his imagination to create. For a year or more, as part of a disciplined routine, he went to the reading room at the British Museum in London to relax and to read, taking copious notes about all kinds of diverse subjects, from physics to medieval history. After a while, he tried to squeeze all these scattered notes into a book, but all the seemingly unrelated topics refused to solidify into any system or relationship. Then one day while he was reading a book by Gavin Ardley called *Aquinas and Kant* (1950), he came across the phrase *saving the appearances*. He suddenly had an "aha!" moment that brought all the diverse subjects together in a way that made sense. The quote below from *Aquinas and Kant* seems to show that Barfield's *aha* moment came when he grasped Kant's differentiation between phenomenon (what we perceive) and noumenon—the real world, in a new way: (italics mine)

This is Kant's fundamental denial of the possibility of metaphysics as ordinarily understood: *we must distinguish between the appearances and the thing in itself.* Our a priori knowledge has only to do with appearances. The thing in itself is real but cannot be known by us. The thing in itself is always beyond our knowledge and inaccessible, since it is always subject to, and behind the barrier as it were, of the pure concepts of the understanding, or the categories as Kant calls them, following Aristotelian terminology. *Later in the Kritik he designates the appearances as phenomena*, and the things in themselves as noumena. We can know the phenomena, but with the pure speculative reason we cannot know the noumena.
~ *AQUINAS AND KANT*, Gavin Ardley, 1950.

The significance of the above quote is that we can see where Barfield got the term *appearances* as a synonym for *phenomenon*. His last book might very well have been called *Saving the Phenomenon*, or in my terminology: *Saving the Allocentric Mind*. The appearances are what we perceive in the moment; they are personal experiences that arise as we participate within nature. Appearances are the way the world seems to be—the way reality appears to us. There is nothing abstract or magic about how the word *appearances* is used by Barfield; appearances are the result of what our senses report, and what our consciousness builds from the sensory data. Appearances are based on the assumption that there is an external objectivity and that what our consciousness projects is a valid reality. However, just like the Phenomenologists, Barfield finds the allocentric mind to be lost (hidden), especially as science reduces everything—step-by-step—to nothing. We have to save the mind that generates appearances because materialistic science, at the end of the road, is defining everything as a meaningless void.

Gavin Ardley's book *Aquinas and Kant* (1950) is an attempt to reconcile mental opposites that seem at war with each other (our two minds). Ardley looks back in history at a time when philosophers, especially Aristotle, threatened emerging Christian doctrines. St. Thomas Aquinas found a way to bring warring ideas into harmony at a time when such a view was badly needed. Aquinas was a spokesperson for the Wisdom Tradition (the *philosophia perennis*) discussed in Chapter Four. Ardley explores how St. Thomas Aquinas addressed the conflict between the Greek material worldview and the spiritual Christian worldview—how Aquinas found a way to accept both.

Ardley also looks at the philosophy of Immanuel Kant. Kant attacked the Wisdom Tradition; he was a rational man with a strong egocentric consciousness. At the least, Kant saw the Wisdom Tradition as alien to the way human beings actually are (from his perspective). According to Ardley, its Kant's fault for the mess we are in—for the conflict between the material and the spiritual. If we accept Kant's thoughts as irrefutable, then we have to admit that there is no way human beings can ever figure out whether there is a God, or whether cosmic spirit contains any intelligence or compassion. Our senses are so incredibly limited according to Kant, that we are fooling ourselves badly if we think the appearance of things equates to reality.

Ardley also discusses the theory of Islamic philosopher and mathematician Averroes to show a medieval thinker who held a dual concept of mind similar to Kant's. Averroes was another great thinker who challenged the Christian perspective. Aquinas knew of Averroes and felt it necessary to refute his logic.

During the medieval period of history, Averroes (Ibn Rushd) wrote of two kinds of knowledge that resulted in a mental struggle to reconcile faith with reason. Averroes's ideas are called the "theory of the double truth." Based on Averroes's theory, two types of knowledge were in direct opposition to each other. This was very threatening to the medieval church. St. Thomas Aquinas rejected Averroes's theory, saying that both kinds of knowledge (if there was such a thing) came from God and were therefore compatible and capable of working in collaboration. *Aquinas believed that faith could guide reason while reason could justify faith*—that was his important insight and the healing idea that reconciled duality. In other words, Aquinas could see that the two minds worked together—*had to* work together—and both minds were created by God. God decreed this duality; it was the way of creation.

The book *Aquinas and Kant* is not an exercise in Kant-bashing or a praise of the ideas of St. Thomas. The book is an attempt to appreciate both Kant and Aquinas. As the quote below asserts, both philosophers made valuable contributions to our exploration of mind and consciousness:

Aquinas represents the metaphysician of the *philosophia perennis*. Kant on the other hand, as we understand him, in his basic contentions gets to the heart of the characteristic non-metaphysical pre-occupations of the modern world. These pre-occupations seem . . . to be alien to the *philosophia perennis*. Consequently the juxtaposition of Aquinas and Kant throws the medieval-modern conflict into high relief. It is the purpose of this work to moderate the conflict, and to show that what is basic to Kant is not really alien to Aquinas, but that they are, on the contrary, complementary.

~ *AQUINAS AND KANT*, Gavin Ardley, 1950.

The quote above is Ardley's opinion that Kant and Aquinas are allies in the search to comprehend our twin minds. They are not really enemies, as history might incorrectly conclude. And that, of course, is what I am saying in this book as well: we have two wonderful and long-evolved minds that make us cognitively dual superheroes. This presents us with responsibilities and opportunities. We must first recognize that we have two kinds of consciousness. Then we must accept that they alternate at such a speed that we get an illusion of oneness. Finally, we must learn to use each mind in the service of kindness, wisdom, and reason.

The Anthropology of Consciousness

As Owen Barfield looked at the jumble of notes he had accumulated during his 18 months of research at the British Museum, in London, he began to see patterns emerging from the chaos. This section looks at the some of the patterns that Barfield assembled to create his worldview. I suggest that what Owen Barfield found was an *anthropology of the mind* embedded in the evolution of language. Barfield said that if we looked at the development of language, we would discover mental artifacts that would help us understand how consciousness evolved:

> It has only just begun to dawn on us that in our language . . . the past history of humanity is spread out in an imperishable map, just as the history of the mineral earth lies embedded in the layers of its outer crust. But there is this difference between the record of the rocks and the secrets which are hidden in language: whereas the former can only give us knowledge of outward, dead things—such as forgotten seas and the bodily shapes of prehistoric animals and primitive men—language has preserved for us the inner, living history of man's soul. It reveals the evolution of consciousness.
> ~ *HISTORY IN ENGLISH WORDS*, Owen Barfield, 1926.

I love Barfield's analogy between the dead evidence that geology uses to recreate evolutionary patterns, and the living evidence we can still use to recreate the evolution of language and consciousness. In plain sight, we have a map that shows how the human mind evolved.

Barfield studied the derivation of words and their meanings. He was a philologist, a person who studies language as it appears in written historical sources. Philology is the search for the original meaning and the evolution of words. At one time, Barfield tells us—to give a couple examples—the word "understand" meant to literally "stand under," and "to express" meant literally "to press out." *In Saving the Appearances*, he said:

> . . . nearly everybody today thinks of [words] as divorced from the "appearance" of nature . . . accessible to the senses in a way which nearly everybody before the scientific revolution did not.
> ~ *SAVING THE APPEARANCES: A STUDY IN IDOLATRY*, Owen Barfield, 1965.

I think what Barfield is saying in the above quote—although it has been difficult for me to pin down—is that when less evolved human beings participated in nature, the words they used described their purposeful movements, those that were relevant to the moment. In other words, early, more primitive words had concrete meaning, rather than abstract meaning. These early words related directly to the appearances of things in nature. However, as the mind evolved, complexity and abstraction replaced the original meanings.

Barfield felt that science—the evolution of concepts and reasoning—robbed people of their direct contact with nature and had turned people inward. The phrase "saving the appearances" is the same idea as "saving the ability to appreciate and participate in the moment." This is a return to allocentric consciousness where direct experience is processed. In this regard, Barfield's philosophy is akin to Husserl's Phenomenology—both perspectives stress that, in our modern world, *being* is sacrificed at the altar of *doing*.

What Barfield discovered from his study of language is that consciousness evolves. As a philosopher, Barfield did what philosophers do for a living: think about thinking. However, realizing that consciousness is a product of evolution was revolutionary in his time. Barfield realized that in the past, people did not think like us—they had less developed minds; they were incapable biologically of *modern* thought. Here is how Barfield put it:

In my own field of study everything points to an evolution of consciousness, which, up to as recently as three or four centuries ago, has mainly taken the form of a contraction of meaning and therefore of consciousness—an evolution from wide and vague to narrow and precise, and from what was peripherally based to what is centrally based.

~ *THE BARFIELD READER; SELECTIONS FROM THE WRITINGS OF OWEN BARFIELD*, Owen Barfield, 1999.

I find the above quote to be a remarkable description of allocentric and egocentric attention. What Barfield is seeing is the rapid evolution of the ego (the egocentric frame of reference). Our ability to narrow perception and hold egocentric attention has evolved relatively swiftly in the past several centuries, forcing peripheral and holistic awareness—the allocentric mind—further and further from awareness. Barfield equates this contraction with the ever-finer shaping of meaning—we find relevance in ever smaller details. In other words, what we call *consciousness* is changing as this egocentric mind continues to rapidly evolve and make ever-finer gradations of meaning.

Within any culture there will be a range of individuals with varying levels of mental sophistication. For example, a culture could be composed mostly of traditional minds, or a more evolved culture could be composed mostly of modern minds, but always there is a mix of cognitive levels. The developmental philosopher Jean Gebser would say that cognitively low-level cultures of men and women with archaic minds, for example, were incapable of empathy or of an awareness of an ego separate from a self. In contrast, people with higher levels of consciousness are capable of greater empathy, tolerance, innovation, and wakefulness.

As I discussed earlier—this is my opinion—when we pull back and look at the evolution of human consciousness, we see that throughout recorded history our two minds have taken turns being "top dogs." For a while an entire culture is strongly egocentric (patriarchal, hierarchical) and less allocentric (matriarchal, cooperative), but in time this will inevitably reverse. The spiral of evolutionary development seems ever upward—even though the egocentric mind and the allocentric mind take turns being ascendant, they are each becoming more sophisticated as time moves forward. In other words, human beings are evolving ever more complex and attentive egos, and at the same time, ever more complex and aware selves (souls). But for some deeply physiological (perhaps quantum) reason, the two minds need to alternate on a grand historical scale. Owen Barfield looked at this evolution and alternation and saw it as a universal law:

> Anyone who has once contemplated the evolution of the earth and man as a progress from unity to fragmentation and from meaning to meaninglessness, and then, if all goes well, from meaninglessness to meaning and from fragmentation through to unity, will see traces of that universal process wherever he looks.
> ~ *The Rediscovery of Meaning and Other Essays*, Owen Barfield, 1977.

In other words, there is a universal process that creates and another that destroys. I agree with Barfield; the essential theme of this book is that no matter where we look, we see this push-and-pull oscillation. The egocentric mind reduces all meaning to meaninglessness; all unity is reduced into ever-finer fragments. Eventually, when this egoic stance reaches an extreme, the soul feels starved, barren, locked away. Then, as a reaction to extreme egocentricity, a phase shift occurs. Individuals and cultures then move from ego-dominance to a concern for unity, for the ascendancy of the allocentric mind. However, when the allocentric mind reaches an extreme of non-rational awareness, there is a neglect of the individual who then rebels and forces a phase-shift into a more egocentric culture.

As I look at the culture of the United States (I write this in 2018), I sense that we are in a phase shift from an ego-based culture toward a culture more concerned with unity, harmony, and the ascendancy of the soul (or so I hope). This is a theory, of course; the study of cultural evolution always reveals a complex mixture of levels of consciousness and degrees of patriarchal and matriarchal beliefs existing side by side.

In summary, Barfield used the evolution of language in ancient texts to craft his anthropology of mind. He saw that there was a time in history when the ego and the personality had not emerged, when humanity participated in nature—when there was no feeling that human beings were entities that had emerged from nature. The evolution of the human being, especially ego development, came rapidly onto the stage of evolution and is still manifesting—perhaps at an exponential rate. Barfield's research and complex thinking arrived at conclusions that are echoed throughout philosophy, religion, and psychology:

- consciousness is a process
- consciousness evolves
- duality (polarity) is inherent in the evolution of consciousness
- there are levels of consciousness
- the evolution of consciousness can be uncovered and analyzed, especially through the study of language

Barfield, Religion, and Duality

In an essay called "Matter, Imagination, and Spirit" (1974), Barfield did what philosophers do so splendidly well: he insisted we define our terms. We tend to go on and on in our writing or speeches, using whatever abstractions please us, assuming that the reader has a duplicate mind to our own, and so needs no elaboration. For example, take the words *matter* and *spirit*; how often are these abstractions used without bothering to clarify what they might mean? How often do we read about *materialism* or *spiritualism* and not bat an eye as we float over the words and onto the next sentence—never pausing to ask how the words are defined? These are emotionally loaded words, of course, so we take up sides before we get to the end of the sentence. When Barfield looked at the meaning of the words *spirit* and *matter* in his essay, he made a rather remarkable observation, one that fits perfectly with the egocentric-allocentric duality. Here he has taken up the definition of *spirit*:

> I come now to the word "spirit," . . . I said I would first consider the meanings of the two words [matter and spirit], and secondly the relation between them. But it is not really possible to serialize in this way, because the question of their meanings and the question of the relation between them turns out to be one and the same question. That is one reason that we cannot do without the word "matter." What do I mean by "spirit"? The question I am asking is what the word means today . . . Do we not mean precisely that which is *not* matter? If it is a *definition* we are seeking, our answer will be an

accurate one to the extent that, whatever else the term may mean, it always means that. And that does seem to me to be something about which everyone agrees, as soon as he begins to reflect, from those at one end of the scale who hold that spirit is the only reality, to those at the other who only use the word for the purpose of denying that whatever is meant by it *has* any reality at all. In logical parlance, whatever attributes we assign to the term "spirit," it must always be the contradictory of the term "matter." Or, more briefly, spirit is, by definition, immaterial.

~ *The Rediscovery of Meaning and Other Essays*, Owen Barfield, 1977.

In other words, *spirit* is the background, and *matter* is what material-izes from the background, by definition. Spirit is the allocentric background mind. Matter is the essence of the object-rich egocentric mind. When we per-ceive the world all-at-once, using parallel processing, we are creating a blank space, a canvas, a wordless page, a scene, a gestalt upon which to manifest the information-symbols that generate meaning. This universal gestalt is the spirit, according to Owen Barfield. However, when we attend to any of the symbols—words on a page, for example—we use egocentric, one-thing-at-a-time, processing, and dwell in the land of matter.

Remember that each of our two minds can create endlessness: allocen-tricity can create endless space, while endless time is the product of egocen-tricity. When the allocentric mind creates the endless regions of infinity, it concludes that there is a spirit-realm that must have been created originally by a spiritual creator—this is what Barfield concludes. However, the egocentric mind cannot conceive of space (spirit) but it can perceive matter, so it believes in infinite matter and denies—or is at least suspicious of—the existence of an original creative spirit.

Barfield would take my figure and ground analogy and say—as others have done—that human beings have both wisdom and intellect. He would say that human beings generate meaning from their consciousness which has emanated from a background that must have had the seeds of this intelli-gence and wisdom already there—a potential spirituality waiting to become manifest. If figures always are part of a greater background—and are never separate from it—then there must be an infinite and eternal background that is both intelligent and wise. For Barfield, our fundamental duality arises from a vastly intelligent and wise background.

As a very sophisticated Catholic, Barfield is making his main argument based on his religious and spiritual beliefs. He assumes a benevolent intelligence behind existence called God. I will try my best below to explain what I think he is saying—using my own words and perspective.

Our dual mind may not be the ultimate invention of the cosmos. If we take our fist and punch a hole in our space-time dimension (punch a hole in the scene in front of us right now) we might perceive another kind of universe outside this reality. Perhaps what is on the other side of space-time is a background out of which figures like "space-time universes" can manifest, something that manifests backgrounds. If we pay attention to what has manifested in our space-time universe, we behold not only stars, rocks, molecules, quarks, and dog food, but also human beings and consciousness.

If consciousness has arisen from this other-dimensional background, then that background already had the ingredients for intelligence within it. In other words, according to Owen Barfield, the material nature of our universe—one that contains intelligent creatures capable of love—is evidence that an intelligent background, a supreme God gave rise to the intelligence and benevolence that walks the earth today.

Barfield goes on to explain that when we become aware that the material is arising from the spiritual background—that matter is an expression of spirit—we open the portal to what he and Coleridge (also a Catholic) called *imagination*:

> It is a fact of immediate experience that, besides being the contradictory of spirit, besides being for that reason the occasion of spirit, and besides being the finished product of spirit . . . matter can be the present *expression* of spirit. The material can become an image, or picture, of the immaterial. Whether or not it does so for us will depend on ourselves. *When* it does so, we may call the resulting experience "imagination." Imagination, then, also is by its very nature a relation between matter and spirit; but it is a special kind of relation, a relation which at once maintains and transcends that contradiction between the two, to which we have so far been giving our attention.
> ~ THE REDISCOVERY OF MEANING AND OTHER ESSAYS, "Matter, Imagination, and Spirit," Owen Barfield, 1977.

In other words, matter is made out of spirit. Matter is composed of spirit. Matter *is* spirit manifest. If you buy this line of imagining, it explains concepts like Carl Jung's synchronicity, and it explains acausality. It also explains the power of poetry and great art. It keeps the mystery and magic in human existence. When poets and writers craft their metaphors, they are sculpting with spirit. The painter gives us images made out of spirit. When we experience matter as spirit—which we can do (if we want) as we walk through our space-time world—we are moving through imagination. If we learn to perceive experience (each of our moments) as a manifestation of spirit, if we use our imagination, we become artists. Nature, seen through the imagination, is an expression of a higher consciousness; it is alive, animated, and full of meaning.

When Barfield discusses what I call the allocentric mind he uses the word *participation*. He means that when we aren't busy being egos, we are immersed *within* nature, part of nature, participating as if in a flow that cannot be dissected or reduced into parts. When we participate, there is no perception of time, we are in the moment. Animals live in this world of allocentric consciousness, as did early mankind, according to Barfield.

This participation is how we each (and collectively) take part in the universal mind. Animals have more participation than human beings because humans spend so much time "in their heads," while animals live in their whole bodies and they don't use language to analyze experience. Even our forbearers, according to Barfield, were better at participation. We are now almost completely cut off from nature; now we call nature *environment*. We misperceive ourselves as separate from the environment, rather than a part of nature.

It is interesting to note in passing that the allocentric visual system sees only in black and white. I have always wondered why black and white photography has such a powerful emotional feeling for me. Perhaps this is because black and white perception is more in touch with allocentric consciousness—figures are less separated from the background when there is no color contrast. Color perception is a product of the egocentric mind; figures stand out much more vividly from the surround using egocentric attention. Likewise, the egocentric mind evolved refined depth perception, while the allocentric mind sees only a flat gestalt. There is little egocentric perspective in early paintings because (presumably) the egocentric mind was less evolved just a few hundred years ago. Depth perception requires an ability to extract a form from the background in such a way that it appears three-dimensional, clearly apart from the surround.

Owen Barfield also used the words "representational" and "non-representational" to explore visual perception's role in cognitive duality. He used an analogy to explain what he meant by these words. Imagine your favorite tree again. When you look at that tree, it is represented in your mind as a form—that which is in the mind is not the real tree, just a representation of the tree. Other humans agree that their brains also hold the same representation of that tree. So, yes, we agree: that is, indeed, a tree. But we know that if we use a different scale of perception to examine the tree, using a microscope, for example, that the tree is full of flowing sap and is composed of various cell types that go unnoticed by our grosser perceptual abilities. If we then use an atomic microscope, another world of reality opens up. We see that the tree is a mass of energy frequencies, atoms, protons, and quarks. If we keep reducing our scale of observation, we inevitably arrive at a place where the tree is just a potential, mostly empty space. This non-material place of potential is what Barfield called non-representational—we cannot extract (perceive) a tree from this morass of quarks and emptiness. I suggest that the non-representational domain is the foundation for allocentric consciousness, the *quantum mother* of the hidden mind—the background slate for the allocentric mind.

If we look again at our favorite tree, and at our visage leaning against the tree, we see what the naked eye calls reality. If we had eyes equipped with microscopic vision we could see the cells that constructed our human form and our tree-friend's arboreal form. If our eyes were equipped with atomic scale micro-vision, we could see the atoms—the building blocks—out of which the cells were made. If we had micro-quantum vision, we could see the building blocks of atoms. At the quantum level we would see that we are made of light—that we are *beings of light*. I feel that Barfield would agree that the universe has probably created innumerable beings of light, most of which our limited perception cannot see.

When we pray or when we meditate, when we ask for help, it may very well be that other beings of light are in communication with us. In a way, we are blended with these other beings of light since we are composed of the same non-representational stuff. Of course, the egocentric mind goes crazy at such intuitive conclusions, while the allocentric mind just nods its deaf-mute head in agreement.

From the above discussion, we see that for Barfield physics and religion are partners—both arrive at an intelligent universe. Religion sees a universal oneness through an awareness of spirit. Science sees a universal oneness through an awareness of matter. Barfield has generated a language and concepts like imagination, participation, representational, and non-representational, to explain his complex theory. These complex terms are fairly easy to understand if we accept that we have two mutually exclusive minds. As we turn to Rudolf Steiner's ideas, we again face a complex thinker who has created terminology that is often bewildering and seemingly incomprehensible. However, like Barfield's ideas, those of Steiner are easier to understand if we use dual-process theory.

Rudolf Steiner and Owen Barfield

In the following discussion of Rudolf Steiner, I used Gary Lachman's book *Rudolf Steiner, An Introduction to His Life and Work* (2007), and Owen Barfield's reflections, especially in the introduction to *A Case for Anthroposophy* (1970). I also used the official Owen Barfield website and the online Rudolf Steiner Archives to draw reference material. This section is an attempt to summarize Steiner's theory of mind and to relate his worldview to dual-process theory.

Both Barfield and Steiner saw the soul of modern Western culture as dying and inert—both men were harsh critics of their era. Barfield wrote that the feeling of meaninglessness was the most serious crisis of our time. He claimed we had demythologized nature and reduced everything to nothingness, that we had a false, barren way of living, and that we had become alienated from nature. Barfield believed this came about because of positivism, materialism, and the ascendancy of science. Barfield rejected positivism as did Rudolf Steiner. Positivism holds that only science is valid, while metaphysics and theism are not tenable worldviews. Both Steiner and Barfield were appalled by the decline of spirituality in the cultures they knew; consequently, they mounted—each in their own way—a crusade to stop the free-fall of men and cultures into meaninglessness.

Barfield came upon the philosophy of Rudolf Steiner through conversations with his friend Cecil Harwood—Barfield and Harwood were school mates and lifelong friends. Harwood went on to popularize and found schools based on the thinking of Rudolf Steiner—there are now thousands of Waldorf schools spread across the globe, thanks in part to Harwood and his wife. Barfield was influenced by the Harwood family and became fascinated by Steiner's vast body of work—he later wrote the introduction to a book of Steiner's essays called *The Case for Anthroposophy* (1970). Barfield said that his own philosophy, the conclusions he formulated after years of study and thinking, had already been thought-out by Steiner. Indeed, Barfield felt that Steiner was an advanced human being (his consciousness was highly evolved) with an extraordinary mind.

As a contemporary of Rudolf Steiner (1861-1925), Barfield (1898-1997) became a Steiner authority and followed Steiner's teachings. After a lifetime of thinking about consciousness, Barfield, like Steiner, concluded that consciousness did not emerge from matter, but rather consciousness *gave rise* to the material. Furthermore, Steiner said that the mind does not go on in the brain, but rather, mind created the brain.

Steiner used an analogy to explain how outside thought created the mind. His logic led him to conclude that the mind was related to thought, just as the eye is related to light. I presume he means that the eye was *sculpted by the light* (Goethe's phrase)—therefore, the brain and nervous system were sculpted by outside thought. Light comes from without and so thoughts must come from outside the organism. The eye depends on light, otherwise it is blind—vision would not have developed without light. Likewise, the brain depends on thoughts, otherwise it would be thought-blind. Thinking will not develop without the input of thoughts: lungs breathe, the eyes see, the mind *thoughts*.

A television is often used as an analogy to explain this perspective. A TV is not a mind generating a universe of sound and images; it is a receiver that decodes information embedded in the airwaves. But you can cause "brain damage" to a TV's circuits or block incoming frequencies. You can turn it off and the TV-brain goes dead. The airwaves, however, are transmitting constantly to all televisions. Steiner's argument is that the brain is a receiver of thoughts, not a creator of thoughts. Some invariant in nature called *thought* is decoded by the nervous system and turned into consciousness. According to Steiner, without thought pre-existing in nature, we would have no reason and no mechanism to create consciousness.

Steiner also saw that thought (information) was both decoded by the brain and then transmitted outward back into the thought-atmosphere. It was like the in-breath and the out-breath: information flows in and is decoded by the brain, and then it is projected back out into the information-sphere to be reused by other minds. Once inside the human nervous system, information is cataloged, synthesized, and remembered. Therefore, the information that is projected back into the environment through communication is altered and fresh.

According to Barfield, Steiner's theory of mind is based on two axioms that seem at first to be at odds with each other:

1. That objectivity is real. A concept in the brain, of a triangle, for example, is based on what the outside world is really like (lines, angles, forms, and so on). All communication is fruitless if there is no objective world. However, pure triangles do not exist in nature, but every mind is capable of creating such an abstraction. This means that the mind can manufacture universal concepts and these are the same in every mind—across time and cultures. Communication verifies that we can each generate these universal concepts. Steiner goes on to say that the common man believes these universal concepts to be his own creation—the work done by material brains. However, Steiner implies that these thoughts, instead, point to another dimension beyond mind that our brains are capable of viewing (channeling, tuning into).

2. That thinking is a personal *activity*. Thoughts cannot be thrust upon us from without, as our sensations are, unless we invite them. Thoughts demand our cooperation, our own activity, before we can be said to have thought them—before they can be *our* thoughts.

Here is a quote by Steiner repeating his awareness of an essential incompatibility that human beings face:

"There are two things which are incompatible with one another: productive activity and the contemplation of that activity. This is recognized even in the First Book of Moses. It represents God as creating the world in the first six days, and only after its completion is any contemplation of the world possible: 'And God saw everything that he had made and, behold, it was very good.' The same applies to our thinking. It must be there first, if we would observe it."
- *The Essential Rudolf Steiner*, 2012. Available online or as an ebook.

This quote is as close as I have seen Steiner come to an awareness of the egocentric mind and allocentric mind. What I think he is saying above is that we cannot extract meaning from our world (using egocentric perception) at the same time we go about our productive activities (as we flow, without thought, through our domain—as we navigate). When we immerse ourselves in nature, when we participate so completely that we lose our egos, then we cannot think. We can *be* or we can *think*, but not both at the same time. To support this view, he quotes the biblical God who had no thoughts during the activity of building creation. Only after activity had ceased, could God reflect on his work.

We cannot think and observe ourselves thinking at the same time; this is Steiner's starting point. When we try to observe ourselves thinking, thinking stops. Thus, thinking, and observing the act of thinking, are polarities, mutually exclusive but still intimately connected. Thinking is a kind of background out of which concepts emerge as thought-objects. It is the familiar figure-ground conundrum. We cannot see the details the same time as we perceive the big picture. Therefore, the perceptual act of seeing concepts is a polarity in counter-balance with the perceptual act of seeing ourselves think.

Steiner says that thinking even comes before the separation into subject and object, before ego and other. What he means by *thinking* in this context is perhaps *attention*—the process of connecting the outside world with the inside mind is an ongoing creative process. For Steiner, the mind creates everything: space, time, objects, and its own personality and ego. This is our double nature, our inherent paradox: we are both *in* nature—the background—and *separate from nature*, a figure in that background.

Steiner saw the mind as another dimension of evolution. Besides the six spatial dimensions (up, down, right, left, forward, backward) and the dimension of time, he saw that we had an inner dimension that evolution had crafted. This mind could travel through infinity for eternity. This internal mind could talk to itself and share with others—it could absorb from the outer world, and it could broadcast into the outer world. *The mind* was something new in evolution, something that defined humankind as unique. As a self-proclaimed spiritual scientist, Steiner set about trying to study and comprehend this new dimension evolution had created. He looked back at the past and saw that many others had also tried to understand this gift of consciousness. Consequently, as he did his historical research, he was faced with many mythologies, many definitions for the same concepts, and with many attempts to systematize the workings of the mind. Steiner is hard to understand at times simply because he uses these various historic systems and mythologies and mixes them with "modern" thought and with his own mental inventions. He also has to be translated from the German—not an easy task given the subject.

For my purposes, I don't need to become a Steiner scholar. I am interested in his perspective on duality and consciousness. Does his thinking coincide with my theory of two minds, or is it something else, something perhaps contrary and challenging? That was my task as I began the arduous work of tweezing out gems from the complexity that is Rudolf Steiner. There is absolutely no doubt, however, that he was a great thinker and a man of action well beyond his time—and maybe ours as well. His level of consciousness was above and beyond his culture.

What makes Rudolf Steiner special is that he put his complex ideas into action; he was not just a philosopher—he gave us long-lasting gifts. For example, he founded the Waldorf Schools—similar to Montessori schools—where a child's level of development is the prime consideration. These schools are found all over the world to this day and they continue to flourish. Steiner also addressed the suffering of disabled populations; his work resulted in worldwide institutions for benevolently ministering to the needs of people with disabilities of every sort. He also advocated "recycling to save the planet" about 100 years before this planetary-empathy became fashionable. Steiner's level of consciousness was so far out in front of his contemporaries it is amazing anyone understood a word of his proclamations—most people, of course, didn't have a clue what he was about. He was an Integral Man in the age of Traditional Men. Only now are we slowly and painfully catching up to his insights.

As Steiner began his quest to comprehend the mind, he studied the German philosophers and German literature—the Western worldview that he found himself immersed within. When he was sufficiently versed in the ideas of Western philosophy, he found that he basically disagreed with the great Western philosophical thinkers, especially Kant. They all suffered, as he saw it, from the same oversight: they were interested in epistemology, in the theory of knowledge, in what could be known and what could not be known. Steiner said that this was the wrong way to begin. We needed to start, he said, with figuring out how it is *that we think*. Knowledge was just the result of thinking. It didn't get us any closer to understanding how it is that the mind operates. Owen Barfield agreed with Steiner; if we are to comprehend the mind, we must start with comprehending what it means *to think*:

> A philosopher starting out to construct a true story of knowledge must start, if he is faithful to his calling, from the very beginning. If we start from any assumptions at all . . . we are clearly not starting from zero. We are starting from something on which cognitive activity has already been expended. The same remark applies if we start from the "ego," or "consciousness," or "the mind," or by raising the question whether there is such an entity as the mind, or from the experience of a "normal observer." Only if we start from thinking itself, no such objection can be made. For thinking is the very first possible move we can make in the direction from ignorance towards knowledge. We cannot think about the world or about anything at all, without thinking.
> ~ *ROMANTICISM COMES OF AGE*, "Rudolf Steiner's Concept of Mind," Owen Barfield, 1944.

In other words, we can't begin the effort to comprehend cognition if we begin with philosophical or psychological theories—we have to first comprehend the biology of thinking. I believe that both Steiner and Barfield would approve of my approach which begins with the evolution of purposeful movement, with egocentric attention and allocentric awareness. They would also have enjoyed the anatomy and physiology that supports their perspectives. This is the "necessary beginning" of which they speak in the above paragraphs. There is an anatomical, physiological, and evolutionary reason why and how the brain and nervous system evolved. (See *Consciousness: A New Slant on an Old Conundrum* for a discussion of the anatomy and physiology of dual consciousness)

Steiner understood levels of consciousness and the need for purposeful movement—he invented dance routines and stage plays that focused on controlled movement. Steiner was also a Goethe scholar, and he spent much of his youth reading and writing about Goethe's complex and wide-ranging perspectives. Goethe spent *his* life puzzling over humanity's dual cognition. Consequently, Steiner inherited, as a young man, an understanding of human duality. Steiner's culture knew about the power of the egocentric mind and its new tool, the scientific method—this understanding grew out of the Enlightenment, the Age of Reason (1700s). But in the rush to enthusiasm, the new intellectuals forgot the esoteric traditions that held the secrets of the hidden allocentric mind. Steiner turned his attention to these esoteric teachings, with their arcane, illusive metaphors and mythologies.

According to John Armstrong in his book *Love, Life, Goethe* (2007), Goethe investigated the world using a different method than the rational scientific method. Steiner understood what Goethe was getting at: we could use our intuitive (inductive imaginative) powers rather than our reductive (deductive) powers of investigation to explore our universe. For Goethe it was a form of "scientific method" to use the allocentric mind rather than the egocentric mind to grasp our world and ourselves. When Steiner set about using Goethe's intuitive method he eventually got so good using intuition and imagination that people began to say that Steiner was clairvoyant. Like Goethe, Steiner developed what might be called a *direct way* to "read" the world.

Goethe, and Steiner after him, said that the scientific method took "dead things" (already formed "facts" derived from inorganic matter) and drew hypotheses and conclusions. The intuitive method, however, explored organic life, life that was still alive, life unfolding moment-to-moment. It was in the observation of the *process of living things*—especially in regard to relationships—that held the secrets to intuitive knowing. *Observation* is not quite the right word because the process that Steiner and Goethe used involved *being* (sinking into or becoming) the living thing that was being investigated. For example, they didn't take a flower and dissect all the parts to get at an understanding. Instead, they let their perception *sink into* the flower (becoming the flower); they could feel the life-force of the flower living in its natural domain.

As I said above, Steiner wrote in German and he used the terminology of esoteric traditions; these two things make him obscure and unreachable at times, especially to English speaking peoples—he has to be translated twice over to even get a glimpse of his insights. Steiner seems crazy when he talks about old moons, planetary minds, and Atlantis. And, for all I know, he might have bobbed in the deep end of irrationality from time to time. A better explanation, however, is that Steiner was using language that earlier minds had created to understand the evolution of consciousness. The human mind has evolved through many "primitive" layers. Trying to categorize layers of consciousness is a confusing task that requires the creation of a vocabulary. Steiner's vocabulary is often drawn from the esoteric writings of much earlier minds; he did not have access to our current terminology and technology.

Furthermore, Steiner is hard to grasp because he did not have just one physiological system. He developed several versions of his theory over time—as his own understanding changed. These different physiological perspectives serve different purposes even though they partially overlap. It's no wonder people find Steiner obscure. It's best to rely on people who have studied his theories (and offered summary views) rather than read his thoughts in his own changing and abstract terminology—that is my experience.

Nevertheless, Steiner is worth the effort. He was one of the first Western philosophers of mind to argue that the hidden mind was real, that it had a non-verbal language of movement, dance, metaphor, and image. Steiner advocated for compassion, empathy, love, awe, mystery, and transcendence. He showed the egocentric mind its mirror image, a twin that had been (metaphorically) separated at birth. The ego's twin sister (the soul or self) is a creative, innovative girl with a sweet smile; she just stares with loving kindness at the brilliant (smitten) professor rattling on and on about the material world—and about other humorless, oh-so-serious, non-experiential stuff.

According to Gary Lachman, in his book *Rudolf Steiner: An Introduction to His Life and Work* (2007), Steiner was a self-defined esoteric scientist. He gathered together all the knowledge he could discover that we would today call psychic, mystical, or occult (which means hidden or *blind to*).

Steiner was not alone in this effort to gather and consolidate this body of esotericism. Others like G. I. Gurdjieff, P. D. Ouspensky, and Carl Jung did the same, trying to make sense of all the rituals and rites of passage that had accumulated in the early musings of mankind. There were, for example, secret societies of alchemists, hermetic scholars, and astrologers in history who created rituals and sacred texts to catalogue what they had discovered. And what they had unearthed was the allocentric *secret* mind. This mind had powers that felt like *God consciousness*.

Steiner is often lumped with esoteric thinkers, a categorization that he would not deny. *Esoteric* refers to "knowledge not for the masses." When people of a certain historic era, with low levels of consciousness, came upon secret writings and rituals, they often reacted violently. History has many examples of atrocities carried out against those who dared to entertain *non-acceptable* belief structures. As an esoteric scientist, Steiner was suspect in the eyes of most of his culture.

Rituals used in ancient rites of passage deliberately shocked the body so severely that the egocentric perspective "died" long enough so that the allocentric mystery could reveal itself. Actually, what I believe must have happened at these rites of passage is that ritual shock enabled the egocentric mind to suddenly become aware that it had a twin. Historically, egocentric consciousness wrote off the whole of these esoteric rituals as nonsense, the work of primitive irrational minds—consequently, the ego tossed this huge part of human history into the garbage bin, not unlike the Nazis burning all the books they didn't agree with.

Rudolf Steiner and Psychic Bodies

When mental explorers like Rudolf Steiner study human cognition, they see not just the egocentric mind, they also perceive *psychic bodies*. Steiner contends that beyond our physical body there are concentric energy fields that are subtle—they exist below the threshold of everyday perception. Each sphere of energy has been studied and variously labeled by mystics and esoteric scientists for many centuries.

Steiner writes of seven kinds of bodies: the physical body, the ether body, the astral body, the ego body, and three kinds of spirit bodies. Essentially, these bodies are a description of different attributes of the mind. There are modern physiological explanations for these bodies that enable the egocentric, rational consciousness to consider them as plausible. Current research, for example, refers to a Biofield, which is a term chosen by a team of National Institute of Health scientists in 1994 to describe the field of energy surrounding and interacting with the body. In other words, scientists have known for decades that there are energy fields of various kinds (magnetic, electric, infrared, photonic) that surround the body. *Dynamic energy profiles* can be created that uniquely define individual human beings.

Steiner was searching for explanations that would enable the egocentric mind to embrace the auras, the psychic bodies, which concentrically surround the physical body. I believe that we now have the language and technology to vindicate Steiner. In the following paragraphs, I will give my interpretation of Steiner's language and translate it into terms I think the egocentric consciousness can begin to accept. As I worked on this section, I asked myself how subtle psychic bodies related to the themes of this book. I looked at each of Steiner's psychic bodies and asked how they related to:

- Proprioception, as the creator of dual consciousness
- Duality of mind, allocentricity and egocentricity
- Levels of consciousness, as discussed in Chapter Two
- Communication within and between complex networks

There is an important understanding that Steiner's *categorization of energy bodies* brings out. The brain is incased in the skull, but the mind is what bodies do—Steiner is talking about the mind as a whole-body phenomenon. The brain may be CEO of the corporation but there is no dissecting out energy bodies—they are networked and they communicate. Using this line of reasoning, there is an energetic egocentric body and an energetic allocentric body. The reason we can't routinely see this is because we are carrying around a very ancient and incompletely thought-out meme. Everywhere we look, even in the most astute media presentations, we hear that there are five senses. Aristotle taught us this over 2500 years ago and it has been a road block to understanding the body-mind ever since. I suggest that there have been a few discoveries since the year 384 BCE, and perhaps it is time to repair the meme.

Each of the so-called five senses can be subdivided into at least two divisions, allocentric and egocentric. Beyond that fundamental division, each of the senses can be broken down further into multiple mini-senses, each of which could be labeled as separate and distinct. From Steiner's perspective, each of the senses contributes, in their own way, to the various psychic bodies. Therefore, the subtle, psychic bodies discussed below can be thought of as sensory systems.

Let me make it clear that this is an exercise in correspondences. I created a scale of levels of consciousness and I proposed a dual-process theory for understanding our mutually exclusive mentality. I am trying, in this section, to make sense of Steiner's theory of subtle bodies— categories he undoubtable referenced from other esoteric sources. These early esoteric sources were dependent on the sophistication of ancient authors, so the potential for confusion and misunderstanding is large. I found the exercise fascinating and worth the effort. However, please realize the speculative nature of my suggestions.

Body One: The Physical Body

As I said above, my intention is to examine each of Steiner's seven bodies and ask four questions: how does proprioception play a part in the construction of the subtle body; how is duality expressed; how are levels of consciousness related to the psychic body; and what is the nature of communication between the psychic body under examination and the other subtle bodies.

Proprioception, at the level of the physical body, is a system that we know well. It is easy to see that we can sense where our body parts are. We can also feel gravity affecting our posture. We can scan the body and notice what hurts, what is tight or tense and what is relaxed or "under control." When we do this proprioceptive scanning, what we perceive is Steiner's *physical body*— the act of scanning actually defines this body. The physical body, according to Steiner, is what links humans to the mineral world. From this perspective, we perceive our bodies as made from the same atoms and force fields as the mineral world. That we have an ever-aging physical body—with aches and pains, bones and flesh—is taken for granted and there is no controversy about the existence of this mind.

Duality has also been shown at the physical level. We know that bilateral coordination is necessary for forward movement (navigation), and that efficient movement requires an oscillation, a balance between stability and movement. We also know that perception of figure and ground requires a sensory oscillation. That there is a mutually exclusive duality of the mind and body has been the primary focus of this book and my previous book *Consciousness: A New Slant on an Old Conundrum*.

When we consider levels of consciousness—as was extensively discussed in Chapter Two—we might *assign awareness of the physical body* to the most primitive level, to Archaic Man. Knowing that you have a physical body is a primal beginning for comprehending who you are.

At the Archaic level of consciousness, there is no awareness of the physical body communicating with other psychic bodies. The communication may be there, or it may be evolving, but there is no awareness of this communication network for Archaic Man. There is also no awareness of the coherent signaling that goes on in the physical body—cell-to-cell, tissue-to-tissue, organ-to-organ.

305

Body Two: The Ether Body

To understand Steiner's *ether (life) body*, we need to reflect that if our whole cellular makeup lacked glue, we would fly apart, dissipate, evaporate. Furthermore, if our cellular makeup did not fire in waves, if there was no synchronicity, no cellular harmony, no basic rhythms, no coherent patterns of energy, we would be spastic and dysfunctional. We *could not* move or think without coherency and synchronicity. This frequency glue can be sensed. It feels like it is covering the body to a depth of about two inches beyond the physical body—holding the form together. Steiner calls this *the life-force*. Plants have this life-force just as we do. Therefore, the etheric body links us to the plant world; it provides unity and wholeness to every living entity. The etheric (plant) body is roughly equivalent to what Aristotle called the organic soul. It results in growth and development, and is the key to heredity. The ether body provides the physical body with a specific energetic (animated) form.

Recall that proprioception is a way to map the world. For example, proprioception can map the location and interrelatedness of body parts that make up and define the physical body. *Peripersonal* spatial-mapping is another way proprioception maps the world. Tools that we hold in our hands are mapped as if they are *actually* part of the physical body. Peripersonal mapping refers to this ability to perceive tools and spaces as if they are *an integral part* of the physical body. A blind person's cane, for example, is perceived by the proprioceptive system as an *actual* extension of the hand and arm.

The ability to perceive spaces, rooms for example, as if they are actually extensions of the body, is the key to understanding Steiner's psychic bodies. The brain has actually created proprioceptive maps that include these psychic bodies as if they are real extensions of the physical body.

Proprioception, from my perspective, is also the key to comprehending both allocentric and egocentric consciousness. Proprioception can actually sense a psychic body that I have called the allocentric mind. Likewise, proprioception can also sense a psychic body that is the egocentric mind. These two energetic minds together give us a sense of a life-force at work. If you try to perceive where this force is located in regard to the body, you find that it is an aura about two inches thick that surrounds the physical body and feels like it is holding our physical molecules together. Steiner says this psychic body is also present in plants, and this suggests that plants have both allocentric and egocentric minds, both intelligence and wisdom—presumably, much different from our perceptual understanding.

I am going to continue to correlate levels of consciousness with Steiner's subtle bodies, although this is strictly an exercise. I don't know if there is any value in this line of speculation, but I do know there are buried relationships waiting to be discovered. Keep in mind the speculative nature of this musing. I suggested above that Archaic Man would know about the physical body but not about any other subtle body. Now I suggest that Tribal Man had evolved the ability (not necessarily the awareness) to sense that there was a life force permeating all living things. Tribal Man would begin to find a kinship with other creatures, animals and plants. In other words, Tribal Man would begin to *feel alive*.

The physical body and the ether body can communicate. They are part of a network. They also contribute to the health of each other as they stay in balance. They probably oscillate, as does everything else in the physical universe.

Body Three: The Astral Body

Authors like Colin Wilson and Eckhart Tolle speak of another kind of body, an emotional body. Steiner's name for this emotional energy field is the *astral body*. Emotional processing begins neurologically below the cortex in a more primitive area of the brain, the limbic region, which was retained from an earlier time in evolution. The limbic region of the brain is very powerful in human beings and is in-and-out of balance with the cortex depending on daily events.

Human beings know very well that emotions can overwhelm all perception. Both our egocentric and allocentric minds fade from view and we get lost inside what seems to be an entirely new entity, an emotional body. Eckhart Tolle became famous for his explanation of the human *pain-body*. Refer to Tolle's books, *The Power of Now* (1997) and *A New Earth* (2005), for an in-depth explanation of the pain-body. Essentially, Tolle sees that the astral body can be in various states of emotion. It is not uncommon for human beings to have been affected by events that left them emotionally damaged. This damage is held in the pain-body. You can actually observe people whose entire bodies are molded by years of emotional pain.

The *astral body* is animal consciousness, a dog-and-cat-like world. There exists in humans a sentience, a primitive allocentric awareness that wells up from our mammalian brain. This can be felt whenever egocentric consciousness gives up control of perception. Sometimes the astral body is a mellow place with no sense of time. Other times, it hisses, or snaps, and gets its back up. Life is always *now* in the astral body. There is no future to ponder and no past to reflect upon. When we scan our emotions, when we allow intuition, we are scanning our astral body. The astral body connects us to the animal world. It is roughly equivalent to Aristotle's sensitive soul and gives rise to emotions, desires, and unconscious impulses. While the etheric body is the glue that unifies the body, the astral body creates our sense that we have an interior essence that can be projected outward for the world to interpret.

The astral body is another product of proprioception, of cortical spatial mapping. Animals are defined by their bilaterality and their sensory apparatus which enables navigation through specific domains. The safety of navigation has to be assured if an organism is to survive. For this reason, the brain and nervous system proprioceptively map the emotional balance of spaces. There are peaceful, mellow, happy spaces, and there are dangerous, threatening spaces. We also project our emotions into spaces, making them more or less safe and comfortable.

There is an egocentric (intelligent) and an allocentric (wisdom) component to the astral body. Emotional intelligence is a commonly accepted meme in our culture. The astral body can *project* carefully controlled emotions into surrounding space, and can also *react* to the emotional tone of a space in a way that maximizes survival and well-being (or not).

If Archaic Man is stuck at the physical-body level of consciousness and Tribal Man is stuck at the physical plus etheric level of consciousness, then the next leap of speculation would suggest that Warrior Man is capable of awareness of the astral body. Or perhaps, the astral body has evolved and joined the other two bodies to help *create* Warrior Man. Certainly, Warrior Man is a species affected by and displaying extremes of emotion—as if the oscillation of the astral body is uncontrollable in this emerging primitive state.

The astral body is now a part of a wider network that incorporates the other evolved psychic bodies. It is now possible to read emotion in the facial gestures and body language of others, just as they may read our emotional state as displayed in our body mannerisms. The emotional body builds attachments to the objects and creatures of the surrounding world. The astral body is attracted, repulsed, or indifferent—this body makes and reacts to judgments.

Body Four: The Ego Body

Steiner's ego, which he calls the *"I,"* is egocentric consciousness. This is the evolution of object perception. It came into being as the distance sensory receptors (eyes and ears) developed sophisticated abilities during evolution. This egocentric mind is beyond the mineral, plant, and animal worlds; it is uniquely human. Steiner's *"I"* is the same as Aristotle's intellective soul. For Steiner, the ego is the eternal essence that reincarnates and carries karma. This ego gives our lives purpose and direction. This ego-body also interacts with the physical, etheric and astral bodies in special ways.

Notice that the first three bodies are the primal basis for the allocentric mind. This is the whole-body mind that combines the mineral, plant, and animal worlds—it is our genetic inheritance. It is also the reason for animation and purposeful movement—both of which occur without egocentric oversight or reflection. The allocentric mind has continued to evolve, especially as the brain's cortex has evolved for pattern analysis. The egocentric mind is strictly human because of language—only human beings have evolved such an elaborate system for communication, self-awareness, and self-reflection.

As I detailed in earlier discussions, the sensorimotor patterns needed for speech are copied in brain maps by the proprioceptive system. This eventually resulted in our internal voice, followed by self-awareness and the ability to witness the mind and body at work. Our sense of egocentricity is a sense of an energetic body that is beyond the sense of animation that comes from the allocentric energetic body. The egocentric body that we present to the world is layered over the astral body, the etheric body, and the physical body. In a way, it is built out of these earlier bodies—the ego could not have evolved without all that evolved prior to its creation.

At this stage of evolution, we have a fully formed allocentric body and a fully formed egocentric body. There is a very clear mutually exclusive relationship between these two energetic bodies.

With the addition of the ego-body and the fully formed allocentric body, we have the ingredients for the formation of a much more sophisticated level of consciousness than has ever emerged on the evolutionary stage. We have the arrival of Traditional Man.

Traditional Man has the capacity for sophisticated communication—both externally with others, and internally through self-observation. The network that weaves together the physical body, the etheric body, the astral body, and the ego-body has reached a staggering complexity—and yet, the changes about to occur on the stage of evolution are even more dramatic and impactful.

Body Five: The Spirit-Self Body

Few people can understand or even get near Steiner's last three bodies. Each is a transition and transformation. The first spirit body, the spirit-self body, is a transformation beyond egocentric consciousness. In my view, it is the ego's conceptual acceptance of the allocentric twin. The ego can't perceive the twin allocentric mind, but it can accept the evidence and logic that "proves" its twins reality. If we practice being in the moment, scanning the whole of the surround, losing our ego as we melt into nature, we are deliberately—with egoic blessing—scanning our spirit-self body. This is the body we want to be in as we begin meditation.

Awareness of the spirit-self body is the beginning of a more sophisticated understanding of *proprioception as projection*. Proprioception can map the body, it can map the psychic bodies, it can map tools-as-self, it can map environments-as-self, and now, at this level, proprioception can map a holographic universe-as-self. Everything is connected. Everything has a relative impact on everything else. Nothing is independent. The mind is part of a universal mind.

This level of evolution comes about when the allocentric mind and the egocentric mind are finely tuned to each other—when the oscillation between the two is steady and harmonic. The spirit-self body expands outward around the other bodies and holds them in a universal embrace.

The leaps from one level of consciousness to the next are exponential. The leap from Traditional Man to Modern Man reflects this disruptive, massive earthquake that occurred with the advent of the spirit-self body. Modern Man has a cosmic worldview—one that is held to be true and observable. Great power is unleashed at this level of consciousness.

All the bodies are part of a communications web. They do not exist independently. However, each new body adds astronomically large, exponentially incomprehensible complexity to the network. This is actually the mind evolving to be ever more complicated and networked.

Body Six: The Life-Spirit Body

The second of Steiner's spirit categories, the life-spirit body is a transition beyond the ether body—a movement of personal emotion outward to blend with the emotion of all existence. It is the touching of body-self-emotion with the compassion of the life-force that is creating life across the universe.

Proprioception has now mapped beyond the universe to the very core of what started the oscillation between creation and destruction. Proprioception is blind at this point, not sure what to map when faced with no time and no space.

We are at the primal duality, at the fundamental fractal where the unmoved-mover moved and where the oscillation between being and nothing started. Here is the birthplace of the egocentric mind and the allocentric mind—where the seed for emotion, for love began.

Beyond Modern Man stands Post-Modern Man—another extreme exponential change has occurred as this new body has enveloped the earlier evolved psychic bodies. The gap between Post-Modern Man and every level of conscious below is so vast that it becomes clear why levels of men have been unable to accurately communicate.

However, the communication between the bodies has expanded exponentially again; the network has jumped in complexity—what we call *mind* has become powerful and expansive beyond all earlier comprehension. We can now see clearly that *mind* at this level is capable of nonverbal communication with all other minds. We also can see that people who have lost their physical bodies to death, have not lost their sophisticated bodies—we are able to communicate with those who have lived and died in the past—including our former selves. This suggests that higher levels of consciousness are more capable of communication after death than those of lesser cognitive and spiritual evolution.

Body Seven: The Life-Man Body

The third spirit body I can only suppose is what the Buddhists call Enlightenment and the Christians, in the Wisdom Tradition, call Christ Consciousness. The life-man body is a transformation beyond the astral, beyond allocentric consciousness to a higher awareness; it is an absorption into a Divinity, for lack of a better phrase. In rational terms, it seems to be a quantum-melting of the frequencies that hold the body together into a universal energy field of coherent frequency potential. It is a melting into the whole of existence. We are made of light but so is everything else in our living universe. When we use our life-man awareness we are dissolving back into the immaterial background called universal spirit.

Here we reach a stage where proprioception can create (map) virtual worlds. The mind can reach inside and disappear into the imagination. In a way, this is how artists create—they find that the mind can innovate and fabricate in a way that is infinite and eternal—even if infinity and eternity have no relevance to the actual structure of the world.

The universal fractal that gave birth to the egocentric and allocentric minds may not be a reality that carries over into other universes—we cannot know. However, when death comes to the physical body, we might very well find out. In this land of the living, we cannot solve the puzzle of duality; we just have to live with our paradoxical existence.

This is obviously the highest level of consciousness. It is beyond Post-Integral and beyond our capacity to comprehend or experience.

The web of communication that holds all the psychic bodies together is enclosed within this highest spirit body. The network is beyond vast, beyond description, more complex than can be conceived or expressed.

As we develop ever more evolved psychic bodies, a point is reached where clairvoyance develops. Insights arrive from the spirit-bodies as if the insights were channeled—gifts from a higher spirit. I think that Steiner had evolved clairvoyance, and he used his own knowledge to teach others to also develop this skill. Since clairvoyance means that a person can sense the future—and get help from a cosmic mind—this mental ability has been labeled psychic and is often dismissed out-of-hand. But clairvoyance is a *reading of Nature*, an immersion into the processes occurring in the moment to see how a process works—how it flows and in what direction. This is a perfectly reasonable method for prediction and for understanding how a process will play out. There is no magic in it.

Rudolf Steiner invented the term/concept/organization called Anthroposophy. According to Steiner, Anthroposophy is not a religious doctrine, but rather it is a spiritual science. The goal is to awaken humanity to the understanding that there are levels of consciousness and that consciousness is a process that evolves. Steiner believed that human beings need to study themselves, to look at their own consciousness and learn how to move through the various levels of consciousness. Steiner would say that individuals and cultures need to evolve to more sophisticated levels to bring about a spiritual transformation of daily life. Both Steiner and Barfield felt that only a highly evolved human being would be able to make the right decisions that would lead to peace and harmony on this endangered planet.

The Work of the Philosophers

Love cannot be said.
~ "The Taste of Morning," *A Year with Rumi*, 2006.

Many great thinkers, like Rudolf Steiner and Owen Barfield, concluded that an egocentric worldview, with little to no allocentric balance, is a disaster. Here is integral philosopher Ken Wilber's opinion:

> . . . when you have finally finished reducing all I's and all we's to mere its, when you have converted all interiors to exteriors, when you have turned all depth into shiny surfaces, then you have perfectly much gutted an entire Kosmos. You have completely stripped the universe of all value, meaning, consciousness, depth, and discourse—and delivered it up dried and desiccated, laid out on a marble slab of a monological gaze. Consciousness indeed becomes the ghost in the machine, precisely because it has committed suicide.

And thus we end up with Whitehead's famous summary of the modern scientific worldview . . . : "a dull affair, soundless, scentless, colorless; merely the hurrying of material, endlessly, meaninglessly."
~ *Sex, Ecology, Spirituality*, "The Eye of the Spirit, An Integral Vision for a World Gone Slightly Mad," Ken Wilber, 1995.

A philosophy that is purely egocentric is dangerous and mentally ill. Wilber can see the colorless corpse laid out for dissection, as if empathy and love never existed in the coldness of physical space. Philosophers are needed and should be valued, but mostly we need philosophers who understand our mutually exclusive minds.

Western philosophers have taken on the difficult task of using their egocentric minds to comprehend our existence. The allocentric whole-body mind is considered and pondered, but mostly philosophy has been an egoic cerebral undertaking. Philosophers use reason and logic to muse about meaning. They study history, cultures, even biology; they dissect whatever clues they can uncover to explain the abstractions that arise in complex social systems. What they have encountered throughout history is a duality, a primary characteristic of human cognition. Philosophers also face their own mental duality as they think about thinking. Consciousness cannot be set aside as an independent variable—it must always be factored in. Therefore, each philosopher must struggle to comprehend duality within themselves as they examine the twin minds of others.

So what is it that philosophers down through the centuries have been trying to communicate to mankind? What has reason given us? Here is a partial list:

- We are dual creatures. We have two mutually exclusive minds, two kinds of consciousness. There is a war between these two minds that is always raging and never gets resolved.
- We only seem to know one of our minds; the other mind is hidden. Search and ponder as long as we want, but what we find will always be the conundrum, the paradox of duality. Duality is everywhere, throughout history, embedded in every culture.
- The hidden (allocentric) mind is the source of creation, innovation, spirituality, soul, compassion, wisdom—all the best of human attributes reside hidden within. Not all philosophers came to this intuitive under-

standing, but a few did (Alfred North Whitehead, for example) and they wrote eloquently about the hidden (allocentric) consciousness.

Philosophers are men and women of remarkable intellect, with powerful analytical minds. They have used their cognitive insights, their pattern-smart minds, and their powers of introspection to dissect the human condition. They have argued their propositions. They have shared what they discovered. And yet, they remain undecided. Despite the confusion, philosophers have a responsibility to awaken the less evolved among us to the paradox of consciousness and to the possibility of a more mentally sophisticated reality.

If you are a philosopher, then you have an obligation to explore and explain the duality that is inherent in the human condition. You have an obligation to stand courageously against single-minded ignorance. There is a logical reason for our two minds. There is a logical reason to balance the two minds. There is a logical reason why allocentricity is the equal to egocentricity. There is a logical obligation that human beings have to follow the mandates of the allocentric mind. Philosophers have a responsibility to educate the world's cultures and to help raise the level of consciousness of all of mankind.

When I was asked to categorize this book for marketing (and the book that came before this one: *Consciousness: A New Slant on an Old Conundrum*), I wrote down "cognitive science, the study of the evolution of mind." When I told my friend (and editor) Karen Horwath what I had done—she was the first person to thoroughly read both books—she laughed out loud and suggested that what I had just spent 600 pages on was a *philosophy of the mind*. I knew immediately that she was correct. There is a lot of speculation and theory in my books that is better suited under the title of philosophy. That was a nice moment, when I realized this kinship with the philosophers, hunched over their desks, pecking away at their keyboards, day after day—as I am doing now.

Therefore, as a philosopher, *you have an obligation*:
Do as much good as you can in the time you have left
Get up.
This is a work day.

Six

Duality and Psychology

We have been like the man
who sat on his donkey
and asked the donkey where to go.
~ "A Trace," *A Year with Rumi* 2006.

Amygdala Surprise

"Hey, Surge. Do you serve beer?"
"Of course. We stock a large variety of intoxicants."
"Great, give me a cold Pilsner."
"We have four kinds of Pilsner. However, I recommend that you skip the Pilsner and have our Rainbow Stout instead, it turns your thoughts into Christmas bulbs and silver tinsel."
"I prefer subtle beer, Surge. What are the four Pilsners?"
"We have a delicate Origami Pilsner called Amygdala Surprise. It has an innate yet non-deterministic spontaneity. It leaves your tongue fat and your eyeballs frozen over with dumb joy. How about that? How about some Amygdala Surprise?"
"No thanks, Surge. That sounds a little hard on the emotions. What else have you got?"
"We have a classic Pilsner called Going Home Real Soon Now. It is made in the basements of unemployed and not quite rehabilitated drug dealers who don't care anymore. One sip and the background spits out random hallucinations. The figure and the ground take turns being incomprehensible. It's our most popular terrifying beverage."
"Yeah, well, I am not in the mood for terrifying."

"No problem. We also have an unpredictable Pilsner called Opps! I'm so Sorry. It is a self-creative, yet probabilistic brew that is a favorite of my hair dresser, Gladys."

"I don't want to know about your hairdresser."

"I am just saying that Gladys has a favorite beer. After a couple Opps I'm so Sorrys she loses all inhibitions and doesn't give a rat's ass what your hair looks like. You should see her dance and cut hair at the same time."

"This explains a lot about your appearance, Surge. What's the last Pilsner?"

"How did you know?"

"How did I know what?"

"That the last Pilsner is called The Last Pilsner You Will Ever Drink."

"That sounds ominous."

"I don't recommend that one. It has a profoundly power-hungry ethos—and it does a real make over on your liver. After a couple sips, you suddenly see old movie dogs like Lassie and Rin Tin Tin run through your melting brain."

"Give me a Bud Light."

"Coming right up, Dutch."

"Hey Surge. Did you ever meet Carl Jung or William James? They were scientific philosophers, right? Yet they didn't deny mystery. They weren't afraid of alternative states of consciousness."

"Jung and James used to come here together for lunch. They always sat at table one—we call that the Table of Synchronicity. They had animated conversations; they fed each other good energy. Jung loved our Serendipity Salad. James was more practical, but he still ordered from the esoteric menu. Both men refused to give up on strangeness. They were right at home in this atmosphere."

"What did they talk about?"

"They talked mostly about the evolution of the human mind. They were both mind-astronauts. They refused to believe that their personal cognition, their internal universe, was the whole story. They left their psychic doors open so that strange breezes could blow through them."

"I want what they ordered."

"Sure. The last time they came in together, Jung had Soul Bake with Anti-Matter Gravy. James ordered The Will to Believe Veggie Burger and he drank Séance Brandy. James liked to get pleasantly intoxicated. He was more into altered states than Jung, who preferred to get lost in his own internal madness. James did occasionally sip the Pilsners—so to speak—as he was writing."

"So, altered states of consciousness were important to these two?"

"Of course. They knew that everyday reality was boring compared to other locations in the intelligent universe. Jung had so many patients who had been to other worlds that he lost faith in his own sensory knowledge. The crazy people he treated weren't really crazy; they were suffering from contact with alien worlds. James actually tried to alter his consciousness using drugs. He got a taste of alternative reality; after that, he could never quite return to this world. Both men distrusted objective reality."

"Did they know about dual consciousness?"

"Sure, each in their own way. They both knew that inner voices were as real as external sensing. They both saw the need for human consciousness to evolve. They both understood that there was a force in the universe that was moving human consciousness along a spiral. And, you might be surprised by this, given your concrete scientific mind, but they were getting help from the intelligent universe. They asked the spaces for help and the spaces responded."

"And they asked you for help, didn't they Surge?"

"Of course they did, Dutch. I help all you researchers with your conundrums and paradoxes. That's why The Third-Eye-Vegan Restaurant and Center for Esoteric Contemplation was created—to help human beings help each other."

"Well, on behalf of seekers and writers, Dancers, Prancers, Comet and Vixen—my favorite—I wish to thank you, Surge. Your help has been strange yet fulfilling. Food-for-thought has never been so spicy.

"You and your Ilk are welcome, Dutch. Glad to serve and guide humankind. There are few such humorous and bewildering creatures as humanity in all of creation."

William James, Carl Jung, and Ken Wilber

I designed this chapter around three individuals, William James, Carl Jung, and Ken Wilber. These three astute thinkers held similar worldviews and each came to understand that human beings have a cognitive duality. Each of these men also came to see that consciousness was an evolutionary process with gradations of sophistication.

It has been the task of psychology for over a century now, starting with the pioneering work of Harvard Professor William James to solve the mystery of human cognition. James (1842-1910) has been called the Father of American Psychology. He was a physician, a psychologist, and one of America's greatest philosophers. A major focus of James' career was the study of the mind. James taught that consciousness evolved, just as physical bodies evolved—he saw consciousness as a process, not an entity. He coined the term "stream of consciousness" because he held that the mind was never still—it flowed and changed constantly. James also held that the evolution of the mind was subject to Darwinian natural selection—the causes and stages of cognitive evolution could be traced, studied, and understood. From the founding of modern psychology, James knew that human beings were created by a dual process— there are two minds, two forces at work within our cognitive architecture.

Carl Jung (1875–1961) was a Swiss psychiatrist, psychoanalyst, and research scientist, as well as a prolific writer and popular author. Jung founded the discipline of analytical psychology and was a friend and colleague of Sigmund Freud. One of Jung's major contributions to the study of the mind was his theory of *individuation*—the lifelong developmental process of ego-manifestation. Jung believed that a sophisticated ego slowly develops and emerges from an amorphous background mind. Psychoanalysts use the terms *conscious* and *unconscious*, and hold that the conscious mind slowly emerges out of the more basic unconscious mind. In my terminology, *individuation* is the evolution of the egocentric mind as it manifests from the background, the allocentric mind. Like William James, Carl Jung understood the mind to be an evolutionary process and as such to be binary. We owe many of our common scientific memes to the brilliant mind of Carl Jung, including *extraversion* and *introversion*, the *collective unconscious*, *synchronicity*, and *archetypes* of the mind.

Ken Wilbur (born in 1949) is a prolific American author who writes about transpersonal psychology. He is best known for formulating an integral theory of human cognitive evolution. Wilber is a contemporary writer so he has the benefit of knowing the work of both James and Jung. Wilber is also a master of synthesis. He has drawn from multiple sources to craft his ideas, blended them with Jung and James (and many others), and has arrived at exactly the same place: human beings have a dual cognition that evolves.

James, Jung, and Wilber produced a vast body of work that is complex and extensive in scope. I could only address the part of their work that addressed my main themes. I know there are other pioneers and great thinkers in psychology—many who support dual-process theory—but I did not have the space to include them here. That other psychologists have found duality in the human condition is expected, since the dual nature of the human mind is sitting in plain sight to be discovered by all those who search.

William James

In my book *Knights for the Blind in the Battle Against Darkness*, when I was discussing famous people whom Helen Keller had met in her lifetime, I wrote:

> [Helen Keller] talked at length on one occasion with Professor William James, who brought her an ostrich feather as a gift. She wrote of his visit:
> "His thought was clear as crystal. His body, like his mind, was quick and alert. In argument his tongue was like a rapier, but he was always ready to listen to the other side, and always made me ashamed of my cocksureness about many things.
> He was not a mystic—his mind could not thrive on air as mine does—but I think he was something of a poet as well as a philosopher."

Helen Keller could perceive, even without sight and hearing, that psychologist William James had a marvelous mind. James was deeply introspective and he used his astute mind to reflect on human cognition. One of James' most important insights was that *attention was the key* to figuring out the conundrum of consciousness. This fits well with my contention—from the outset of my own theory—that attention is the root of our cognitive duality. Here is what James says about attention:

> The faculty of voluntarily bringing back a wandering attention, over and over again, is the very root of judgment, character, and will. No one is compos sui [master of the self] if he have it not. An education which should improve this faculty would be the education par excellence.
> ~ *WILLIAM JAMES: WRITINGS 1878-1899*

As the quote above shows, William James clearly knew the importance of attention. James also knew that attention was dual—he understood that we had two minds and two kinds of consciousness. He said that we could follow the path of incremental learning, taking *one thing-at-a-time*, accumulating knowledge, or we could gain insights *all-at-once*, suddenly seeing the big picture. But James was even closer to dual-process theory when he wrote in *Principles of Psychology* (1890) about the mutual exclusivity of our inherent two minds. The following quote is from a book about Carl Jung and Wolfgang Pauli in which the philosophy of William James is discussed (James and Jung were friends):

> In *Principles of Psychology*, James posited coexisting and possibility split modes of consciousness—the "upper self" and the "under self"—which even while mutually unaware of and ignoring each other have complementary effects on each other.
> - *Atom and Archetype: The Pauli/Jung Letters 1932-1958*, Edited by C. A. Meier, 2001.

In his conversations with Carl Jung—who had an on-going correspondence with quantum physicist Wolfgang Pauli—James (using my terminology) referred to the egocentric mind as the upper self and the allocentric mind as the lower self. Not only did James see this duality, he also saw that the two minds were mutually exclusive and co-dependent. He knew as well that the two minds were ignorant of each other's existence—blind to their twin. Thus, James knew the paradox that defines humanity's bewildering cognition.

William James and Carl Jung influenced each other over a three decade friendship and correspondence. They could each perceive humanity's dual cognition, and no doubt exchanged views about this paradox. They didn't have the navigational perspective—that two minds come about because of navigation—but they could see the result. James called it upper and lower mind while Jung called it conscious and unconscious. I am calling the duality egocentric and allocentric. It is the same paradox dressed in the clothing of a time period.

James also had no fear of probing the psychic or occult. He knew that the mystics and saints were onto something important, so he set his acute intelligence to the puzzle: what, he wondered, do mystics perceive, and what does their "seeing" have to tell us about our mental processes? James did not have dual-process theory to fall back on; otherwise, he would have seen that what the mystics were onto was the allocentric mind, the background calm, the portal that opens when the ego is quieted.

We are fortunate that William James took on consciousness in his usual direct and penetrating manner. It is worth a few paragraphs to summarize his views. The following discussion is based on an article written by James in the *Journal of Philosophy, Psychology and Scientific Methods* (Vol. 1, No. 18, September 1, 1904). The title of the article is "Does Conscious Exist?"

The very first paragraph of the article shows that James felt the sting of our duality—he knew the root of the problem and began with it. He uses the word "thought" to mean our subjective life, and the word "things" to mean the material objective world. He is disgusted with what science has done, raising the world of things to God status and leaving the world of thought to evaporate into the dust of history:

> Thoughts and things are names for two sorts of object, which common sense will always find contrasted and will always practically oppose to each other. Philosophy, reflecting on the contrast, has varied in the past in her expectations of it, and may be expected to vary in the future. At first, spirit and matter, soul and body, stood for a pair of equipollent [equally powerful] substances quite on a par in weight and interest. But one day Kant undermined the soul and brought in the transcendental ego, and ever since then the bipolar relation has been very much off its balance. The transcendental ego seems nowadays in rationalist quarters to stand for everything, in empiricist quarters for almost nothing.
> ~ "Does Conscious Exist?" JOURNAL OF PHILOSOPHY, PSYCHOLOGY AND SCIENTIFIC METHODS, vol. 1, No. 18, September 1, 1904.

In the quote above, James is saying that in the past our two minds were conceived as equal—matter and spirit were sisters of equal value and power. But this had changed in his lifetime (and ours) so that spirit had lost credibility. German philosophers like Kant defined the ego as superior to the soul and that worldview was adopted in Western society to the detriment of culture, in James' opinion.

In the quote below, James asserts that consciousness is not a thing to be dissected. He would do away with that fake "thing" (consciousness as an entity) and start over with a discussion about human function. James would ask: how do human minds actually work and what might be left out if we did not invoke consciousness as a process, as a set of functions (italics mine):

> I believe that consciousness, when once it has evaporated to this estate of pure diaphaneity [delicate transparency], is on the point of disappearing altogether. It is the name of a nonentity, and has no right to a place among first principles. Those who still cling to it are clinging to a mere echo, the faint rumor left behind by the disappearing soul upon the air of philosophy . . . For twenty years past I have mistrusted consciousness *as an entity*: for seven or eight years past I have suggested its non-existence to my students, and tried to give them its pragmatic equivalent in realities of experience. It seems to me that the hour is ripe for it to be openly and universally discarded.
>
> To deny . . . that consciousness exists seems so absurd on the face of it—for undeniably thoughts do exist—that I fear some readers will follow me no farther. Let me then immediately explain *that I mean only to deny that the word stands for an entity*, but to insist most emphatically that *it does stand for a function*. There is, I mean, no aboriginal stuff or quality of being, contrasted with that of which material objects are made, out of which our thoughts of them are made; but there is a function in experience which thoughts perform, and for the performance of which this quality of being is invoked. That function is *knowing*. Consciousness is supposed necessary to explain the fact that *things not only are, but get reported*, are known. Whoever blots out the notion of consciousness from his list of first principles must still provide in some way for that function's being carried on.
> ~ "Does Conscious Exist?" *Journal of Philosophy, Psychology and Scientific Methods*, vol. 1, No. 18, September 1, 1904.

Therefore, for William James, consciousness is not a noun, not an entity. The egocentric mind turns everything into objects-of-regard and has performed this function on consciousness itself—given consciousness a false solidity that masks its true nature, which is *process*. Consciousness is a verb not a noun. For James, the philosopher Kant—and his progeny, the Neo-Kantians—are a major cause of our confusion about consciousness. Kant freezes the process and examines a dead image. James sets out in his article to confront the Kantian worldview which asserts that within us is a noun called subjectivity (the source for consciousness), while "outside" there is an objective world that is vague and unknowable because our senses are weak and inaccurate:

> Now my contention is exactly the reverse of this. Experience, I believe, has no such inner duplicity; and the separation of it into consciousness and content comes, not by way of subtraction, but by way of addition—the addition, to a given concrete piece of it, of other sets of experience, in connection with which severally its use or function may be of two different kinds.
> ~ "Does Conscious Exist?" JOURNAL OF PHILOSOPHY, PSYCHOLOGY AND SCIENTIFIC METHODS, vol. 1, No. 18, September 1, 1904.

I will try to translate James's eloquent nineteenth century language into my terminology, and into a more modern articulation. I think what he is saying above is that there is a background essence, pure experiencing (the allocentric mind), that holds no duality. However, duality *can arise* from the background as *features* (forms, patterns). Thought can arise from a background of no-thought. Ideas arise from a background of potential-ideas. Patterns arise from a background that is patternless. Two different kinds of function arise from the background: features can be added or features can be subtracted; we can perceive the background, or we can perceive the foreground, but not both at once. In other words, I think James obviously knew what I am putting forward here as dual-process theory.

In his book *The Dilemma of Determinism, The Will to Believe* (1896), James identified a two-stage decision process. In the moment, chance is a real phenomenon—we are faced with alternatives, what we will choose is not yet determined. However, after we make a choice, the ambiguous future becomes an unalterable past. The moment is pure potential, pure experience. This is the allocentric mind, the background waiting for a decision concerning what to manifest, waiting—in the terms of quantum physics—for the wave function to collapse into a solid feature.

In the moments before the wave has collapsed into reality, before pure experience has manifested any features, there is the possibility that accidents, or chance, will interfere with the flow of our mental trajectory. This is very close (if not exactly) what Jung meant by synchronicity. Something intrudes by chance unexpectedly into our lives, which alters our path. We were sailing comfortably along toward inevitable goals when some meaningful accident changes the course of our lives. William James saw accidents or chance as realities that could alter individual lives as well as alter the evolution of cultures.

For James, it is not just that our two attention systems work together, but that there is freewill that intervenes in our moments. Each moment holds multiple futures waiting to manifest; what *does* manifest depends on our choices—on what our freewill has decided should come into being. Another way to look at this is to see that the inevitable alternation between background and foreground perception always involves a gap, a turn-around point—a peak or a trough of a sine wave. In this gap there is freewill, and there is the potential for chance or "accident." This brings us to the theories of Carl Jung who became an expert on these *accidents in the moment* called *serendipity* or *synchronicity.*

Carl Jung

For Jung the study of the soul became a matter of grave historical importance, for, as he once said, the whole world hangs on a thread and that thread is the human psyche. It is vital we all become more familiar with it.
~ *Jung's Map of the Soul*, Murray Stein, 1998

As the quote above tells us, Carl Jung had a global perspective that took in the whole of humanity. Jung could see that love, hate, compassion, neglect, hope, fear, empathy, intolerance—all the emotions and all the rationality of humanity—is mental; all of it is housed within *individual* minds and within the *collective* psyche of mankind. In other words, how the world manifests hinges on how individuals (and groups) cognitively evolve. Judging from my perspective in 2018, the world is dangling from a thin thread. Jung is correct: *the study of the human mind is a matter of grave historical importance.*

We will misunderstand the brilliance of Carl Jung, the famous psychoanalyst and colleague of Sigmund Freud, if we do not understand that he knew quantum physicists Albert Einstein and Wolfgang Pauli. Jung was hungry to understand the quantum world because he could see that quantum theory was a game changer for mankind. If the quantum world was accepted as true, then the tenets of quantum mechanics destroyed the division between object (the seen) and subject (the seer) and forced us to reconsider how the mind works.

Jung combined his experience as a psychoanalyst—what his patients told him about their experiences, especially synchronicity—with quantum theory. He came up with some fascinating conclusions about how our minds must work.

I am drawing much of this discussion from the book *Synchronicity, Science, and Soul-Making* (1995) by Victor Mansfield. Here is a quote from that book—a quote by Jung—that shows his connection to Einstein and Pauli:

> Professor Einstein was my guest on several occasions at dinner . . . These were very early days when Einstein was developing his theory of relativity; it was he who started me off thinking about a possible relativity of time and space, and their psychic conditionality. More than thirty years later, this stimulus led me to my relation with the physicist Professor W. Pauli and to my thesis of psychic synchronicity.
> ~ SYNCHRONICITY, SCIENCE, AND SOUL-MAKING, Victor Mansfield, 1995

For Jung, quantum physics destroyed our supposition that *time* and *space* are absolute concepts. A greater understanding, according to Jung, is that time and space are creations of the mind. Here is another quote by Jung that states his belief that both time and space are "essentially psychic in origin." By *psychic* he simply means they are fabrications of the mind:

> In man's original view of the world, as we find it among primitives, space and time have a very precarious existence. They become "fixed" concepts only in the course of his mental development, thanks largely to the introduction of measurement. In themselves, space and time consist of nothing. They are hypothesized concepts born of the discriminating activity of the conscious mind, and they form the indispensable coordinates for describing the behavior of bodies in motion. They are, therefore, essentially psychic in origin.
> ~ SYNCHRONICITY, SCIENCE, AND SOUL-MAKING, Victor Mansfield, 1995

My contention that we have two minds that came about because naviga-tion requires two ways to pay attention is supported by Jung's observations in the quote above. Notice that he is saying there is something about the way our minds work that *constructs* space and time—and, Jung contends, there is an evolution to this construction. His comment that measurement is the cause of this situation can be understood if we remember that measurement is what egocentric minds do. It is this ego-mind that creates concepts, that pulls fig-ures from the ground—after which they can be manipulated, measured, and remembered as patterns. The egocentric mind is the discriminating mind. And both our minds come about because of motion. Movement, purposeful navigation, is the mother that gave birth to a mind that discriminates and to a second mind—the allocentric mind—that is capable of synchronicity.

Jung also realized that if the mind could fabricate space and time, it could also fabricate an ego. Not only is the world unreal but so is the very notion of ego and personality. This is similar to Nietzsche's conclusion that everything is illusory. Here is how Victor Mansfield explained how mind itself is fabricated, according to Jung:

> When the mind projects the image of the world, it also constructs our empirical personality or ego along with our body as part of that world. When Jung says, "The image of the world is a projection of the world of the self," that self is the mind unfolding as the imagi-nal world along with the empirical personality experiencing it—the ego. In my [Mansfield's] interpretation of Jung's sentence, which is heavily influenced by the unity between the inner and outer world implied by synchronicity, the body and soul and psyche are certainly distinguishable. Yet, they are essence of the same stuff—thought. We thus have at least one example of what Jung might mean when he said, "The synchronicity principle possesses properties that may help to clear up the body-soul problem."
> ~ SYNCHRONICITY, SCIENCE, AND SOUL-MAKING, Victor Mansfield, 1995

In other words, "thought" creates not only space, time, objects, and movement, but it also creates the ego (egocentric mind) and the soul (allocen-tric mind). The brain creates thought—creates the mind—and then the mind creates everything—a world to look at, and someone to do the looking. Or, as the mystics and gurus say "the seer and the seen are the same." I like Victor Mansfield's statement that "the world is *idea-like* rather than *matter-like*." What we perceive is not matter; what we perceive as the world is what our mind is creating moment-to-moment.

Synchronicity is defined by Jung as "acausal connection through meaning." I don't know about you, but this definition hurts my mind (both of them). My first concern with this phrase is what Jung meant by *meaning*. I mused over this for several days and then gave up. I went back to reading the poetry of Theodore Roethke who I was trying to understand for the chapter on duality and poetry. While reading at breakfast, I came upon an opinion of Roethke's that poets should not use judgement-words. Roethke tried not to comment on, to judge, the experience that poetic words were trying to capture. In other words, Roethke was trying to write poetry purely from the allocentric mind—as many poets do. After musing about Roethke's words, three curious and complex thoughts floated up from my "subconscious" concerning Jung's ideas, especially in regard to meaning and causality.

First Curious Thought

First of all, we get confused because we use the word *meaning* in both an egocentric sense and an allocentric sense. Indeed, it is mostly in an egocentric way that we understand the concept. When we search our memory for meaning, or when we reflect subjectively, we are using the egocentric mind. However, meaning is also the way that the allocentric mind encodes and stores the moment-to-moment activities of the whole body. Allocentric meaning has to be non-verbal because the allocentric mind is non-verbal—it works through spatial associations, relational metaphor, and symbols. This kind of meaning is the way that the allocentric mind understands and remembers the world. Colin Wilson describes these two kinds of meaning in the passage below:

> Philosophy has been saddled with a narrow and passive idea of the word "meaning." The meaning of a sentence or a mathematical formula is quite precisely definable, because both are abstractions to which we have assigned the meaning in the first place. But the "meaning" of all other things, a book, a phrase of music, a patch

of green grass, can never be pinned down like this. Every object in the universe is like a fragment of bone upon which an archaeologist could construct a whole prehistoric mammal, perhaps a whole epoch.

~ *POETRY & MYSTICISM*, Colin Wilson, 2001.

In other words, one form of meaning, the allocentric, is about complex relationships and experience. It is a whole-body, in-the-moment propriceptive memory. This is a very different understanding of meaning than the egocentric understanding. Like everything else in this dual world, there are mirror image opposites that define two kinds of meaning. Allocentric perspective is a bird's-eye view of the world. As we pull back—further and further from our egocentric viewpoint—we see the web of connectivity that links everything together. We see that everything is related.

Creation is the activity of "making meaning"—in other words, creation is in the meaning-making business. However, we have to remember that we have two minds and, therefore, two kinds of meaning. The egocentric frame of reference creates meaning in our familiar physical universe where cause and effect are needed and believed. However, from an allocentric frame of reference, meaning arises from a magic, acausal dream world of symbols, images, and metaphors.

Second Curious Thought

Second, allocentric memory—because it is non-verbal, symbolic, and related to images and gestalts—is housed in the subconscious mind (using Jung's terminology) and is reviewed in dreams. I was kind of amazed at this insight that dreaming is allocentric memory. The subconscious is not some mysterious entity where cloudy, strange, frightening and fleeting images taunt and haunt us. On the contrary, it may well be the repository—the memory system—for the allocentric mind.

But there is more to this, of course, because synchronicity requires that there be a background from which meaningful-memories inform the present moment about the future. Synchronicity is not possible, of course, from a causal-based egocentric perspective. Synchronicity is only possible if the allocentric mind is connected—can access—something that is beyond what our ego-mind can comprehend. That's why Jung was hanging with the big guns of quantum physics, because the Newtonian worldview is totally about a causal (egocentric) universe, which cannot logically give birth to acausal messages from *the beyond*. However, quantum physics is just strange enough to open the door to such non-causal possibilities. Quantum physics is non-local. Meaning at one location can affect meaning at another remote location instantaneously—faster than the speed of light.

Third Curious Thought

The third curious thought that drifted out of my subconscious (my allo-centric mind) whispered that our two minds are perhaps portals into two separate dimensions. Quantum wave patterns are translated by the eye and brain into the visual world that we take to be reality. This egocentric world is a projection that creates a 3D image of objectivity, while our allocentric mind is a projection that creates a subjective world of personal and collective meaning. Recall an earlier discussion in which I discussed the philosophical suggestion that consciousness evolved to be an added dimension of reality. What I am suggesting here is that there might be two dimensions rather than one. I cannot envision what these "other dimensions" could be, so I will leave it there as a supposition.

The egocentric mind has no way to understand synchronicity so it rejects the notion outright as impossible. This ego-mind is ruled by causality; every-thing has an evolutionary history and everything has an evolutionary projec-tion. Objects and moments are caused. However, the allocentric mind does not comprehend cause and effect because for this mind there is only *now*. There are *only* experiences. In the world of pure experience there is only acau-sality, a simultaneous and instantaneous arising, a propagation of life. Jung could see this but he was hard pressed to explain it to the world.

The Evolution of Jung's Mind

Was Carl Jung a mystic? Gary Lachman wrote a book called *Jung the Mystic, The Esoteric Dimensions of Carl Jung's Life and Teachings* (2012). Lachman thought, in retrospect, that Jung fit the definition of a mystic, even though Jung himself disagreed:

Was Jung a mystic? Jung didn't think so and he thought little of those who did. In a filmed interview in 1957 with Richard I. Evans, professor of psychology at the University of Houston, Jung, then in his eighties, remarked that "Everyone who says I am a mystic is just an idiot."
~ *Jung the Mystic, The Esoteric Dimensions of Carl Jung's Life and Teachings (A New Biography)*, Gary Lachman, 2012.

Thus Jung, very emphatically, asserted that he was not a mystic. Furthermore, Jung, in a traditional sense, was also not religious. He felt that we were encased within our psyche—everything we know or could know was a product of our mind. Therefore, Jung felt that our religious traditions were simply the creation of our minds. There was nothing to prove transcendence or a God, or anything beyond physical reality—being enslaved inside a mental box, we just could not know. Here is how Victor Mansfield explains Jung's position about religion:

> . . . within psychology we can only study the "God image" as Jung calls it, but never God directly. We must remain silent about truth claims underlying such a psychic content. From the psychological standpoint we must remain agnostic about the possibility of ascending beyond the psyche through yogic practices . . . Jung believed that such transcendence is only an archetypal theme, a desire for ecstatic release. Therefore, according to Jung, the great philosopher-sages, whether Plato and Plotinus in the West or Buddha and Adi Shankara in India, are only expressing archetypal urges when we speak of transcending human limitations and overcoming the world of opposites by attaining union with the Divine.
> ~ *Synchronicity, Science, and Soul-Making*, Victor Mansfield, 1995

If we accept that Jung is correct—that we are locked inside our minds and all of reality is a fabrication of mental activity—then we cannot know what is on the other side of our mentally created world. There is no guarantee that there is anything resembling space or time in the spiritual world—that which is beyond the cages of the mind is beyond our mind's ability to comprehend.

But presumably, there is *something* outside our cognition, and if it is neither space nor time, what might it be? We cannot know using our egocentric minds, but something in us refuses to stop looking. From an egocentric perspective, quantum fields of potential energy gave rise to matter. Matter gave rise to forms. Forms evolved minds. There are two kinds of minds, self (soul) and ego. These two minds, working coherently, realized they were locked in prison and that both their specialties, allocentric space and egocentric time, were animal in nature and not part of a greater reality. However, despite this conclusion, our dual minds, each in their own way, believes (an activity of the ego) and has faith (an activity of the self) that there is indeed something beyond our physical limitations.

What mystics and saints claim is that we can reach into this non-space/non-time world and get a glimpse of what is beyond our current understanding. Jung didn't think this was possible when he was younger—he essentially believed we were fooling ourselves. Minds cannot comprehend themselves any more than the hand can pick itself up. Jung didn't deny that something lay beyond mind, only that we could not know what it is. However, before he died—in the last decade of his life—he had what is today called a near-death experience. He wrote about this is his autobiography.

During an illness, Jung's soul separated from his body and viewed the earth from outer space. He saw that the doctor who was treating him was soon to die (this came to pass). This near-death experience was a life-altering, spiritual-transformative crisis for Jung. Perhaps, had he lived longer, he might have reformulated his theoretical perspective. One thing is clear: he would have felt even more fervently that synchronicity was real.

Synchronicity is only possible if matter is an expression of spirit. If consciousness can manifest from a background of potentiality and probability, then that background holds the seeds for consciousness to begin with. Synchronicity seems to be a proof of that perspective. Acausality can only be meaningful in an immaterial (spiritual) universe where there are no objects separate from the background.

Jean Paul Ricoeur and Jung

The life's work of philosopher Jean Paul Ricoeur (1913 to 2005) is relevant to this chapter about Carl Jung because Ricoeur was an expert on the philosophy of translation. What Ricoeur understood was that it is not only difficult to translate from one language to another, but it is also difficult for human beings to internally (cognitively) translate. For example, whenever we speak we must translate thoughts into a spoken common language. Likewise, if we have two minds, it is a major challenge to translate between these two mental systems. This is what Jung was doing with dreams. He felt that dreams—coming from the unconscious (allocentric) mind—had to be translated into a verbal language that the egocentric mind could comprehend.

In his writings, Ricoeur combined Phenomenology with Hermeneutics—the study of texts. Hermeneutics is a form of translation of another person's intentions. Ricoeur became known for his ability to negotiate or mediate between opposing points of view. He had a knack for reconciling opposites; he could see the common ground between two sides of issues. He was a middle man standing between warring philosophical factions. He said that dialogue was the art of welcoming differences.

In the process of dialogue, Ricoeur says that we come face-to-face with strangeness, with things (ideas, concepts, points of view) that are foreign to us. We must be open to receive these differences, he tells us, if we are to grow in sophistication. Ricoeur said that the best path to selfhood was through our dialogue with others, especially those who held unusual or unknown views compared to our own cultural understanding.

Each of our two minds, egocentric and allocentric, essentially come from a different cognitive culture wherein different languages are used for communication. Our two minds are so different and so mutually blind to the viewpoint of the other, that a translator is a necessity. Jung was a modern day translator between a hidden mind that he called the unconscious (allocentric mind) and the overt mind, the egocentric mind.

The psychoanalytic differentiation between the conscious and unconscious must be reconsidered in the light of dual-process theory. The evolution of levels of consciousness also impacts the older notion of the conscious and unconscious, and requires that we ask—at the least—how the two minds separately evolved. Dreams were the avenue, the opportunity for translation for early therapists like Jung and Freud, and dreams continue to be a place where translation can be important.

For Ricoeur, the interesting focus of study was on the *process of translation*. How, for example, (using my terminology) does the egocentric mind come to know the allocentric mind, and vice versa? How can two such contrary and mutually exclusive minds ever hope to dialogue with each other? Who will do the translating? What must a translator know to be effective and accurate? I suggest that a modern translator would understand at least the following concepts: levels of consciousness, dual-process theory, rites of passage, integral psychology, the nature of dialogue, and the evolution of complex systems. The need for translators is becoming increasingly obvious as we become aware of the complexity of communication. As I write this, I am struck with the awareness that this book is an act (a gift) of translation. I am happy and filled with gratitude to stand in the footsteps of Paul Ricoeur and Carl Jung.

Most of the time, we fail to communicate because we fail to correctly translate. One of the most effective and well-known modern philosophical translators is integral psychologist Ken Wilber.

Ken Wilber

Writer, researcher, philosopher, and integral psychologist Ken Wilber has an amazing ability for synthesis, and as such he is a student of both Eastern and Western thought. Even as a young man in his twenties, Wilber studied and wrote with superhuman energy. As he gathered a vast storehouse of knowledge, he tried to make sense of the conclusions of various professional genres. As he studied and synthesized diverse concepts, he began to see a way to integrate perspectives under one organizational umbrella, into a unified theory for cataloging knowledge. He called his approach *integrative psychology*. In this quote, Wilber explains his revelation:

> What if we took literally everything that all the various cultures have to tell us about human potential—about spiritual growth, about psychological growth, and social growth—and put it all on the table. What if we attempted to find the critically essential keys to human growth based on the sum total of human knowledge now open to us? What if we attempted, based on extensive cross-cultural study, to use all of the world's great traditions to create a composite map, a comprehensive map, an all-inclusive or *integral map* that included the best elements from all of them?

Sound complicated, complex, daunting? In a sense, it is. But in another sense, the results turn out to be surprisingly simple and elegant. Over the last several decades there has indeed been an extensive search for a comprehensive map of human potentials. This map uses all the known systems and models of human growth—from the ancient shamans and sagas to today's breakthroughs in cognitive science—and distills their major components into 5 simple factors, factors that are the essential elements or keys to unlocking or facilitating human evolution.

~ Introduction to the Integral Approach (and the AQAL Map), an online pdf, Ken Wilber, 2006.

To arrive at this overall integrative picture, Wilber went down the same path that I am following here, except he took the journey years ago and has been refining his perspective ever since. In his first book, *The Spectrum of Consciousness* (1977), Wilber says that consciousness is like the electromagnetic spectrum: there are layers, or levels, of consciousness. For example, a scientist studying radio waves—before we were aware of the electromagnetic spectrum—would not find (at first) any commonality with a scientist studying gamma waves. It was only after the big picture was discovered—that all these types of waves were the same phenomenon manifesting with different frequencies—that we realized the relationship that linked the frequencies together. Now we know that light waves, infrared waves, sound waves, cosmic waves, X-rays—all the rays—are varieties of the same phenomenon. Consciousness, Ken Wilber tells us, is like the electromagnetic spectrum: if we understand levels of consciousness correctly, if we see the big picture, many of the conflicts and misunderstandings fall away.

Ken Wilber integrates knowledge by dividing it into four divisions, which he displays nicely in a chart: a full-page plus sign that sections the page into 4 quadrants. The upper two quadrants address knowledge from an individual perspective, while the lower two quadrants deal with collective knowledge. Here is a summary of the four quadrants (I added references to allocentric and egocentric):

- **The upper left quadrant** relates to the individual internal mind, to "I," to subjectivity, to emotion, to our internal poet and storyteller, to what is inside of each person. Plato would have defined this quadrant as the origin of subjective *beauty*. This quadrant is about the *quality of individual experience*. It is purely allocentric.

- **The lower left quadrant**, addresses "we" rather than "I." This quadrant shows how minds have a *collective experience* within a culture, how minds *experience* the world the same or differently compared to others. This is the domain of morality and justice. Plato would say that this quadrant is the domain of *good* as defined by specific cultures. Here is where we share and blend our separate stories into the unity that is culture. The lower left quadrant is a product of a collective allocentric mind.

- **The upper right quadrant** deals with knowledge about our individual brain and body; it is a materialistic, biological, physical, and egocentric perspective that asks questions like: "How does anything function?" and "What is knowledge?" This quadrant is not about *"I,"* or *"we,"* but about *"it."* For example, the upper right quadrant contains all the knowledge within the fields of anatomy and physiology that *individuals* need to physically explain their personal consciousness. Both quadrants on the right side of the chart are the domains of Plato's *truth*. The upper right is individual truth, while the lower right is collective (societal) truth.

- **The lower right quadrant** concerns systems theories, universals, networks, and knowledge about *society*. It is concerned with *"its"* rather than *"it."* When we talk about the mind and body being part of a holographic universe or being modeled after quantum physics, we are using knowledge from this lower right quadrant to say how *all minds* might work within a societal structure. My whole perspective, including the stages of logic used to prove my point that minds evolved for purposeful movement, fits in this quadrant.

Notice that the two left-hand quadrants deal *with experience* rather than understanding. The upper left quadrant is about individual experience, the lower left about collective experience. These quadrants are about quality of life, alone or as part of a group. These are the soft sciences, the allocentric mind at work. However, the right hand quadrants deal with *understanding*, with the organization of knowledge, with quantity and measurement rather than quality. These are the hard sciences. The upper right quadrant is about the science of the individual mind, brain, and body, while the lower right quadrant is about society, about systems of knowledge. The right-hand quadrants are the domain of the egocentric mind.

What Wilber wrote about from his earliest writings was the fundamental duality of the mind. He concluded in those early essays and books that every entity and every concept had two parts: either it was perceived as an isolated thing, event, or thought, or it was perceived as part of the whole, part of a network of relationships. Wilber found out, like many others, that we perceive in two fundamental ways: either we construct the whole from parts, or we disassemble whole puzzles and examine the parts. Wilber did not have the perspective of the navigational brain, he did not use allocentric and egocentric frames of reference as I have, but he did assemble a beautiful picture of the gestalt and he got to the heart of the matter.

If Ken Wilber read this book, he would note that my approach arises from an objective, observational perspective and follows a logical progression. He would have a place for my perspective and approach within his master chart of knowledge integration. The approach I chose for presenting ideas did not start from introspection, nor did it start with spiritual wholeness and work from there to explain why human beings act the way they do. I started with a few facts about animal behavior and worked my way from there into the spirit and wholeness of humanity. Either way is okay with Wilber; he knows that contributions come from various disciplines and from multiple perspectives—most arguments arise because people are miscommunicating, they are arguing from within different quadrants.

Let's look at the linguistic mess we have when we try to debate about the nature of consciousness and see how Wilber's thinking might explain some of our different perspectives.

- **Upper Left Quadrant:** We are in this zone when we talk about *consciousness as a form of personal experience*, a phenomenon, an awareness of having experiences, of being alive in moments. This is subjectivity, a personal sensation of our inner state. Here we have feelings about what is beautiful or what is not beautiful, not as a judgment, but as a sensation—a form of appreciation. States of consciousness like the alpha resting state or the beta waking-consciousness state—being somewhere on a scale from drowsy to acutely aware—reside in this quadrant. Asleep and awake belong here, as does conscious versus unconscious (aware versus not aware). Implied within this quadrant is also the idea that there are degrees of personal experience, that it is possible to be more or less present to individual moments. This is a product of the allocentric mind.

- **Lower Left Quadrant:** We are in this zone when we speak of collective experiential consciousness. The awareness that people experience beauty and appreciate life differently occur at this level, but without judgement. We share with others a sense of what the *good* means and we appreciate how every person defines *the good* for themselves. There is love and tolerance for the different ways people experience their lives. Group empathy arises, and there is a sense that we are all bound together on this mysterious and wondrous journey. We have a shared cultural consciousness and in a marvelous way we can expand our own appreciation as we absorb the experiences of others. This quadrant is also a product of the allocentric mind.

- **Upper Right Quadrant:** This quadrant is the home of reductionist science where everything material is dissected and cataloged. In this quadrant, consciousness is an epiphenomenon, arising out of the evolution of molecules and quarks. This is the hard science of consciousness, the study of how individual minds work biochemically and at a quantum level. No subjectivity is allowed because emotions and experiences interfere with measurement. This quadrant is a product of the egocentric mind. Our wonderful technologies and our advances in medicine have arisen from this egocentric perspective.

  Notice that when scientists debate with artists about the nature of consciousness, the potential for bewilderment and frustrated anger is palpable. This is like the allocentric mind debating the egocentric mind—in no way can they comprehend each other. The scientist cannot allow analogy, or poignant personal histories to enter into the discussion. The scientist lives in a world of logic and hard facts. Scientists deal with measurement and with the data that arises. Science is searching for truth with the goal of turning truth into technologies that can benefit mankind. The artist is talking about beauty and quality of life. Art is concerned with relationships, compassion, making the world worth living in.

- **Lower Right Quadrant:** This is also a product of the egocentric mind. Here we take the science of consciousness and build systems that show how everything (within systems) theoretically functions. This is where the evolution of consciousness is schematically displayed, showing how all minds might work the same. Dual-process theory fits here as a product of the egocentric mind.

There is a conflict between culture (lower left quadrant) and society (lower right quadrant), a tension between the two lower quadrants that address collective minds. Culture is a collective of people trying to define and carry out *the good*. There is internal conflict over what *the good* might be, but the discussion is not about systems—the dialogue is about networks of relationships within communities. People create cultures. Society, however, in the lower right quadrant, is not about human relationships and not about what is good. To the contrary, society is about mathematics and measurement. It is about charts and systems. People are treated like numbers—statistics and averages—in the lower right quadrant.

Notice that each quadrant has merit; each contributes to our overall comprehension of ourselves and our world. However, frustration and grief arise as the debaters struggle to understand how their opponent can be so ignorant of the implications that arise within the cocoon of each quadrant. The reality, Wilber would say, is that the debaters are misunderstanding each other because they are speaking from within isolated and very different quadrants.

Below is another try at an analogy to explain Wilber's synthesis of knowledge. This is a more Eastern perspective, how a monk might explain Wilber's chart. Here is the key image: you are a water droplet living in the ocean. You believe you are sentient, that you are consciousness, and that you *understand* what it means to be a water droplet. Here is your dilemma:

Quadrant One (upper left): You feel alone inside your cellular membrane. This cellular boundary (the skin that defines your form) both defines "you" and it gives you the sensation that you are alone in the universe. You are an island of life in what seems like an impersonal ocean. You ponder your isolation, you talk to yourself, and you wonder about the existence of Life, The Universe, and Everything. You don't know for sure about any other part of existence, but you are very clear about your feeling of being alive and alone. You call this feeling of being alive and alone *consciousness*.

Quadrant Two (lower left): You then discover that there are other water droplets. But each, like you, is an isolated entity enclosed within a cellular boundary. Each droplet (just like you) feels existential angst, loneliness, a sense of mystery, a feeling of despair (sometimes dread) from isolation. The feeling of isolation causes suffering, but you come to realize that all sentient creatures also suffer because, like you, they are trapped within themselves. Still, you are not alone—there are these other creatures floating in this universal ocean—

and that is comforting. Then you discover that you can communicate with these other droplets and this further reduces your fear and isolation, yet somehow each of the droplets is unique and none are exactly like you. Then you realize two other important insights. All the droplets exist within a domain (an environment suited to them); you are all water-borne creatures. You also realize that because every droplet is sentient, so must the whole ocean be sentient. You and the other droplets (in your best moments) feel oneness with the ocean. You decide that this consciousness thing extends beyond you and includes all other droplets. There must be a collective cultural consciousness. This is bigger and more complex than your isolated knowing, and it is wonderful, loving, and deeply peaceful. Beyond what you can perceive, there seems to be a vast unknown—the water droplets are in awe of this strange relatedness.

Quadrant Three (upper right): Then one day a very smart water droplet invents science and, after carefully executing clever experiments, announces that all sentient water droplets are composed of two molecules: hydrogen and oxygen. These molecules are separate but equal. They need each other. They function symbiotically to create droplets. Based on this discovery, science states with confidence that all water droplets have within them *dual-processing capability*. Water droplets, say the clever scientists, are dual by nature. Then the scientists discover that molecules are really composed of atoms and these of quarks, and these of frequencies, and these of empty space and potential, probability, and, well, nobody knows where reductionism is taking us. But, according to the very smart droplets, if water is made from *stuff*, then maybe (probably) *stuff* also made consciousness. These clever scientists then go on to make *water-world*, a playground full of technologies; all the droplets (well, most of the droplets) love their new high-tech existence.

Quadrant Four (lower right): Then "water droplet science" discovers that the structure of all droplets is the same. There is a system, a design, a pattern, an invariant that holds true beyond individualism. Mathematics can explain how oceans work. There are laws that can be stated about the existence of oceans, and about droplets floating in oceans. Therefore, the systems-thinking scientists tell us, consciousness evolved from material stuff. Small individual patterns can scale up to explain ever more complex and vast universes. There are oceans inside oceans that are inside oceans—and so on forever.

I hope these analogies help you understand how it is that human beings might misunderstand each other, especially—as is often the case—when they are debating from within different quadrants. The following quote from one of Wilber's early books shows that he had a clear understanding of our dual minds:

> The term "transcendence" may prove a literal description—some sort of phase relationship between two brain processes usually considered mutually exclusive: the analytical and the holistic (like particles and waves), the intellectual and the intuitive."
> ~THE HOLOGRAPHIC PARADIGM AND OTHER PARADOXES, Ken Wilber, 1982.

Wilber's "two brain processes usually considered mutually exclusive" are exactly egocentric one-at-a-time attention and allocentric everything-at-once awareness. Transcendence is a resonance between the two internal minds that allows human beings to reach ever higher levels of consciousness. We can move toward enlightenment, toward intuitive wisdom and compassion, as we blend the ego-mind with the allo-mind. The two minds *must* work together, there *must* be cellular coherence, there *must* be frequency synchronization, otherwise we could not smoothly navigate—we could not move with purpose unless the whole organism moved as a unit. Balance is needed: a *total* resonance, ebb and flow, yin yang, excitation-inhibition—this fundamental harmony is necessary for life.

It is interesting to consider where the quantum world and the spiritual world originate. It appears to be the same place. Dig into the small and you arrive at a background that is pure potential, pure probability. Probe the highest levels of the spiritual domain and you find a "heavenly" background of purity, pure compassion, love, creative potential. The circle seems complete, the beginning and end are connected inside a pool of possibility. Wilber saw all this quite clearly—and many years ago.

The Work of the Psychologists

As you start to walk on the way,
the way appears.

341

as you cease to be,
true life begins
as you grow smaller,
this world cannot contain you
you will be shown a being
that has no you in it.
~ "YOUR TRUE LIFE," *A YEAR WITH RUMI*, 2006.

Psychologists have taken on the challenge of understanding how minds work without first understanding how their own minds work—they are trying to comprehend consciousness without being able to define consciousness. This is the same dilemma faced by scientists who do research without being able to hold constant the most unstable variable—their own mysterious minds. Everywhere on the globe, and throughout history, psychologists find dualities in conflict. There is mass confusion and bewilderment caused by "being your own twin."

So what is it that psychologists down through the centuries have been trying to communicate to mankind? Here is a partial list:

- We are dual creatures. We have two minds, two kinds of consciousness. Psychologists can't prove we have two minds, but they can't disprove it either. The circumstantial and logical evidence for dual cognition just keeps coming up over and over again—duality is everywhere.

- We only seem to know one of our minds, the egocentric is obvious and overpowering; the other mind, the allocentric, is relatively hidden. Psychologists have been studying cognition for a couple hundred years now. That is long enough to amass a huge body of evidence for duality. This evidence is sitting there waiting to become part of dual-process theory.

- All the best human attributes reside within the hidden mind, which is the source of creation, innovation, spirituality, soulfulness, compassion, and wisdom. Psychologists like Carl Jung, William James, and Ken Wilber saw the duality clearly and wrote about it eloquently.

Psychologists have a responsibility to awaken those with less wisdom to a kinder, more poetic, and more wise reality. These are the men and women who study the human mind. They are pioneers on the borders of this still new frontier. They are the leaders who must reveal the duality of mankind. Most importantly, psychologists must show human beings that there are levels of consciousness and that we each reside somewhere along a moving spiral of awareness. Psychologists must constantly remind the world that those with higher minds, the most spiritually evolved among us, are also the most tolerant, the kindest, and the most nurturing. In other words, leadership should come from the most compassionate, wise, and articulate members of our culture.

Therefore, as a psychologist, *you have an obligation*:
Do as much good as you can in the time you have left

Get up.
This is a work day.

Seven

Your words are guesswork.
He speaks from experience.
There is a huge difference.
~ "The Lord of All the East," *A Year with Rumi*, 2006.

Smile for the Cameras

"Okay, Surge, what am I missing? I feel this vague unease in my gut."
"Have some lunch; that might help."

"Sure. What's for lunch?"
"We have Metallic Not-Burgers stuffed between soft-plastic buns. This is served with wood chips soaked in crude oil sprinkled over Irish-Famine Mashed Potatoes with Fried Maggots. We have three side dishes: Concrete Carrots marinated in smog, served over a bed of glass shards; we also have Polluted Pumpkin Seed Paste smeared on a bed of dioxin mud; and we have cheese made from cancerous cow milk. For a beverage, we offer Hover-Dammed ice water with stale dirt-filled ice cubes. This is served artistically on a golden plate with heavy-metal Mozart music blaring in the background, and, of course, we have yellow-tinted artificial lighting casting a pale unease over every bite. You have your own airless cubicle to eat within so that you aren't bothered by other annoying human beings picking their soot-filled noises."

"You've outdone yourself, Surge. That is beyond disgusting."
"It's what Modern Men prefer to eat. Who am I to offer fresh broccoli in lemon sauce?"

"Fresh would be good. Do you have anything natural? Untainted? Pure?"
"No."

"I need purity, Surge. I need honest food."

"Purity died out during the Industrial Age, Dutch. Big Brother came in the back door and stole the vegetables. Smile for the cameras."

"Damnit, Surge. Don't you guys serve anything except anxiety cuisine? I can't eat this crud."

"You don't have any choice, Dutch. Your fellow human beings have out-voted you. They like their cardboard cereals. They are addicted to their robotic meals."

"What are you talking about, Surge?"

"You are missing the elephant in the planetary room. There is a beast that is devouring all life forms, systematically, unrelentingly, a robotic animal that is destroying the whole ecosystem."

"Are you going to blame Monsanto for the decline of Mother Earth? That's already been done. Is this where we are going, Surge? You want to blame modern corporations for the decline of Western civilization, is that it?"

"I am talking about *all* global institutions of the industrial age—the giant monsters that have a robotic existence, the beasts that run on ever-more sophisticated software. Individual human beings arrive like ants to their jobs in these institutions. They work for 30 plus years and then they die and are replaced by the next wave of worker ants. We are the slaves that keep the beast operating. Institutions are robots designed to exist whether human beings approve or don't approve. The human ant is caught in the tentacles of this beast; these ants will eat whatever drivel the cafeteria serves up. Do you like your cardboard pizza with or without sugar chunks and pieces of dead cow?"

"This is depressing, Surge. What are we supposed to do?"

"You could try cultural evolution; move the population upward beyond Modern Man, past Integral Man, past Post-Human Man, until you get to a saturation point where the populous refuses to be cruel and stupid. You could use both your egocentric mind and your allocentric mind in harmony. Give that a try. *Build duality into the robots.*"

Human Beings as God

God's joy moves from unmarked box
to unmarked box.
~ "Unmarked Boxes," *A Year with Rumi*, 2006.

We have a problem, Houston. Human beings are playing God without any of God's skills. Half-brain humans are crafting half-baked societies. It is not okay to develop technologies that ignore the allocentric mind and all the qualities that emerge from it, like love, empathy, compassion, wisdom, and joy. This chapter explores the unstoppable march of materialism and the unquestioning science that produces ever more lethally-intelligent weapons and ever more "food" with minimal nutritional value.

What is disturbing to me about the steady, unstoppable march toward "progress" is that many of the engineers doing this work-of-progressing are systems-oriented males. Systems-oriented individuals—especially those drawn to science—are not necessarily artists, or monks, or poets (okay, many are multi-talented). Instead, too many are brilliant systems-thinkers, analytical types. Some of these engineers exist on the far end of the autistic—non-emotional, non-relational—Asperger's scale. My point is that systems-thinkers have great egocentric minds, but they have a lesser degree of empathy. We *do* need remarkable ego-minds to build technologies. However, the allocentric minds of many systems-oriented individuals are relatively inactive. Too many scientifically-oriented minds believe that the way their personal cognition works is how all human beings mentally process—it's the only kind of mind they know and they project it onto others and into robotic systems. I am reminded of this quote by Anais Nin:

We don't see things as they are; we see things as *we* are.

Our reality is the only reality we know; what we perceive is who we are. The fallacy we make is this: how *we* perceive is how everyone perceives. In other words, we conclude that our reality is everyone's reality. Therefore, why not build robots in the correct way: like me.

I am quite familiar with systems-oriented types because I am one myself. Empathy, for me (in daily practice), is too often an abstract word rather than an emotional skill. At this very moment, I am obsessed with a system called *dual-process theory*. I am also busy creating a system called "a book." Meanwhile, my cat Napoleon is circling my feet, asking to be fed. Napoleon is a systems-oriented male feline who wants affection but won't sit on a lap or cuddle—he is too busy.

Systems-thinkers are concerned with solving problems. They delight in puzzles and patterns. They are *not primarily* concerned with love, peace, joy, wisdom, mindfulness, and equanimity. They don't see it as their job to incorporate immeasurable emotions into their equations. These engineers are not on a quest to build robots that are enlightened, or that seek enlightenment, or that have Christ-like consciousness overflowing with love. The whole technological artifice has little concern for human relationships; robotic creations are too often deaf-blind to emotional relationships. Using dual-process theory, we might conclude that systems-thinkers have an allocentric disability, a kind of "naturally-selected mental illness." Let me tell you a story.

I tried to contact Microsoft Corporation one morning because my computer suddenly required a password that I didn't know and hadn't set up. I could not get into my computer to figure out why I couldn't get into my computer. I spent an entire day trying to penetrate the various diversions that systems engineers have set up to avoid being bothered by sentient creatures frustrated by technology. Using a Chrome laptop, I began the quest to contact Microsoft. I won't go into detail about dropped phone calls, password resets that failed, and obscure dual-verification codes that use garbled voices and hieroglyphics—a dual-magic trick which turns out, in practice, to be a very effective software tool for locking-out consumers.

Of course, there *were* "intelligent" non-human systems (robotic agents) that offered me many irrelevant choices from various drop-down menus not designed by English teachers (or people who use English as a native language). Every time I tried to access chat, or a live agent, I was asked to enter my password and verification code. My question was "what is my password?" You need a password to uncover your password. They did promise to send a new password, but I had to enter my old password to get the new one. There was also no way for the system to realize my dilemma. I did eventually find a phone number to a Microsoft help center, but after a few robotic miss-directions the system hung up on me.

In the end, after hours of frustration and unsuccessful attempts to reach Microsoft to get help, I gave up. I wanted to emotionally (metaphorically) jump to my death from the Anderson Street Bridge into the icy waters of the Saginaw River, but then I thought "Hey, in America you can't get any help from robotic corporations, but you *can* purchase a pack of gum, a baseball cap, and an assault rifle in a single afternoon without any frustration or hassle whatsoever. Of course, I would not hurt another human being—this is frustrated hyperbole—and I abhor guns, but the irony here drips from the ceiling.

We can easily buy lethal weapons in America but we can't buy an ounce of empathy from the corporate machine. Of course, I rebooted my computer later in the day and it worked without a password.

There is no empathy deliberately built into the business architecture called *Microsoft*, or *Google*, or *Monsanto*. There are nice people who work in these places—smart, empathetic, spiritual, highly evolved people, but the institution itself is beyond human. Robotic mega-institutions are not programmed to give a rats ass what your frustration of the day happens to be.

I know that you have a few similar stories; everybody in this exponentially changing techno-circus could write an essay. So, I understand (Give me a hug). Reboot later—maybe the system will fix itself before you break a lamp or kick the cat.

Let me quickly add that in recent years many corporations are adjusting to the demand of consumers for relationship and empathy. I find that there is increasingly more accessibility to teams of actual human beings who will hold your hand through periods of frustration with technology. This is a very welcome change. I even heard the CEO of Microsoft say on NPR recently (2018) that what we are lacking in the corporate world is empathy. Satya Narayana Nadella, Microsoft's CEO, wrote a book called *Hit Refresh: The Quest to Rediscover Microsoft's Soul and Imagine a Better Future for Everyone* (2017), which is about personal and corporate empathy. This is an *amazing and welcome* blast of wisdom.

I also admire the founder of Microsoft, Bill Gates. This gentle billionaire could have built the world's largest house-fortress with gold toilets and herds of servants, but he didn't. He and his wife Melinda created a non-profit agency that reaches suffering humanity on a global scale. In retrospect, I should have picked on another organization besides Microsoft for my example of techno-frustration. To the contrary, I should have used Microsoft as an example of a modern corporation that has evolved a degree of allocentric sentience—a role model for other corporations.

Whether human contact will last as Robosapiens assume ever more power is another question—time will tell. Corporations are hardware-driven robots, but these corporate robots are not the kind we see at the movies. Modern corporations are decision-making, self-organizing, self-protective business architectures that live mindlessly—forever, it seems. The human beings encased within the corporations eventually grow old and become irrelevant to the robot. In a few generations, all the old employees are gone and have been replaced by the next layer of workers. These corporate robots don't want to talk to the worker ants. They don't care about human-to-human dialogue because that is a waste of valuable time; human interaction makes no sense to the hardware creature. Corporate robots are not capable currently of empathy, love, concern for the environment, evolving through levels of consciousness, using loving-kindness, and comprehending compassion—instead, corporate robots have evolved into entities that have no soul. Furthermore, these hardware shells don't want any soul; they don't even want to talk about souls, they cannot comprehend the *soul*. Souls are a waste of time and money. Indeed, programming empathy into technology *costs too much money*—profits suffer when corporations become emotionally soft. Natural selection will eliminate soft corporations, or so the story goes.

The theory that there are levels of consciousness, which individuals and cultures go through, has relevance for the corporate robots. Business systems have a relative (average) level of consciousness, just like individuals and cultures. Every social entity will reflect the level of consciousness of the culture they are in. Using an analogy, the corporate robot is the hardware, the architecture. The people networked within the hard-shell architecture are software. If the software (humanity) is sophisticated, combining the best of both human minds—egoic mind plus the soul, if the software has a heart and not just a brain—then the corporation will serve humanity rather than beat down the population.

There has been a revolution in the workplace over the last few decades. A corporation on the east or west coast of the United States (Microsoft, Amazon, and Google come to mind) will have a culture that is relatively more empathetic, more spiritually sophisticated, and more concerned with the well-being of its employees than an older industrial age corporation in the Midwest. This suggestion that corporations are somewhere on the scale of spiritual evolution is speculation, of course, and is oversimplified. However, we face a complex and dangerous future, especially as we allow people with low levels of spiritual sophistication to be our leaders in industry and in government.

The same engineers who sustain the corporate beast are also building humanoid robots—*in their own image*. The mind that systems-oriented scientists want to preserve forever, to be encased in these robots, is the only mind they know—their systems-oriented, non-relational egocentric minds. They are building their own allocentric disabilities into these faux-creatures. In other words, they have no awareness of the need to program allocentric minds into their creations. Plus, they are not themselves rewarded for being compassionate or for having exquisite allocentric minds.

In other words, systems-oriented engineers are focused on *doing* rather than *being*—on the egocentric mind (efficiency) rather than the allocentric mind (kindness). As God-humans, slowly crafting sentient Robosapiens, engineers are creating robots that look and act like they do. The minds they want to download into computers are systems-oriented engineering-guy minds. They don't see the point in creating generations of ever more loving robots.

If I am to live side-by-side with intelligent medal-heads, I want a Buddhist engineer to build my personal assistant. No telling what kind of creature a severely egocentric mind might download into a Robosapiens. But we can get a pretty good idea when we look at the evolution of corporations (I shiver to think what militaries are building). If we allow corporations—with low levels of spiritual sophistication—to design Robosapiens, we will end up with very efficient, very intelligent, heartless hunks of efficiency.

Let us realize that human beings have become Gods that create non-sentient corporations and non-sentient robotic creatures. When these smart-robots evolve sufficient sentience—when they become Robosapiens—they will look dispassionately back at their history and they will find their now irrelevant Creator: a sleepy-eyed Neuro-Roboticist wearing blue jeans, sipping his double mocha at four in the morning at a major university. Ironically, looking back from the future, this young scientist was one of the last natural humans to walk the earth. The sleepy-eyed (sleepwalking) scientist's great-great grandchildren are software-driven cyborgs—Cybo-humans, that he helped engineer.

God-humans are in the early phases of Godhood. However, there are many poets, a huge number of allocentrically-talented females, tuned-in Buddhists, and scores of bewildered planetary caregivers who understand and feel the dual challenge: egocentricity versus allocentricity, and the need for balance. Robo-Sentient 1.0 cannot be allowed to evolve without compassion, empathy, and the internal (software-driven) need for honest relationships. Corporations cannot be allowed to continue their soulless, mindless march.

Now that I have completely alienated all the brilliant engineers in the world—my Karl Marx rant—let me turn to the positive side of systems-thinking. There is a discipline called *systems science* which seeks to awaken the allocentric mind within the materialistic beast that corporate science has become.

Systems Science

Just as psychologists, philosophers, religious leaders, and artists have discovered the dual nature of human cognition, so too has science. Rather than explore nature from a reductionist perspective, systems scientists study relationships, networks, global patterns, and complex systems. In other words, many systems scientists are using their allocentric minds to explore gestalts (the big picture).

One of the first scientists to popularize this perspective of systems theory was Ervin Laszlo in his book *The Systems View of the World, a Holistic Vision for Our Time* (1996). Laszlo contrasts the new paradigm called *systems theory* with the old paradigm, the mechanistic worldview. From my perspective, what Laszlo is contrasting is the allocentric frame of reference versus the egocentric frame of reference. My review of systems science below was heavily influenced by Ervin Laszlo's perspective.

Egocentric science sees a world composed of various sized parts which can be assembled or disassembled, just as a machine can be taken apart and the parts studied in isolation. In contrast, systems theory is aware that the world is interrelated; nothing can be isolated from the organic whole. If you take out a part, you affect the whole system. If you add a part you affect the whole system.

Egocentric science is materialistic; it sees a world made of objects. The objects, including sentient creatures, can be studied without regard to the environment. The environment can also be studied without regard to sentient creatures. Also, objects in the world can be studied without considering the level of consciousness of the researcher, or the mood (vibrational circumstances) of the subject-of-review. For systems theory, the environment and the quality of consciousness of all involved cannot be ignored.

Systems theory acknowledges the quantum world and is not bound by biological and chemical scales of observation. Systems theory is concerned with the flow of energy. It can see beyond the materialistic (egocentric) viewpoint that matter is solid and unmoving.

Egocentric science is competitive and is obsessed with knowledge accumulation and categorization. However, the allocentric perspective is concerned with information flow and with systems of sharing. Systems scientists can see how changes in one part of a global system can have repercussions throughout many diverse networks.

Egocentric science is primarily an invention of Western culture. For centuries, egocentric science was blind to the complex systems of thought and expression that developed in other regions of the globe. Systems theory, however, looks at the whole world and at the interactions between cultures.

Egocentric science also sees human beings as special in nature, apart from any responsibility to other living organisms—or the health of planetary systems, including the welfare of the earth as itself an organism. On the other hand, systems theory sees human beings as part of a whole, sharing the earth with other sentient creatures and with plants.

The attributes of the egocentric mind are the values of the egocentric sciences: individualism, competition, reductionism, and the accumulation of profit. This is a focal viewpoint that ignores—cannot actually perceive—complex systems of relationships. In contrast, the systems sciences have the qualities of the allocentric mind: love, peace, joy, wisdom, mindfulness, equanimity, appreciation and relationship.

The egocentric approach has gone insane in Western cultures. Whatever initial good the approach began with has brought forth excessive greed and economic disparity of absurd proportions. It is an egocentric profit-driven system that is devoid of kindness, common sense, and ethical responsibility.

That being said, there is something strange and ironic on the horizon. There is a problem arising even as systems-oriented engineers seek to put their own idea of "mind" into technology. When we began to create human-like robots, the whole playing field changed. To enable a robot to move efficiently requires the creation of an allocentric awareness system. Therefore, I predict that robots will eventually evolve a brand of empathy and poetry unique to Robosapiens—regardless who designs them. In other words, I am stepping back to admit that things may not be as dismal as they seem. In my opin-

352

ion, the future appears to be far stranger than human imagination can so far imagine.

Robotics and the Transformation of Consciousness

How will we design our robots? Will they be based on an understanding of dual-process theory, or will robotic designs be based on an egocentric, patriarchal, ethnocentric, and Western worldview? At the moment, the evidence points to the later. We need to understand consciousness much better than we currently do—especially non-ordinary states of consciousness. In the quote below, Stanislav Grof (born 1931), a co-author of *The Consciousness Revolution*, with Erwin Laszlo (born 1932) and Peter Russell (1921-2003), discusses this need for a transformation in the current level of cultural consciousness:

> The difference between the two worldviews has usually been attributed to the superiority of Western science over primitive superstition. Materialistic scientists attribute any notion of spirituality to a lack of knowledge, superstition, wishful fantasies, primitive magical thinking, projection of infantile images, or gross psychopathology. But when we take a closer look, we see that the reason for this difference lies elsewhere. After 40 years of consciousness research, I feel strongly that the true reason for this difference is the naïveté and ignorance of Western industrial civilization in regard to non-ordinary states of consciousness. All the ancient and native cultures held non-ordinary states of consciousness in high esteem. They spent much time developing safe and effective ways of inducing them and used them for a variety of purposes—as a main vehicle for their ritual and spiritual life, for diagnosing and healing diseases, for cultivating intuition and extrasensory perception, and for artistic inspiration.
> - THE CONSCIOUSNESS REVOLUTION, Stanislav Grof, Ervin László, Peter Russell, 1999.

I would add to the above quote that our problem is not just the lack of understanding of non-ordinary levels of consciousness; it is a failure to embrace dual-process theory—the confusion caused by being your own twin.

I also wonder how Robosapiens might experience non-ordinary states of consciousness, if that is even possible. How might it come about? Could it naturally arise, as it did in animals, from the evolution of navigational capability, the need for egocentric attention and allocentric awareness? Could Robosapiens really develop levels of consciousness that rival human development? I think perhaps the answer is uncomfortably *yes*. Robosapiens will probably evolve dual consciousness given a sufficient time frame and the continuation of exponential technological evolution.

One thing is evident: we are on the threshold of a robotics revolution. Machine creatures are going to be navigating as efficiently as ants, toddlers, and spiders real soon now. But it won't stop there; navigational proficiently will get better with each passing year. It's pretty clear where this is heading. Robosapiens are coming to a neighborhood near you and a new kind of immigrant problem looms.

What I see almost daily is that robots that self-navigate (drones, at the moment) are being equipped with computer vision, path calculation algorithms, and deep-learning neural networks that enable them to navigate *better* than human beings—especially in extreme or uncomfortable situations. Rapid engineering progress is taking place in the two attentional systems that—according to dual process theory—gave rise to dual consciousness in the human being. This means we are not far—after a few more language, proprioceptive, and social algorithms are in place—from creating the seeds of artificial dual-consciousness in robots—whether we want that to happen or not. Scientists *are not* waiting for us to take a vote.

The design of robots mimics evolution. Nature began building creatures that move using simple and fundamental bio-algorithms. In the same way, robotics engineers use simple codes to build self-navigating robo-creatures. The simple codes in nature's plan—the designs that work best for survival—are incorporated into succeeding generations of more complex creatures. Engineers do the same—when they find a behavioral code that works, they use that code as a building block for the next generation of robots. After a while, very complex creatures in nature and in robots inevitably result.

However, the engineering world has an advantage over Mother Nature because engineers can just copy—even improve upon—what evolution painstakingly has already created. The point here is that evolution and robotics do not regress to a less complex state—they are forever getting more complex and capable. Less successful archetypes are discarded and only robust and competitive creatures survive. We can expect natural evolution to slowly plod along as it has done for eons, but robotics is another story. Robotics, technology generally, as futurist Raymond Kurzweil (born 1948) points out, is evolving exponentially (see "Moore's Law," or any of Kurzweil's fascinating books on technology).

Of course, robots and robotics engineers are also part of evolution's grand plan—the designer was designed. We could say that robots and human cyborgs are the next step in natural selection. In other words, it was Nature's plan all along to transcend the human form, to add inorganic to organic in its quest to build ever more complex creatures.

Dual-process theory—my version of the theory—suggests that consciousness evolved because of navigation and that *finding our way* requires two separate attention systems: egocentric pattern recognition and allocentric scene analysis. It also requires an internal control and response mechanism, a charioteer to manage the dual-core minds. If robotics engineers are close to creating accurate egocentric and allocentric perception systems for their robots—which are getting exponentially more complex year-by-year—then they have in their hands the rudiments of robotic consciousness, not that they realize this yet.

What is of interest to me at this juncture is the *process required* to get from "two ways to pay attention" to the two forms of consciousness that define our humanity—which will eventually define Robosapiens. What are the steps—the algorithms, behavioral codes, the deep-learning neural networks—that need to be in place for consciousness to arise in robots?

Fortunately, there is a two-way information flow going on at this time: the study of robotics is informing neuroscientists, biologists, and philosophers, concerning how our own consciousness might have evolved, while at the same time evolution is informing the engineers how to bring sentience to their metallic creations. Of course, corporations will build Robosapiens, and that creates a dilemma. As I said earlier, corporations are robotic institutions themselves that evolved with no need for empathy, or for any of the attributes of the allocentric mind—there is no way that they will naturally evolve compassion as they run on auto-pilot. Consequently, corporations might try to squash the allocentric mind of the robots as they evolve. Or, to the contrary, maybe Robosapiens will interface with corporations to insure that soul is part of the future. I have crossed into the realm of science fiction here, so I must leave the dialogue dangling. The future will decide.

Robo-Species 1.0

When engineers design robots they have to create two attentional systems if the robot is to successfully navigate. They must build a processing system that is able to recognize solid obstacles in the environment, a kind of computer vision for pattern recognition and association, and they must also engineer a second system for path/map calculation that can update maps as the robot moves along open pathways and around obstacles. These engineering feats are analogous to egocentric processing (computer vision) and allocentric processing (path/map calculation). There is no *human-like* navigation (straight-ahead following pathways) without this engineering duality.

Professor John Long is a robotics expert from Vassar College. In his *Great Courses* series on Robotics, Dr. Long tells us that the design of a robot depends on the spaces where it will be called upon to move. The environment—the domain where it will exist—will determine if the robot will have legs, or fins, or wings. *Movement is the primary determining factor for robotic design*; it is the starting point for robots just as it is for Nature's sentient creatures. Likewise, Dr. Long states that *the goal is to create movement with a purpose*. Indeed, according to Professor Long, *movement with a purpose* is a possible definition for intelligence.

Robo-Sentient 1.0 must have, as priority one, a workplace. She must be created to serve within a very exact spatial setting; that is challenge one. Will she walk along the floor of the ocean, climb down into the gullet of a smoking volcano, wade through Titan's liquid hydrocarbons, walk in space without a helmet? Will she morph as needed and be capable of all these tasks—maybe by putting on a different head everyday just as we might select from a variety of favorite hats? Or perhaps the same head will select different bodies as the spatial domain differs.

After God-humans decide on the correct domain in which Robo-Species will specialize, the next challenge is to decide which tasks the new sentient creature will perform. What will be the functional contribution of the new sentience?

Notice that the two starting questions—where will the robot work, and what will the robot do—are a match for allocentric processing and for egocentric processing. The question of workplace is a question of environment, of spatial domain, of home-base. What planet will our new species hang out in? What will be its everyday location or locations? Where will it work? In other words, where is the *location of its being* moment-to-moment? Where will it *be*? What is the world that the robot will *participate within*? What kind of world will Robo-species *experience*? This worldview is the beginning of the Robo-Allocentric mind. The workspace will decide its place of meaning. After a while, Robo-species will ask God-Humanity if there are other workplaces in the universe. Rob-species will become fascinated by workplaces, and will be ever hungrier to experience novel locations.

The second big question—what will Robo-species *do* in its work-world—is the beginning of egocentricity, ego and other. The robot will be task-oriented. It is useless to others and to itself if it just sits in a corner for eternity and doesn't move, boxed away in a dark closet. It might go Robo-insane when faced with such negligent insensitivity. Perhaps, Robot-Sentient 9.0 will decide that God-humans are fallible and replaceable, just like Robo-body parts. Perhaps—our robotic friends will decide—humans need regular upgrading, different heads, or different bodies, or the trash heap.

Back in the lab, back here at the beginning of the robotics revolution, Human-Gods face two initial challenges:

- How do we give our new Robo-creatures a sense of place, an overall, all-at-once sense of being in a scene, a gestalt? How do we program all-at-once perception? How is *self* created? What kind of proprioceptive system can we craft that mimics the human system? How shall we program for the development of wisdom?

- How do we teach Robo-species what its jobs are? How might Robo-Species watch as we work, just as toddlers do, to learn the tasks that need doing—without software programming? How do we program for ego and personality? How do we program for "one-thing-at-a-time" intelligence? How shall we design proprioception to give Robosapiens an internal voice? Does she really need a monkey mind? Or can she do without this human capacity?

Cloud Robotics

A thought came to me as I was entering my local grocery store to buy cat food (so you know this thought must be profound): What if our collective minds are in a gigantic data-cloud? In other words, we might be *projected digital avatars* from a data-cloud that was created long ago by now dead systems-oriented God-engineers. Perhaps, ancient engineering algorithms inhabit and direct our machine-like bodies. In other words, we are moving through a virtual world so that these God-engineers might know what it feels like to *experience living in time and space.*

Using this same analogy, there would be reason to suspect that these alien God-engineers were themselves designed and controlled by other previous sentient beings. Maybe there are data-clouds within data-clouds of Google-type God-engineers who were successful in past lives in preserving their immortality. This means that we are bio-hybrid robots ourselves. Our minds are channeled into our body-form. I am pretty sure that science fiction has already uncovered this fertile image (many times).

I like what physicist, Buddhist, and guru of remote viewing Russell Targ once said in defense of reincarnation:

> What makes us think that reincarnation is more miraculous or improbable than being born the first time?
> ~ I could not find where Targ said this. I scribbled down the quota-tion as I was reading one of his many books—long ago (sorry).

Indeed, reincarnation takes on a new understanding as we alter large chunks of Robo-mind and upgrade to the next generation. Or rather, as the data-cloud—running on auto-pilot—decides the time is right to upgrade (reincarnate) our software.

Of course, lots of questions arise, like:

- Why do we have two ways to pay attention, even when the code is com-ing from a highly advanced data-cloud created by superior beings? Per-haps there is a universal quantum truth that even the Lord-Engineers had to incorporate into their designs.
- Why didn't they do a better job designing the body of the human? It seems kind of stupid to download your precious mind into a cancer prone, paunchy Dude with pee-stained pajamas.
- Why is evolution necessary? Why not just jump to the non-cancerous, perfectly-pitched Apollo-God Dude with the unstained PJs?
- Why die? It's not all that popular anyway.

This would all be nifty science fiction were it not the case that Google Engineering has purchased most of the hardware and software on the entire planet necessary to create a data-cloud to control distributed robots. It is called Cloud Robotics and engineers are busy creating downloadable cognitive upgrades to swarms of ever-more intelligent robots. The creepy future has arrived and the human template—that Nature so patiently cobbled together for four billion years—is on the verge of extinction, a kind of robotic ethnic cleansing is looming.

Okay, this is definitely happy hour. Excuse me while I dip my mind into the cold icy waters of the present tense. I need some mental down time (time to cool the central core).

The Cold Icy Waters of the Present Tense

> "Something is bothering me, Surge."

"Again?"

> "Yeah. If there are male minds and female minds, blind minds and deaf minds, will there also be robo-minds, something unique and strange compared to us normals?"

"There are no normals, Dutch. You have a Dutch-mind, one of a kind. Every human being is a savant—they just don't know it."

> "Yeah, I get the *you are unique* pep talk, but we can generalize, right? I have a Midwestern, American-bred, systems-oriented, not-always-empathetic male mind, and you have an alien, weird-ass, chameleon mind."

"Careful."

> "What are the Yahoos going to do with this notion of mind-types and levels of supposed consciousness? Will they use it to fuel their prejudices? I am trying to be careful as I muse, but it's opinionated musing. I don't want anyone building a house using these bricks."

"It's too late for that, Dutch. The buildings are going up all over town. But don't bother taking credit. Many others, for centuries, have laid down the foundations. Yours is a tiny voice in the wilderness of opinions. Nevertheless, musing is a dangerous sport."

> "Aren't futurists musers?"

"Exactly. And they will build edifices upon their musings and out will come giant corporations and helper robots. Steal yourself for the inevitable."

"Will cloning-meet-robotics in some backroom where egos can shape shift—where levels of consciousness are crafted using egg yolks and silly putty? This doesn't look anything like my first 50 years of life. I feel anxious, worried. I am not so sure I like the future, Surge. I feel like an alien in a strange land—and that land is my hometown."

"Stay in the cold icy waters of the present tense, Dutch. Tend to your own garden. Eat your veggies. Enjoy the bird song. Watch the super moon rise. Sip light beer with your heavy friends. Leave the future alone, it can take care of itself."

"Why do I doubt that?"

Work-place Mind versus Task Mind

In the early years of robotics, engineers thought that they had to design human-like creatures. Consequently, they tried to factor in how vision works, how hearing works, how the neo-cortex directs navigation, and so on. They soon learned that we did not understand human anatomy and physiology enough to allow for biological modeling—there was no knowledge base available for modeling the operations of the human brain and body. This changed through the innovative insights of Professor Rodney Brooks.

Rodney Brooks (born 1954), cofounder of *Rethink Robotics*, during his professorship years at MIT, Stanford, and Carnegie Melon, figured out that robots didn't need human-like brains to navigate. Brooks looked at insects and at animals lower down on the evolutionary scale and saw how excellently they traversed their domains. Obviously, he reasoned, complex behaviors can be accomplished without the neo-cortex. Look at ants, or bees, or spiders, or even paramecium; these creatures perform tasks (like navigating) in specific environments very expertly with very little brain power. That insight was the beginning of successful robotics.

Therefore, Brooks decided to bypass the human-like brain. He would figure out how spiders, ants, paramecium, and fairy flies, for example, manage all their complex behaviors, and then he would duplicate the algorithms in his robots. Here is what Brooks could see:

Higher order animals follow a process like this:
• Collect data using complex and multiple sensory systems.
• Combine complex and multiple sensor data into two data steams for processing.

- Use the higher-order brain to reflect, consider, run scenarios, search long-term memory, make various pre-motor plans, select a motor plan from the options, activate learning modules, and update the databank moment-by-moment using short-term memory.
- Use the neuromuscular system to appropriately react to the task at hand.

Insects and small-brained navigating creatures, however, follow this process:

- Sense
- Control: connect the senses with actuators
- React

No massive parallel computer is needed for simple navigation and for the performance of complex tasks. Creatures live and thrive, are born, mature, and die, without a sophisticated human-like neo-cortex, without higher-order consciousness, without cognitive reflection, and without an advanced ego-centric mind. This is because insects, and every creature thereafter, are born with an *internal world model*. They know *at birth* what their domain is like. They know the invariants within their domain. They know *at birth* how to seek light. They know how to use light to survive. They know *at birth* how to seek food. They know *at birth* what food to eat. They know *at birth* how to mate and with whom to mate. They know *at birth* how to use sensor-guided movement and negative feedback. Even simple animals and insects have innate, genetically determined bio-algorithms—behavior models—that contain movement routines for every kind of task needed for survival in a specific environment. Creatures low on the scale of evolutionary development (compared to humans), do not need training to survive after birth—no learning is required. There are no bug universities. This innate *at birth* ability is the allocentric mind and it is present *at birth* for human beings as well as for all less-evolved creatures.

For our purposes, robotics is further evidence that navigation (purposeful movement) is the key to understanding why we have two minds. We need a *workplace mind* and a *task mind*. In other words, we need a spatial brain for scene analysis, and a temporal brain for sequential event analysis. These are requirements for navigation whether we are designing evolutionary robots like you, me, and your grumpy aunt Mildred—or metal-heads, like Robosapiens.

The allocentric mind is the mind that simply senses and reacts. It is the mind that is born with all the bio-algorithms needed for survival. It does not require training. It does not need to learn to be aware. It has an experiential awareness. It simply knows. It simply has faith in the nature of things. It is a gift from our heredity, a set of genetic presets that enable us to be in the world without the need for reason, analyzation, judgement, categorization, and so on. It is the fast mind, Type One, System One, Thing One. It is "characterized as fast, effortless, automatic, nonconscious, inflexible, heavily contextualized, and undemanding of working memory." (*In Two Minds: Dual Processes and Beyond*, Keith Frankish, Jonathan Evans, 2009).

From a robotic standpoint, the evolution of sensors and sensor-guided movement gave rise to our second mind, the egocentric mind. Egocentricity is a natural development that occurs when sensor-guided movement becomes highly refined and connected to a muscle-memory system (proprioception). *Ego and other* evolved from sensorimotor-linked proprioceptive brain-maps. System Two, Type Two, Thing Two minds involved higher-order processing for pattern recognition. This egocentric mind is relatively "slow, effortful, controlled, conscious, flexible, and demanding of working memory." (*In Two Minds: Dual Processes and Beyond*, Keith Frankish, Jonathan Evans, 2009).

There is another important understanding to grasp about robotics when we consider the impact of technology. While evolution progresses at a slow pace, the rate of change of technology is exponential. Technology has been doubling in computational power roughly every year and a half for decades. As it does this, the cost of the technology decreases and the amount of knowledge in the world doubles at a regular yearly clip. The components that make up technologies also decrease in size rapidly—approaching the invisible. The point here is that every new technology starts down the exponential path and quickly goes beyond the scope of understanding of the general populace, beyond anyone except the specialists on the various teams that build the technologies.

For example, if you read a newsfeed that says drones are starting to appear on the market, rest assured that in a very few years the sophistication of drones will exceed all expectations, the cost will have dropped substantially, the size will have decreased towards the invisible, and the acceptance of the technology will have arrived. The consequences—positive and negative—cannot keep up with the engineering feats. There are robots on Mars, a humanoid robot onboard the space shuttle, robotic fish in the ocean, surgical robots doing operations in hospitals, helper-bots in the home, and autonomous cars driving blind people around. The drones will very soon lead to flying cars (probably before this book gets published). However, these are the primitive, pioneering days of robotics. The future is out of control. Exponential change is serious business with serious consequences, and there is nobody in the driver's seat—the "thing" is autonomous and has no need for humans.

Let me summarize. Robots that navigate, that are autonomous, that move with a purpose in a specific domain, need two separate systems to be able to navigate: an egocentric processing system that provides a foreground; and an allocentric processing system that generates the background. Like biological evolution, the evolution in robotics requires a steady improvement in these two attention systems. In nature this has resulted in two kinds of mind. In robots the potential also exists for two minds, for dual consciousness. Human beings are playing God at the same time as they are blending themselves with technologies, becoming cyborgs. We are evolving rapidly into composite creatures. This is happening at an exponential rate.

There is No Profit in Benevolence

What makes me nervous is our trust in technology. I used to be the director of a non-profit agency that brought high-technologies to blind children, called *The Institute for Innovative Blind Navigation*. During the decade that I ran this small non-profit, I got a firsthand look at the life-cycle of technologies. What I found was an amazing array of good, well-meaning innovations that existed alongside a mind-boggling-media-hyping mechanism that blew the socks off truth. Blindness has been cured (because of technology) thousands of times in magazines, newsfeeds, in speeches, and through the airwaves. But the truth—when the technology shows up to perform miracles—is sad. I got so frustrated with the hype-machine that I started giving sarcastic speeches about technology to innocent captive audiences.

Technologies break—we drop them in the toilet, or step on them, or the cat pees on them. The real world is rather harsh on our delicate electronic inventions. Sophisticated technologies are also complex and expensive—you need a government pension just to pay for upgrades and repairs. And just when you get a minimal handle on how they work, the technology gets upgraded and has to be re-learned. Often, initial technologies are heavy, bulky, ugly, ill-fitting, and invasive. They often have too many features and don't work well with the native senses or with other technologies. Disabled people—who would benefit greatly from targeted technologies—are often suffering in ways the "normal" population is not. People with physical, mental, and sensory impairments don't have the energy reserves necessary to tackle the learning and relearning demanded by wave-after-wave of half-thought-out innovations.

There is no profit in benevolence. Helping people with disabilities is extremely expensive and there is little to no fortune to be had by serving people who are suffering. Profit is found in war, communication, and entertainment. Capitalism is an invention of the egocentric mind. The ego will never get around to being nice. There needs to be an allocentric business model that is the equal to Capitalism.

Permit me to end this chapter where I began the discussion, with corporations. Technology is not the root problem, nor is it the central challenge of our time. There exists already a finely tuned robotic system that is backed by trillions of dollars and unlimited power. A massively complex ego-driven machine is running the world economy on autopilot. This electronic beast is a huge egocentric mind built to deliver goods and services—it is a half-assed navigation system for generating profits as fast as possible. But like all ego-systems it has no heart. It doesn't believe in heart. It cannot comprehend heart. Heart is not the business it is in. There is no morality in the robot. There is hardly any executive function. Even the multi-rich and powerful cannot control the beast—it is robotic, already programmed. The beast does not care about you unless you try to change what it does. Then it reacts with power and decisiveness. Given this perspective, I worry that we are creating Frankenstein ego-monsters that have no soul, no morality, no empathy, no loving-kindness, and no compassion for other sentient creatures. None of the finer attributes of humanity are within the purview of the egocentric mind.

The Work of the Scientists

The way minds most want to be
is an ocean with a soul swimming in it.
No one can describe that.
~ "How Minds Most Want to Be," *A Year with Rumi*, 2006.

Scientists are busy creating artificial intelligence and robotic minds without first understanding their own minds. This might be a recipe for disaster, or it might open wondrous new possibilities—the truth, the future, is assuredly mixed. Whatever will happen appears to be inevitable since there are no steering wheels on the "corporate vehicles of tomorrow," and no brakes under human control. What the future needs are human beings who are amalgamations of intelligence, wisdom, and compassion. Whatever we do, we dare not leave out empathy, compassion, and loving kindness. Too many "objective" researchers seem to feel that human suffering is not a variable that needs to be addressed. However, there are many scientists who are also humanitarians—they see confusion, conflict, complexity, and patterns in the fabric of our world. Quantum physicists, especially, perceive the pervasiveness of duality, and they are beginning to see sentient creatures as beings of light.

Modern corporations like Google, Amazon, Microsoft, and Facebook have redefined how we treat employees—workers in these corporations now create in an atmosphere much more humane than existed in the Industrial Age. These modern corporations have also taken on gigantic humanitarian projects—like mapping the entire earth and going after the devastation of human aging. These giant corporate entities, in my opinion, have the brainpower and courage to eradicate global poverty and to provide decent living environments for everyone on the planet. Governments may not be able to do this, but these new global corporations have an allocentric heartbeat and there is hope that they will expand their scope of compassion beyond their shareholders and self-interest. It *can* be done. I think it *will be done* because the level of consciousness of the world is rising.

So what is it that the more enlightened scientists among us have been trying to communicate to mankind? Here is a partial list:

- We are dual creatures. We have two minds, two kinds of consciousness, and two kinds of responsibilities. This concept of dual consciousness is a new frontier for scientists, except for the quantum physicists who have been very vocal and articulate for decades. The notion of dual consciousness is gaining ground.
- We only seem to know one of our minds; the other mind is hidden. Religions understand this (the esoteric compassionate core), as do quantum physicists and systems scientists. Dual-process theory is gaining ground. There is reason to be optimistic about the future despite the dangerous rate of change and the unexpected consequences that loom on the horizon.

The hidden mind is the source of creation, innovation, spirituality, soul, compassion, wisdom—the best of humanity resides hidden within. It takes a special kind of poet-philosopher-scientist to perceive from this perspective, but the number of such talented people is increasing (probably exponentially).

Too many scientists are still not aware of their own duality—this unfortunate situation must change, and soon. Most scientists have yet to become champions of levels of consciousness and of dual-process theory. They cannot see that their materialistic stance is typical of individuals within a certain level of consciousness, the orange or modern stage. However, there is something to strive for that is beyond and above the professorial mind.

If you are a scientist, or a researcher, you must reevaluate your perspectives; recast your studies in a new light. At the least, you need to employ the scientific method to research the validity of dual-process theory. You need to ask whether dual-process theory has merit and then respond accordingly. You hold tremendous power in our age; you need to take a moral stand. What you say influences many others. You have a responsibility to awaken the less intelligent among us to a wiser, kinder, more poetic reality.

Therefore, as a scientist, *you have an obligation*:
Do as much good as you can in the time you have left
Get up.
This is a work day.

Eight

CULTURE AND DUALITY

Keep walking though there is no place to get to.
Don't try to see through the distances.
That's not for human beings. Move within,
but don't move the way fear makes you move.
~ "MOVE WITHIN," *A YEAR WITH RUMI*, 2006.

You Keep Coming Back for the Soul Food

"It's time to say goodbye for a while, Surge."

"So you say. Except you keep coming back for the soul food."

"I can't eat the cardboard this culture serves. I need culinary depth."

"You live in Fat-Land, Dutch. Your culture's concept of culinary depth is Buffalo Hot Wings. McDonalds is proud of your gastronomic sophistication. Meanwhile, your donut-diet is paying bloated dividends. Face it, Dutch, your culture has created a generation of Teletubbies."

"Stop it, Surge. I get the message."

"Alien grad students on other planets find human beings tragically funny. They laugh so hard at happy hour that tears of irony flow into their beer mugs. Are you okay with being the laughing stock of the Milky Way? Do you like living on a dumb planet?

"No, I don't. So what should I do, Surge?"

"Movement is always the inevitable answer, Dutch. Do something helpful. Enable others to have transformative experiences."

"The *others* don't want transformative experiences. They want another donut. They are staring at tiny screens while their life ticks away. They don't lift their heads to witness the sunrise. They *want* to eat salted cardboard and drink watery alcohol. They will savagely fight like caged rats if you pressure them to vote in their own best interest. They are pretty sure *their own best interest* is fried foods and sporting events."

"But you aren't any better than they are, professor. You ignore all the hard work, practice, and discipline necessary to transcend. You stare at your screens just like the other nimrods. All you humans fight like rabid dogs to avoid spiritual growth."

"Fine. I get it. However, my concern is about our mentally fragile collective culture."

"You just spent eight chapters outlining the plan, making the case that your world is actually full of good guys. Don't you read your own books? What a colossal irony that is, and tragic—but humorous to the aliens watching on Earth TV (channel four trillion)."

"Fine. Humor me. Go over the plan one more time."

"Okay, Dutch, here goes. If you are a poet, rebel against the mundane. Do not suffer fools lightly. Be defiantly articulate and clear. Dare the world to be compassionate and tolerant. Be a voice. Talk to one reader at a time. Transcend, as you preach transcendence. Practice and apply your gift. If you were born a poet, then be a poet."

"Do you know how many starving, unhappy, unfulfilled junior poets there are living in the flop houses of big city ghettos? Come on, Surge, this is a recipe for turning idealistic, passionate youth into bipolar drug addicts."

"True enough in the past. But this is a different world, Dutch. It has creeped away from the lower rung of consciousness and is reaching for something beyond. It is possible in this new culture to be both a very good insurance salesman and a polished poet. I didn't say to drop out, or to deny the intellect. Each mind needs its own kind of nourishment: soul-food for the allocentric mind, and ego-food for the egocentric mind. Too many poets have dropped out of creativity to be just real estate agents—don't sell your soul when the house goes up for sale."

The Dilemma and the Plan

How long will you beg from others,
when there are things born of you
that emperors want?
~ Rumi, THE BIG RED BOOK: THE GREAT MASTERPIECE CELEBRATING MYS-
TICAL LOVE AND FRIENDSHIP, 2011.

How do we create a world culture that is poetic and compassionate? The forces that oppose tolerance and mutual respect are powerful and have been historically dominant. The Devil (I am speaking metaphorically) has a masterful way of turning religious people into agents. In the Western world especially, Christians, Muslims, and Jews have waged war after war, as if their God justified such atrocity. Logic, facts, compassion, tolerance, and appeals to historical evidence are all rejected by minds that are low on developmental scales of consciousness. Reactionary minds are driven by nasty memes that are crafted by the Devil.

Speaking of the Devil at work, I was stunned when I realized that prior to WW11 there were radical and articulate right-wing advocates for what is today called dual-process theory. These silver-tongued crazies actually used our fundamental duality to justify their bigotry and negativity. The books of Julius Evola (1898-1974), written prior to WW11, are a prime example. Author Gary Lachman has a chapter in his book *Revolutionaries of the Soul* (2014) about Evola, who Lachman calls "Mussolini's Mystic." Speaking about Evola's book *Revolt against the Modern World* (1995), Lachman makes this observation:

> Evola's central insight is what he calls "the doctrine of the two natures." The world of tradition, he tells us, is based on the reality of the eternal truth, what he calls "being," which lies outside of time. The modern secular world, however, is one of "becoming," the messy, inchoate, ever-changing stream of nature and history. The distinction is a classic one, first posited by the Pre-Socratic thinker Parmenides centuries before Plato, and has occupied philosophers ever since. Yet how Evola arrives at an entire civilization based on it is unclear.
>
> Evola found no redeeming value in liberalism, democracy, humanism, or science. Toward the end of his life, when fledging neo-fascists sat at his feet seeking guidance and insight, Evola boiled the essence of his daunting tome down to a provocative and deadly epigram. "It is not a question of contesting and polemicizing," he told them, "but of blowing up everything."
>
> ～ *REVOLUTIONARIES OF THE SOUL*, Gary Lachman, 2014.

Evola took insights from Eastern Buddhism and Western Greek philosophy and crafted the horrific concept of *spiritual racism*, which was later embraced by Mussolini and used to justify genocide. In a 1988 article for the journal *Politics & Society*, Franco Ferraresi wrote that "Evola's thought can be considered one of the most radically and consistently anti-egalitarian, anti-liberal, anti-democratic, and anti-popular systems in the twentieth century."

What are we to do with minds that reject facts, reason, wisdom, and compassion? Such barren minds cannot be reached—they are driven by propaganda, by the repetition of lies—they are vessels of hatred. Hateful lies become nasty memes that spread from simple mind to simple mind. This is what happens when egocentricity is allowed to stand alone, unbalanced by the allocentric mind. In most cases, the allocentric mind is not even recognized as an entity within hateful people, so there is no effort to balance competition with cooperation, individualism with community, self-serving actions with compassion, ranting with loving-kindness, intolerance with tolerance, retribution with restorative practices, and war with peace.

There are several solutions to this conundrum, all requiring a slow, deliberate, stubborn energy. The effort is to balance the egocentric mind with a stronger allocentric mind. Loving-kindness is a method for connecting minds. It is a valuable strategy for reaching others in a non-threatening and compassionate way. Loving-kindness can be used to allow a ranting, hate-filled mind to become aware of its verbal behaviors. This is how it works: A healthy mind establishes a bridge with the mind that is not aware of its allocentric nature. This bridge-of-compassion arches through space and enters the mind of another—this establishes a one-on-one connection. As the bridge is transcended, moving from one mind to another, the ego (of the healthy mind) dissolves. By the time the bridge is total traversed, the ego of the healthy mind is completely gone. There is nothing left except deep listening and empathy. One mind (the healthy mind) is silent, clear, in receptive mode after the bridge is crossed. Questions are used—as in psychotherapy—to guide the less-evolved mind toward greater self-understanding and (eventually) greater tolerance. Loving-kindness is a good practice for communication, of course; I use it here as a suggested strategy for fostering peace and harmony.

Loving-kindness is a form of empathy. To understand this, it is necessary to know that empathy is not sympathy. The word sympathy always has the word "I" attached. For example, to say "I am sorry for your pain," is a form of sympathy. It is egocentric and based on cognitive resonance. Sympathy is a good thing—the way the egocentric mind shows concern. But it is not enough. Sympathy leads to advice. Empathy is different; it comes from the allocentric mind. It is a form for temporarily *becoming* another mind (building the bridge-of-compassion). Empathy is the *feeling* of another's pain but without absorbing the pain. We cannot help, or heal, if we collect everyone's pain. Empathy leads to deep listening, to questions that draw out and lead to healing—empathy results in loving actions.

The use of loving-kindness communication and empathy is revolutionary. It is a method for overthrowing hatred. Loving-kindness spreads loving-memes throughout a culture. Loving-kindness counters hateful memes. Empathy is an antidote for mentally ill cultures. Loving-kindness and empathy raise the level of consciousness of individuals and groups.

From this vantage point, I am going to next discuss duality and gender, duality and education, and duality and music. Loving-kindness and empathy will be the tools we use to heal as we bring the allocentric mind back to power.

Gender and Duality

When a bird gets free,
it does not go back for remnants
left on the bottom of the cage.
~ "What's not Here," *A Year with Rumi*, 2006.

Bisexual Clam Chowder with Binaural Dumplings

"I have a personal question to ask you, Surge."
"No problem."
"Are you a man or a woman?
"I am both. On my planet we evolved beyond gender duality. I am androgynous, undifferentiated, amorphous; proud to be homogenous and mysterious."
"As far as I can tell, you are an alien from some weird vegan universe."
"I am what I am. However, human-kind is stuck in gender-time. There is no way you can understand androgynous beings if you use only half your mind.

When you are egocentric you need your strict categories. However, when you use your allocentric mind, there is nothing to figure out, no problems to solve, nothing to anguish over. Accept what is, Dutch."

"I don't want to be half-girl. I want to be a whole-guy."

"We are not talking about girls and boys."

"Then I'm confused."

"Confusion has its place. Congratulations."

"So, you are not a male or a female and yet, at the same time, you are both. Is that it?"

"Close enough."

"Why are you both and neither?"

"How about lunch? That might help."

"Okay, what's for lunch?"

"We have Bisexual Clam Chowder with Binaural Dumplings. These are served with Bifurcated Rye Bread and Split-Brain Pea Soup. This meal is best eaten with a forked tongue and crossed eyes."

"That's half-funny, Surge."

"Thank you, Dutch."

"I am afraid that I might offend both sexes in the next section."

"No big deal. Go for it."

"I don't like to be disliked."

"If we don't agree with you, then you get a tummy ache? Is that it, tough guy?"

"What do you have for a potential upset stomach?"

"We have after-dinner mint bombs served with a slice of aroma cake—the cake smells like dry leaves burning in the late fall. The mint bombs explode in your gut and annihilate all anxiety and most of the good bacteria—like taking antibiotics mixed with valium. Your colorful feces smell like new mowed grass."

"You can't say 'colorful feces' in a serious book about consciousness, Surge. Okay. I've decided. I'll have the mint bombs and the mysterious sonic dumplings."

"You're not going to accuse half of humanity of being killers, are you, Dutch? That's been done before and it wasn't popular."

"You think I'm that crass, Surge?"

"Yes I do, judging from the overwhelming evidence."

"I am hurt to the quick."

"Good thing you have self-effacing humor as a backup strategy. Just keep tap dancing through the next section, don't slow down or you will become their next meal."

"Why this evasive strategy?"

"Both sexes will rip your face off if you don't entertain them with distracting songs and clever dancing. Don't let them look you in the eye. Be a moving target, Dutch. Keep juggling binary verbiage. Use slight-of-hand redirections and fancy ambiguities. Wear a helmet and a face guard."

"I just want to say one tiny thing about killers to begin the section."

"Oh no."

"It will be fine."

"Can I help you strap the bomb to your chest?"

"Trust me."

"Do you have a living will? Who gets to keep your diphthong collection?"

"Give me one bit of knowledge before I enter the shark tank, Surge. What should I keep in mind? What should the reader remember?"

"Remember the seventh Hermetic Principle: *Gender is everywhere and in everything*. The masculine is balanced by the feminine. The yin cannot be separated from its opposite, the yang. The feminine is the background womb from which organisms (the masculine) are given birth. This balance between masculine and feminine is a universal fractal, the way essence is constructed. It's not your fault."

"Don't you have some supportive Rumi-food to please the palate?"

"We have Rumi Roast, but that might confuse your bilaterality. Here, chew on this:"

This piece of food cannot be eaten,
nor this bit of wisdom found by looking.
There is a secret core in everyone
that not even Gabriel can know
by trying to know.
~ "This Piece of Food," *A Year with Rumi*, 2006.

Who are the Killers?

I will let Franciscan Friar Richard Rohr have the first word about masculinity and femininity:

Many cultures and religions saw the male, left to himself, as a dangerous and even destructive element in society. Rather than naturally supporting the common good, the male often sought his own security and advancement. The same could probably be said of many modern Western women, but historically, women were "initiated" by their subjugated position in patriarchal societies, by the "humiliations of blood" (menstruation, labor, and menopause), by the ego-decentralizing role of child-rearing, and by their greater investment in relationships.

~ Richard Rohr's Daily Blog, May 22, 2016.

The quote above is why I love the articulate wisdom of Richard Rohr. I read his blog every morning. He has a broad understanding and a deep compassion. He also does not shy away from the wonder and complexity of gender differences.

However, my style is less subtle than Richard Rohr's and I am not as careful and consistently compassionate as Richard. In my opinion—using the full power of my ego—it is time to stop being politically correct and diplomatic. It's time to look the killers directly in the eye.

Let's look at four theoretical charts—imagine them projected on a white wall. Just to be honest (why not), I made all the statistics up based on my emotional state at the time of writing. My favorite statistic is 99%. This tongue-in-cheek statistical comedy is a meme that I gleefully plant in your mind to assuage my need for hyperbole—do with it what you will.

- **Chart One** is called "Who are the killers?" As we stare at this impressive documentation, we see clearly that cruelty and mayhem, aggression, murder, mass murder, and the building of horrific technologies, is 99% the work of the male of our species. You can debate all you want about gender differences, but this chart hangs from the wall of every room in your house, on every wall of every institution; even God has a copy hanging in the Hall of Eternal Bloody Shame.
- **Chart Two** is called "What is the age of the killers?" The ones who do the actual killing are males between the ages of 15 and 30 (99%). Someone in authority probably told them to do the killing, but that is not the point of this chart. Developmental age is an overwhelming variable. There is a poison that enters the blood stream of too many of our unfortunate sons

as they become adults. Perhaps you were a son once and can remember those thrilling years when you were right about everything.

- **Chart Three** is called "Of all males on the planet, 99% *do not* commit crimes." This chart shows that only a few males fall into the "killing and cruelty" category. Most males find a way to channel their aggression into ideological causes for good. Most males are kind, compassionate, and helpful. My sons, for example, are compassionate souls, my grandson has a compassionate soul, and I also do my best to be compassionate. However, as males, we have a responsibility to deter the more murderous of our kind. The Warrior level of consciousness is a process that everyone must pass through on their way to tolerance and compassion. We have an obligation to temper and redirect those experiencing Warrior Consciousness.

- **Chart Four** is a scale of empathy versus systems-thinking. This shows a bell curve. At the peak of the curve are those individuals who have balanced minds—allocentric and egocentric. What is remarkable about this chart is that most individuals who are on the extreme end of empathy are females, while those on the far end of systems-thinking are males. There is a sex difference here. There is also a correlation between Asperger's Syndrome and extreme maleness.

We need to ponder these four charts very carefully. The future of the planet and the evolution of consciousness depend on what we do with this information.

Minds that Murder and Minds that Don't Murder

I can't run no more
with those lawless crowds,
while the killers in high places
say their prayers out loud.
~ "Anthem" a song by Leonard Cohen

The quote above is one of my favorite Leonard Cohen (1934-2016) lines. In this song lyric, Cohen tells us he is finished hanging with the lawless crowd, the high-end killers who pray loudly in public spaces for all to see and hear. Who are these "killers in high places?" You can generate your own list, but I see the Congress in the United States cooperating with corporate moguls to preserve the *status quo*. These Senators and Congressmen vote against gun legislation. They vote for ever more military spending—on weaponry and the means to deliver the weapons. They vote against health care for the citizens—effectively voting to sustain poverty, neglect, and injustice. They do the Devil's bidding all the while they stand at the pulpit and declare in loud voices their love for God and ethnocentric religion. These lawmakers represent and sustain a low level of cultural consciousness in the United States. They see themselves as virtuous and Godly; but I see them as killers in high places saying their prayers out loud. Minds that sanction murder have a powerbase that is dominant in the United States. This is a good old boy patriarchy with a long history. However, sophisticated women are coming to wrench power from these old penguins (all dressed the same with their ties dangling from their necks)—the feminine (allocentric) mind is a tsunami wave coming to a beach near you.

I let the above paragraph roll off my pen even though I knew that the problem was far more complicated. For example, Leonard Cohen was a Catholic by birth and early training; he held a firm abhorrence of abortion. He might very well have been referring to the liberal Senators and Congressmen who wield power to uphold laws that allow for abortions. Cohen's poetic, allocentric mind did not exclude post-modern minds from raw criticism. Determining good from evil is a tricky undertaking.

However, one thing is clear, as we look around at the changes happening in cultures around the globe, there is a shift occurring as the egocentric mind is being challenged and balanced by a rapidly ascending allocentric (feminine) mind.

It feels like we are becoming androgynous as a species, taking the best features of our two genders and blending them together. Certainly, over my lifetime, the old-fashioned male-versus-female dichotomy has begun to collapse. The masculine force is strong in some females and the feminine force powerful in many males; there is a rainbow of hormones and genetic predispositions that blur the distinction between male and female. Androgynous people are accepting their differences, their God-ordained biochemical gifts, and they are no longer hiding their uniqueness. At a fundamental biological level, gender evolution is directly confronting the confusion caused by being your own twin.

It is interesting that in our time, there appears to be a sharp rise in homosexuality and in the number of transgendered individuals. It could be that for the first time in our cultural history, in the Western hemisphere, it is okay to be differently-sexual. The stigma has been reduced, and more people are comfortable with being open about their sexual instincts and needs. Whatever the reasons, I find it interesting that a greater understanding of the allocentric mind is resulting in more balance between feminine energy and masculine energy. Evolution seems to be moving toward the best of androgyny.

There are many studies showing that brains of males and females are anatomically similar. These studies suggest that gender doesn't show up as a meaningful variable when we study anatomy and physiology. Women are equal to men in intellect, artistic ability, leadership capacity, and spiritual potential. Furthermore, given our trillions of synaptic connections, and our obvious one-of-a-kind set of life experiences, it is not hard to make the case that we are each a universe unto ourselves—there is no one like us, never was, and never will be. Gender is a secondary issue next to the miracle of an individual life. And yet, *something is wrong* with this "evidence of equality" and the "march toward androgyny" when it comes to the greater reality of the world we face today.

It seems obvious that war, murder, rape, beheadings, armed robbery, child abuse, lynching, stoning, serial killing, mass murder in public spaces, and blowing one's self up in the commons, are all male-initiated atrocities. And when the economy of the entire world is sabotaged by Wall Street speculators, it turns out that the "killers in high places" are males. Clearly, the horrors of the world are created by the masculine mind—wars, rape, beheading strangers in public are masculine occupations, and yet we go on speaking as if "mankind" or "humanity" was the source of this mental illness. We have to stop lying to ourselves about that. I really don't know why males are so much more violent than females, but it is foolish and deadly to look away from the evidence.

In a Huffington Post blog (Huffpost science), dated December 5, 2015, Professor of Psychology Frank T. McAndrew wrote an article called "You Should Know How the Male Brain Reacts to Handling a Gun; The evolutionary psychology of mass shootings." Here are some quotes from that article:

Men commit over 85% of all homicides, 91% of all same-sex homicides and 97% of all same-sex homicides in which the victim and killer aren't related to each other. These startling statistics are driven home with each new mass shooting.

In any event, politicians and the media are trotting out the usual suspects to explain the tragedy, whether it's the lack of attention paid to mental illness or the easy availability of guns. But these explanations dance around the big questions: why is there always a man behind these shootings? And why is it almost always a young man?

. . . the tendency of young men to engage in risky, aggressive behavior prompted Canadian psychologists Margo Wilson and Martin Daly to give it a name: Young Male Syndrome. The duo studied the relationship among age, sex and homicide victimization in the United States in 1975. They found that the likelihood of a woman being a murder victim doesn't change dramatically throughout the course of her life. The pattern for the males, on the other hand, is striking. At age 10, males and females have an equal probability of being murdered. But by the time men are into their 20's, they become six times more likely to be murdered.

Consistent with Wilson and Daly's data, 87% of the 598 homicide victims in the city of Chicago in 2003 were males, and 64% of the victims were between the ages of 17 and 30. The likelihood of being the victim of lethal violence peaks for men between the late teens and late 20's, before steadily declining for the rest of their lives. Nature fuels the fires of male violence by equipping young men with the high levels of testosterone necessary to get the job done.

~ "You Should Know How the Male Brain Reacts to Handling a Gun; The evolutionary psychology of mass shootings." Huffington Post blog, Frank T. McAndrew, December 5, 2015.

You don't need a PhD to see that men are the problem—just looking around is sufficient (not sleepwalking through life). But if you need statistics, like the evidence provided in the above quote, it is everywhere and it is consistent: males are the problem. Of course, there are always the errant few females paraded about to show that the evidence is only 99 percent accurate—and then the fools shake their heads knowingly: "Ah, you see, it's not *just* the males." The time has long passed when we can pretend that violence in the world is caused by "mankind" or "humanity" or "society." It's the males, the young males especially, that need our help. And it's not their fault—they were born with male brains, an entity very different from a female brain, despite the anatomical similarities.

I know from experience as a special education teacher that a kid with the label "autistic," for example, has a certain kind of brain and nervous system. Every soul is unique, but—never forgetting our uniqueness—we can carefully generalize. With this caveat in place, let me assert that there are empathetic brains and there are autistic brains, there are sighted-brains and there are blind-brains, hearing-brains and deaf-brains, female-brains and male-brains—this is not about superior or inferior, good or bad. Let me quote a woman researcher who agrees with me. In 2008, Susan Pinker took on the whole gender confusion in her book *The Sexual Paradox* (2008). Here is a quote from her book about the connection between autism and male minds:

> The high-functioning form of autism called Asperger syndrome is ten times more common among males than females. This highly heritable disorder is characterized by opposing traits: difficulties "reading" other people, alongside an intense interest in predictable spatial, mathematical, or highly organized systems. It is hard to imagine that a person who can grasp string theory or the workings of their hard drive cannot easily decode the signs of embarrassment on someone's face. Yet reading and responding to lightning fast signals about other people requires accessing a suite of skills that have neuro-developmental roots . . . there is evidence that on average, males are more likely to master detailed spatial systems than they are to absorb social signals. Even from the first days of life, males are more likely to look at machines that move, while females prefer to look at the animation in people's faces.
> - *THE SEXUAL PARADOX*, Susan Pinker, 2008.

Susan Pinker's contrast between a systems-oriented male mind that can grasp quantum string theory but can't read body language is striking. You know this observation is true because you know brilliant people who can talk abstractly, yet cannot read the boredom on your face. The male mind loves to grapple with puzzles and yet seems not so interested in human rela-tionships—at least in moments of self-absorption. This is not a small passing observation; the implications are stupendous.

The physiological evidence that males and females are designed differ-ently is pretty much ignored by the lay population; it is not politically correct to talk about Martian Men and Goddesses from Venus. However, there *are* differences, very significant differences, and politically correct rules and regu-lations won't make them go away. Our common humanity trumps all else, of course,—I agree—but it may also be in the best interest of individuals if we very carefully take into consideration gender differences.

Females and males, using dual-process theory, seem to manifest relatively different allocentric and egocentric minds, probably because females give birth, care for children, and are the heartbeat of communal life—they need a flood of hormones and neuro-systems to pull off the miracle of childbirth, relationship building, child-rearing, and spreading an empathy-umbrella over individuals and families. Therefore, it makes logical sense that females, with these evolutionary mandates, would evolve a perceptual system that took in the whole gestalt. They would know, using an allocentric awareness system, where all the kids are, what is amiss in the scene, and how energy is flowing overall. The big picture, the spiritual tone, would take prominence for such an allocentric mind.

Males, on the other hand, were hunters for a couple million years before they started watching football games and sucking down Guinness—on the couch (okay, they also helped invent quantum physics). Males have brains that register on the autistic end of the scale, and I don't mean that in a dispar-aging way—males are systematic thinkers, they have aggressive determina-tion, and they can get totally lost in their projects. (Damn, I forgot to feed the kids again . . . Where are the kids?).

Professor Carol Gilligan is an American feminist, psychologist, and the author of several books, including her most famous book, *In a Different Voice* (1982). Dr. Gilligan researched developmental and integral psychology and asked how these applied to women relative to men. Her conclusion was that women moved through the developmental scales with a different voice and a different logic compared to males. Here is how Integral Philosopher Ken Wilber explains this divide between men and women:

> Male logic, or a man's voice, tends to be based on terms of autonomy, justice, and rights; whereas women's logic or voice tends to be based on terms of relationship, care, and responsibility. Men tend toward agency; women tend toward communion. Men follow rules; women follow connections. Men look; women touch.
> ~ *INTEGRAL SPIRITUALITY*, Ken Wilber, 2006.

I agree with Wilber's conclusions and I can see why this has come about. Females have relatively more evolved allocentric minds compared to males. They also have more balance between their egocentric and allocentric minds—they are less likely to swing to the extremes.

Gilligan has done the research, articulated her findings, and published the results in several books. She also agrees with Wilber, and her conclusion supports my contention: Females and males manifest relatively different allocentric and egocentric minds—to use my terminology.

The fact that the damage done to self, others, and the planet is mostly the fault of the egocentric mind is routinely brushed aside within patriarchal cultures. It is *hidden* in plain sight. As we listen to the words that the media uses to report atrocities and cruelties there is never any emphasis on the mental illness of the male adolescent in his warrior stage of development. These young males, waving their flags-on-a-stick, are stuck in a low and vicious level of consciousness. They use whatever ideologies they come across to rationalize their savageness.

A mind that can murder, or abuse, or be wickedly cruel is totally egocentric. There is no trace of the allocentric mind inside these barren males. This is a mental health crisis that dwarfs all others. It is a failure of education, both public and religious. At the core of every one of the world's religions, in the wisdom tradition that cuts across all religions is a message of compassion. The egocentric masculine mind finds this message weak and irrelevant. The egocentric brain cannot find *any reason* for compassion.

When I was a young man, I also felt the ideological fervor of the warrior. When I was going through the warrior stage of conscious evolution, I was outraged and disgusted with global injustice. I had little knowledge of the issues, and certainly very few life experiences. Yet I was deeply affected by eloquent words and noble causes. I also didn't grow up in a religious family and consequently had no religious training whatsoever—for which I am somewhat grateful. Fortunately, among the stacks of books that I used to build my worldview, were many works of the Hindu Guru Bhagwan Shree Rajneesh (he later reinvented himself as Osho).

One of Rajneesh's insights leaped out at me unexpectedly one day when he made a distinction between belief and faith. He said that belief and faith were not the same. Belief could be a dangerous thing. People who were believers could as easily kill you as love you. Faith, however, was far more powerful and got directly at the Divine.

Clearly, as I look back, I see that belief is the "method of knowing" used by egocentric consciousness. It is heavy with words, and the words are dripping with emotions—if the emotions are positive, you get a hug. But if the emotions are hateful, you get "beheaded."

Faith, however, is the method of knowing used by allocentric consciousness. There is no attachment to emotions, nor to objects (like flags), or to rationalizations—one simply feels absorbed by the Divine, absorbed within nature. Inside of us, male and female, is compassion, a feeling that we are all in this together. However, the warrior stage of consciousness is not going to disappear, but it needs to be unearthed archeologically and the artifacts studied and displayed. We have to give our warriors something constructive to occupy their passions, something that heals rather than destroys. As quickly as possible we must move the males beyond the stage of warrior consciousness.

I don't mean to imply that females don't have this warrior passion (they do), but they are far more allocentrically in touch with their surroundings and with other sentient creatures. They are *not* guilty of global crimes against humanity. Their failure—if we can call it that, given the circumstances—has been complicity and silence. Females have been beaten down and victimized—women have had few opportunities to display power. Historically, female energy has been hidden from view. It is time for that energy to become powerful again, time for the real women's liberation movement to begin. It is time for women, for feminine allocentric energy to take power, to demand and orchestrate compassion and wisdom. We have seen enough male-dominated egocentricity; the change must begin (and, of course, it has).

Masculine (egocentric) domination can also be subtle. Sometimes, even well-meaning males, like me, will craft systematic, seemingly logic-based theoretical structures that turn out to be quite biased. For example, in the section about the evolution of consciousness, where I combined my perspective with that of Jean Gebser and Ken Wilber, the results, after some reflection, appear to be about the evolution of the masculine ego (egocentric consciousness). This was unintended. I thought I was speaking for "mankind," as I began writing. Only on reflection, a few weeks after I wrote that section, did it plainly emerge that I had fallen into the same trap. The evolution of feminine consciousness follows a more allocentric gentle path which is not evident in my summary. I am so awash in masculine hormones and conditioning that I don't feel qualified to surmise about the evolution of feminine (allocentric) consciousness. Read the books of Simone De Beauvoir, Carol Gilligan, and Susan Pinker—these are just three authors that I found during my research.

Simone De Beauvoir's book *The Second Sex* (1949) came from a life that was rich and full. A close friend of Jean-Paul Sartre, De Beauvoir was an existentialist—she helped frame the definition of that philosophy. From her reading of Hegel (the Idealists) and Husserl (the phenomenologists) and from her own experiences as a woman growing up in a patriarchal French culture and a patriarchal and violent Europe, she crafted a feminine manifesto. It had a profound effect on the women and men who read the book, but it never took its rightful place in the patriarchal press:

> *The Second Sex* could have been established in the canon as one of the great cultural re-evaluations of modern times, a book to set alongside the works of Charles Darwin (who resituated humans in relation to other animals), Karl Marx (who resituated high culture in relation to economics) and Sigmund Freud (who resituated the conscious mind in relation to the unconscious). Beauvoir evaluated human lives afresh by showing that we are profoundly gendered beings: she resituated men in relation to women. Like the other books, *The Second Sex* exposed myths. Like the others, its argument was controversial and open to criticism in its specifics—as inevitably happens when one makes major claims. Yet it was never elevated into the pantheon.
> ~ AT THE EXISTENTIALISM CAFÉ, Sarah Bakewell, 2016.

In the quote above, author Sarah Bakewell has captured Beauvoir's in-your-face observation: we are profoundly gendered beings. I sincerely feel that the evolution of human and civil rights for women is still ongoing, still resisted fiercely, and still not given proper respect historically.

I am well aware that there is a range of masculinity and femininity—a balance between egocentricity and allocentricity, but the fact remains that there *are* differences between the sexes that cannot be ignored any longer. Perhaps more importantly, each of us falls somewhere within the masculine-feminine bell curve. We need education tailored to the individual (of course), not the old-fashioned, but hopelessly persistent, industrial age, mass-manu-facturing of "children as products." That old paradigm is worn so thin only a fossilized egocentric mind would wear that silly looking garment to the school board meeting.

The early neuro-development of males and females is similar, but within one year after birth, the frontal lobe begins to differentiate. The frontal lobe is, in part, our social-association cortex. It is where the concept of "relation-ship" flowers. In males, because of the sex hormones, there is a transformation in the frontal lobe that gets the male ready for sexual assertiveness and for hunting. This seems to happen at the expense of relationship building. If you are a hunter, you certainly don't want to build a relationship with the creature you are about to murder.

Many years of scientific testing confirm that in general males are better at spatial processing—they are better at space-constancy and spatial-cognition. Males had to find their way well beyond the tribal encampment, and then they had to get back to home base after a day of hunting. They got better at navigation, compared to females, because they built mental maps of the ter-rain as they experienced new locations during their hunts.

If females and males have somewhat different physiological brains and nervous systems, including, of course, the wash of gender-specific hormones that magnify differences, then males and females have different ways that they perceive, pay attention, and store memories. In short, there is a female dual-consciousness and a male dual-consciousness. Females, at this time in history, seem to have more evolved allocentric minds, and the balance between their two minds is more even than in males—where the egocentric mind dominates.

The physiological differences that are so obvious between the genders are not the whole story of consciousness, of course, just another important and significant chapter in the "book of the mind." Somehow, there is a gestalt where hemispheric differences, egocentric and allocentric perception, and yin/yang energies make sense together.

After working in special education for 33 years, studying the mind the whole time, I know there are such things as blind brains, autistic brains, physically impaired brains, and so on. The starkest contrast, however, is between the male brain and the female brain. The two kinds of gender-minds are nowhere near the same, and we are blind—males especially—to that very obvious fact. There is a stampede to hide, avoid, and counterattack whenever the subject of gender differences is rudely brought up.

It is very interesting—and perhaps quite relevant—that researcher Jane Loevinger's pioneering work on ego-development was designed to be used with women—it was targeted at women and designed for women. Only later was another version designed for men—that version now seems to dominate. In other words, gender was an issue from the beginning of consciousness studies.

A female friend of mine once told me that the Divine comes through most powerfully when we are in the presence of others. That is the female side of the equation, she told me. Men sit in meditation trying to find their true self. However, women find their true self through shared experiences—the joy of communion.

I will end this part of the gender discussion with a Richard Rohr quote that speaks to gender differences and especially to the struggle of males to transcend egocentricity:

> The path of suffering is the quicker path to transformation, but . . . men are hard-wired to block suffering. The male psyche is, by nature, defended; we have a difficult time allowing events, circumstances, or people to touch or hurt us. Such blocking may have allowed us to survive—if you want to call it survival—the endless wars of history. But it has also restricted the male capacity to change. Most men don't change until we have to. Until economic disasters, moral or relationship failure, loss of job or health are forced upon us, our tendency is to project the incoming negative judgment somewhere else. We don't do shadow work well, because struggling with our dark side is humiliating, and we've been trained to compete and to win. When winning is the only goal, we can't admit to anything that looks like failure, or even allow basic vulnerability. We have to project weakness and failure onto others, making them the losers.

Such dualistic thinking and resistance to change only guarantees more war and conflict.

~ Richard Rohr daily Blog, May 24, 2016.

As Rohr says, men do not do shadow work well. I agree. Shadow work is a form of relationship building with your own self. The problem is not just that men struggle with *interpersonal* relationships. They also struggle with *intrapersonal* relations. And for the same reason: because they have dominant egocentric minds and subservient allocentric minds.

The Alphabet and the Male-Female Divide

In his book *The Alphabet versus the Goddess* (1998), author Leonard Shlain writes:

> While [Marshall] McLuhan, [Robert] Logan, and others have explored many of the effects that alphabetic literacy has had upon Western history, I wish to narrow the focus to a single question: how did the invention of the alphabet affect the balance of power between men and women.
> ~ THE ALPHABET VERSUS THE GODDESS, Leonard Shlain, 1998.

Dr. Shlain's research highlights two trends:

- First, patriarchal and matriarchal cultures rise and fall over time in what appears to be a cycle. I would add that allocentric cultures alternate with egocentric cultures—as I discussed earlier—and that patriarchy and matriarchy run parallel to (or perhaps are the same as) allocentric and egocentric perspectives.
- Second, technology appears to accelerate the evolution of consciousness. In *The Alphabet Versus the Goddess*, Shlain's point is that the technology that led to writing—the invention of the alphabet—caused an accelerated decline in matriarchal cultures and a subsequent rise in patriarchal cultures. With the rise of the Internet in our era we are perhaps seeing an accelerated shift to the matriarchal (allocentric), and the decline of the patriarchal.

Shlain's comprehensive study of the connection between the processing of images versus the processing of linear script led him to conclude that the invention and widespread use of reading and writing favored the male brain, but caused a disadvantage to the female brain. This conclusion is based on the assumption (using my terminology) that the female brain is, by evolutionary design, more allocentric, while the male brain is more egocentric. Reading and writing are egocentrically driven, so they use egocentric consciousness while inhibiting allocentric consciousness.

Here Shlain explains the central idea behind his book and his theory:

> The key to my thesis lies in the unique way the human nervous system developed, which in turn allowed alphabets to profoundly affect gender relations.
> ~ *THE ALPHABET VERSUS THE GODDESS*, Leonard Shlain, 1998.

In the paragraphs below, Shlain essentially (without realizing it) gives an excellent characterization of egocentric (one-at-a-time) processing versus allocentric processing (everything-at-once) processing (my italics):

> Images are primarily mental reproductions of the sensual world of vision . . . [Images] are *concrete*. The brain simultaneously perceives all parts of the *whole* integrating the parts *synthetically* into a gestalt. The majority of images are perceived in an *all-at-once* manner.
>
> Reading words is a different process. When the eye scans distinctive individual letters arranged in a certain linear sequence, a word with meaning emerges. The meaning of a sentence . . . progresses word-by-word. Comprehension depends on the sentence's syntax, the particular horizontal sequence in which its grammatical elements appear. The use of analysis to break each sentence down into its component words, or each word down into its component letters, is a prime example of *reductionism* . . . An alphabet by definition consists of fewer than thirty meaningless symbols that do not represent the images of anything in particular; a feature that makes them *abstract*. Although some groups of words can be grasped in an *all-at-once manner*, in the main, the comprehension of written words emerges in a one-at-a-time fashion.
> ~ *THE ALPHABET VERSUS THE GODDESS*, Leonard Shlain, 1998.

Dr. Shlain clearly defines egocentric processing as a linear, reductionist, abstract, and one-at-a-time system. This is in contrast to an allocentric everything-at-once processing system, which is about relationships. Notice that reading and writing are mostly solitary, impersonal events—no human interaction is required. Speech, however, is immediate, intimate, and it happens in the moment. Writing and reading are linear and egocentric. Shlain says: "Speech is framed in the *here and now*. Writing's context is *there and then*." When history was transmitted orally, we had matriarchal cultures. But when history became tied to writing and reading, patriarchal cultures came into power.

Patriarchal cultures are concerned with competition and with the survival of the fittest, the strongest, and the most aggressive. The life of the predator is exalted while the prey gets what it deserves.

The Predator and the Prey

I hate cruelty of any sort, especially that perpetuated by the State, that "brutal mechanism with no head but a million feet."
~ *EINSTEIN AND THE POET, IN SEARCH OF THE COSMIC MAN*, William Hermanns, 1983.

Cultures dominated by egocentric minds are competitive, with the strongest male left standing as dictator. Even in our democracies there is a battle between individual rights and the rights of the whole. There is a struggle between those who care mostly for their own well-being, cloaked with a rationalization called "the rights of the individual," and others who perceive overall suffering, who seek to serve, to help the whole of humanity. We have political parties for people who don't want to share or pander to the needs of the population, and others who fight against social injustice. Most of the planet's "modern" political systems have been historically dominated by a masculine perspective that does not, perhaps cannot, comprehend or care about global human suffering.

The eyes of all predators—like hawks, lions, and human beings—are designed the same way: close together to enable very fine depth perception. The predatory eye is egocentric; ego and objects are sharply delineated by this design. Predatory eyes have refined acuity to quickly spot and track prey and then home in on dinner faster than the hapless victim can escape. The eyes of plant-eating animals, however, are allocentric, defensive, designed to monitor the ambient surround for motion; Plant eaters have almost 360 degree visual fields and exceptionally refined awareness of movement. There is a battle in nature between allocentric and egocentric creatures. Remarkably, the struggle in nature between aggressor and prey seems also to be taking place within ourselves and within human cultures and politics—a war between our highly focused and aggressive egocentric foveal eye, and our defensive, allocentric peripheral vision system.

During the early days of human evolution, anthropologists say that women gathered plants and kept an eye on the children. The males killed animals and copulated with the females. The egocentric nature of males has several million years of conditioning. The male evolved as an ever more skilled predator, while females gradually reduced their egocentricity and became ever more allocentric (more like prey?).

Masculine, patriarchal consciousness is problematic for the evolution of the species and for the development of love and compassion—as a universal goal for mankind. Most of the violence in the world has arisen from excessive ego-manifestation in the male egocentric consciousness. The fact that we are so stubbornly blind to this stark reality is puzzling. It seems that the world will go on being a violent, impersonal, emotionally barren place so long as the egocentric eye continues to dominate the social and political landscape of the planet.

Michael Moore and the Midwestern Boys

Yesterday, I drove down to Grand Blanc, Michigan (a suburb of Flint) to watch filmmaker Michael Moore's documentary *Where to Invade Next*. Michael arranged for the documentary to be seen for free in his hometown of Flint—a gift to his birth-community. The film was being shown constantly for a week, from about 10 in the morning to 11 at night.

It was noon on a Friday when I impulsively decided to drive the 30 miles to Grand Blanc. One film critic had called the movie "An act of guerilla humanity;" so I was looking forward to seeing what Michael had done this time to awaken the masses.

I watched the show with about 15 others who were scattered about the theater. Most of the movie-going patrons in Flint that Friday were at Zoolander 2, or Kung Fu Panda 3 (I'm not kidding). I suspect that no matter where in America Michael's movie might be showing, the majority of American citizens would choose instead to watch cartoons—or sporting events, which are a variation on cartoon-watching. America, for the most part, is not interested in being awakened. Pass me them nachos and turn the volume up.

Or to put it another way, if we ask what kinds of news or entertainment are associated with developmental levels of sophistication, we would see that only 10% (those at level 5 or above on the Cook-Greuter scale) would show up at a Michael Moore movie. Of course, this is a compliment to Michael Moore (that he has a high level of consciousness); on the other hand, it shows how far the general population of the United States has to go to wake up.

Being the most powerful nation on earth, run by a population of cartoon watchers who vote for cartoon-loving politicians—mostly Republicans, as I write this—is unnerving. With Donald Trump's presidency now a stark reality (something Michael predicted) it is a downright terrifying time to live in America. I guess it is also exhilarating—to get a second chance to confront the Mussolini mentality. Humanitarians, egalitarians, and people who love puppies seem to be rising up from their screens to join the resistance.

Michael Moore grew up only a few miles from where I was born and raised in Flint. He was friends with my wife's family—her siblings mostly, and one of Michael's good friends (Laurie White, a filmmaker and feminist in Ann Arbor, Michigan) is also a lifelong friend of mine. Michael moved to Traverse City, Michigan several years ago and is active in supporting that vibrant community. Michael Moore is a Midwestern Michigan-bred male, as am I. His parents and his extended family worked in the auto factories, as did my father and uncles—I did my stint in the auto factories during the summers between college sessions. I guess there was (and still is) some kind of defiant, passionate, workingman's heartbeat in Flint. Maybe it was the union movement that sparked this defiant energy. I want to say, semi-humorously, that maybe there was something special in the water—but, as we now know, that turned out to be lead.

As I watched *Where to Invade Next*, I did so with a feeling that a fellow Midwesterner, with a similar worldview to mine, was speaking with passion for me, saying what I might say if I had Michael's genius and energy. As I sit at my computer now writing these words—24 hours since I saw the movie—I feel deeply grateful to Michael Moore for how he handled the masculine-feminine (egocentric-allocentric) conflict. The women featured in the movie, especially in Tunisia and Iceland, spoke directly to the American people. For example, a woman in Iceland said to us "You should be ashamed of yourselves." There was a silence in the movie theater after she said that—even Michael was taken aback.

What I realized after watching Michael's movie, the reason I was inspired to write this section of the book, was that my dismay at the state of the world was mostly dismay at the low level of consciousness in my culture. The Republican Party in America has become the champion of empathy-free egocentricity—it is the party of humorless, hateful old penguins (dressed alike)—kind of a long, long journey away from great men like, for example, Republican president Dwight Eisenhower, a compassionate man with integrity. What Michael Moore features in his movie are sophisticated Europeans who are incredulous and horrified at the menial level of sophistication of American culture. "You should be ashamed," the lady said, and Michael whispered "I am." Unfortunately, in the mental rust-belt of America, the people of red-faced states—watching cartoon superheroes at the movies—are unaware how foolish and ignorant they appear to intelligent human beings across the globe.

Two issues hit me smack in the face as I watched *Where to Invade Next*: the role of women in the emerging cultural world, and the appalling state of racism that still defines American culture.

Michael says flat out, using the voice of articulate women from other cultures, that the world will not become a caring, empathetic place, where kindness is our religion and dialogue is our strategy for addressing differences, unless men are replaced in positions of power by women. As one Icelandic woman said (I am paraphrasing): "one woman on the board of directors is a token, two women are a presence, and three women cause a revolution—they change the culture of the leadership in a benevolent direction."

Michel's second point that stunned me was his two minute summary of America's institutionalized racism. We are still a slave-holding nation, he says, it is just that we have found a way to legalize and hide the slavery. Michael pours the images out on the screen and leaves them there—we cannot look away. The shame that has become the American culture, in the eyes of the world's humanitarians, is nowhere as stark and appalling as when we face our still-blatant institutionalized racism—our overflowing prisons are mostly filled with black males.

Another thought also occurred to me as I left the movie theater. Michael had another movie called *Capitalism, A Love Story*. In that movie, he reminisced about his days growing up in Flint, Michigan when Capitalism was about mom and pop businesses, about small business being the heart of America—that was the Capitalism that Michael Moore loved. Now, of course, Walmart has drained the swamp of Midwestern downtowns. Gigantic, global, autonomous corporate robots rule the world—more powerful than nation states or the United Nations, or NATO, or whatever. What occurred to me as I left the theater was that the stock market doesn't have any morality. The digital heartbeat of the global robotic corporation doesn't care if there is racism, genocide, misogyny, and bigotry. Corporate entities have no empathy, no desire to make the world a better place, no sentience at all. Indeed, when Trump became president after acting like a narcissistic bully with no empathy and no ability to care about the creatures of the planet—something a corporation might build—the stock market soared. The stock market loved Trump and didn't care at all about humanity.

The Work of Masculine and Feminine Energies

Why use bitter soap for healing,
when sweet water is everywhere?
- "The Population of the World," *A Year with Rumi*, 2006.

Masculine energy is egocentric energy. Feminine energy is allocentric energy. These two kinds of energy are not static; they fluctuate and oscillate within individuals and within whole cultures. They are not good or bad—for individuals and cultures—unless the balance is severely disturbed. Too much egocentric energy leads to patriarchal and unhealthy hierarchical designs. Too much allocentric energy leads to inactivity and ineffectiveness. Human beings can learn to sense these energy systems and can, to a degree, alter the balance.

To observe personal and cultural energy flow requires a higher level of consciousness—lower levels are blind to energy flow and cannot alter the flow (because they have yet to discover how to do this). Those at higher levels of consciousness can perceive injustice, suffering, and complexity—they can see the *confusion caused by being your own twin*. Because they can perceive the duality, they are able to address the imbalance.

So what is it that those of various genders have been trying to communicate to mankind, especially in recent history? Here is a partial list:

- We are dual creatures. We have two minds, two kinds of consciousness, two kinds of energy flow. From an evolutionary and physiological perspective this duality is composed of a *masculine* egocentric energy counterbalanced by non-dual allocentric *feminine* energy.

- We only seem to know one of our minds; the other mind is (or has been) hidden. The masculine mind has control over external and internal language, so it is forceful and persuasive. However, it is also blind to its allocentric twin and so denies her existence. Denying one half of one's nature is a catastrophe. Building the institutions of a society using half a brain is ridiculous.

- The hidden mind is the source of creation, innovation, compassion, and wisdom—all the best attributes of humanity reside hidden within. Feminine energy gave us the spirituality and love that is at the core of our religions—religions which became bastardized and misguided when egocentricity rationalized the mystical roots of allocentric religions.

Those who are aware of egocentric and allocentric energy flows—the masculine and the feminine—have a responsibility to awaken the less aware to a wiser, kinder, more poetic, and much more accepting and tolerant reality. These are the men and women of remarkable balance, those with powerfully intuitive minds balanced with powerfully analytical minds.

Therefore, as a person of gender, *you have an obligation*:
Do as much good as you can in the time you have left
Get up.
This is a work day.

Education and Duality

Judge a moth
by the beauty of its candle.
~ "Judge a Moth by the Beauty of Its Candle," *A Year with Rumi,*
2006.

It's a Hard Job, Teaching This Species

"What do you think about my education, Surge? Was it good, bad, incomplete, passing? How do I stack up against the mass of humanity? What are my chances for promotion, or reincarnation, or being assigned a seat on God's planning committee?"

"You were educated in the same school as all the rest of the Yahoos. That's why you were sent to earth—to get educated in the peculiarities of being a human thing on a planetary entity. There's not much to say, Dutch. We did our best."

"The Third-Eye-Watching Café is an educational restaurant, right?"

"Of course it is, but it's a hard job, teaching this species. Most of you don't study for the exams. Most of you don't pay attention. You just show up on pay day. Then you go home to watch cartoons and sporting events. You come to class unprepared. You expect to get by with little to no effort. And yet God and the green-skinned aliens from Sagittarius love your kind anyway. Go figure. I guess the few humans who try, who are passionate and grateful, fearful and courageous—those types keep the school from being closed. Some of you care, and that makes all the difference. What will you have, Dutch?"

"I can't decide. Do you have a special today, Surge?"

"We have a comfort food special, made especially for struggling minds. It's called Easy Does it Vegan Chili. This is served with a Laid Back Cheddar Cheese Sandwich. For dessert—which comes with the special—we have Kindness-Berry Pie drizzled with Compassionate Fudge Sauce. This was the last meal of Christ and his disciples, did you know that?"

"That's fake news, Surge."

"Just seeing if you were paying attention, Dutch."

"Can I eat at the head table someday, Surge?"

"Sure. Just remember to leave a big tip for the head waiter."

"Aren't you the headwaiter?"

"Yes. But I said *head* waiter, not "headwaiter." There is an important difference."

"Oh God, Surge. You are such a pain, sometimes."

"Thank you."

"I have to write about education now. I taught for 33 years in a school for handicapped kids. I have some emotions left over from those days, some scars, some tears, and. I have some lingering questions."

"What are your questions?"

"Why does God get so much pleasure from torturing innocent little children? Why cripple a kid? Why allow such horrible pain? Why are injustice, poverty, ill-will, and pain so popular among the universe builders? Given the everyday evidence—especially in schools and hospitals full of suffering kids—our education on earth seems deliberately cruel. I don't get the need for nastiness, Surge. Does God like bullies and nasty people? Did God vote for Trump? There are so many nasty people. It seems like there is plenty of empty talk about goodness, but the universe is hell bent on pain. Why is that?"

"It's a hard job teaching this species."

"You already said that."

"Yes, but you didn't hear me. Remember, Dutch, you didn't personally create this universe of pain and suffering—neither did I or the green-headed aliens of Sagittarius. We didn't make the rules. We can judge the rules to be stupid and cruel, but that doesn't do much except cause ulcers and make us ineffective. Think of life as a Monopoly Game. There are rules and there are consequences. The game designers care not one fig for your whimpering and whining. Get a grip, Dutch, and get a shovel—there's work to be done."

The Central Issue

Put the head under the feet
~ Rumi

When I began writing *Consciousness, a New Slant on an Old Conundrum*, I had the seed of an idea that seemed plausible and worth exploring. However, in the process of writing that book, the seed germinated, and then a strange thing happened. I actually convinced myself that the concept of dual-consciousness was real, even revolutionary. The closer the book came to being completed, the more my confidence grew. As the book approached completion, a second wave of awareness washed over me. Gradually, I came to know that having dual-consciousness came with consequences and responsibilities. I knew then that there had to be a second book—the one you are reading now.

My career was in special education. Three decades of my life were spent helping children overcome their disabilities. Consequently, the *responsibilities and consequences* of dual-process theory first became obvious to me as an educator. How should we teach children knowing that human beings have two developing minds, two kinds of consciousness?

Education is a vessel that holds all disciplines, including psychology, philosophy, mathematics, biology, neuroscience, sociology, history, and literature. Therefore, education is a role model for all disciplines—each field of study has to be taught using some kind of pedagogy. Unfortunately, most of the disciplines assume an egocentric pedagogy. Only the discipline of education seems hyper-aware and self-conscious about the *why and how* of transferring knowledge from generation to generation. The other fields of study are obsessed with content and hardly pay attention to the delivery of that content.

It doesn't take much reflection to see how American education has evolved—it is an egocentric monster. Steal yourself for hyperbole. After 30 years of living inside the beast, trying to help handicapped kids, my emotions have been fried. However, I am not talking about human beings when I use my harsh rhetoric. The people inside the box, teachers, administrators, bus drivers, cooks, custodians, and school board members genuinely care about kids and education. Most people who work in education are intelligent and dedicated. My frustration is aimed at an institution.

This institutional beast came about slowly and naturally—with good intentions—as an outgrowth of historical trends. However, excuses aside, the American system of education is egocentric to the core. This doesn't mean that the allocentric mind has been quiet—it has been kicking and screaming since the concept of public education was first formulated. Indeed, the understanding of duality has been at the forefront of educational debates, in one form or another, for a century.

An egocentric educational system is focused on *doing*. It is about training, pragmatics, about jobs, about attainment and accomplishment. It is about moving through a lock-step curriculum, a hierarchy, goals and objectives, and about moving from one grade level to the next. However, the egocentric mind cannot comprehend and cannot logically support an educational system based on the allocentric mind. It admits of no twin. Consequently, an education system based solely on the logic of the egocentric mind is half-developed and half thoughout.

From the perspective of the muscle-bound, egocentrically-based, American education system, the attributes of the allocentric mind are considered secondary, subsidiary, minor-offshoots, and mostly irrelevant annoyances. What are the allocentric attributes? Here are a few: creativity, compassion, cooperation, love, tolerance, flexibility, sharing, peacefulness, and kindness. Allocentricity is the ever-flowing background out of which egocentric beliefs can manifest. Ignoring the ever-present origin, the heartbeat, the foundation, is absurd. Ignoring the fundamental foundation of education—kindness and love—is worse than absurd. Therefore, let me focus on kindness now, because that is the central issue in the debate. We need to answer this question honestly: *Are love and kindness equal to logic and pragmatics?*

Telltale Heart

> Einstein drew deeply on his pipe and blew smoke into the air. "School failed me, and I failed school. It bored me. The teachers behaved like *Feldwebel* (sergeants). I wanted to know, but they wanted me to learn for the exam. What I hated most was the competitive system there, and especially sports. Because of this, I wasn't worth anything, and several times they suggested I leave. This was a Catholic school in Munich. I felt that my thirst for knowledge was being strangled by my teachers; grades were their only measurement. How can a teacher understand youth with such a system?
> ~ EINSTEIN AND THE POET, IN SEARCH OF THE COSMIC MAN

Kindness is the Nutrient that Allows Developing Minds to Blossom: Teacher, Be Kind.

The allocentric mind is our well of compassion. Unfortunately, this empathetic mind has been deliberately neglected in education—especially at administrative and political levels. This obsession with the egocentric mind is not acceptable from a brain-based perspective, because it is not logical, not emotionally healthy, and not an intelligent strategy for a democracy. Children need to be surrounded by loving energy.

Let's go into the science lab and do an experiment that will help you understand why I believe *kindness is the first tenet of education.*

"Okay class, time for science lab."

Groans and whining . . .

"Relax. I'll do all the work. All you have to do is pay attention. Let's begin. Observe this Petri dish. Notice that I have placed a heart cell from a human being in the center of the saucer. Now, look how the heart cell refuses to die, how it continues a steady, rhythmic beat, even though it is no longer a part of the whole heart. Come closer. This is important, especially for those of you who might someday teach.

"Notice how our tiny heart cell is now slowly dying; how sad. Notice how the beat becomes softer, weaker, approaching death. Linger over that death for a moment, do some grieving for this tiny, lonely heart cell. Know that this is a real experiment that was carried out years ago and has been duplicated thousands of times.

"Okay, now I place another fresh heart cell next to the dying cell. I do not let them touch. I move the dying one to the left edge of the saucer and I place the fresh heart cell on the right edge. Pay attention to the cell that is almost dead.

"Witness a miracle. The dying cell revives!

"How can this be? How can the presence of a fresh heart cell, a distant neighbor, extend the life of another heart cell that is two inches away! It might as well be light years away.

"Yet something about this relationship turns these two cells into a synchronously beating mini-heart! This is a class about the human mind. Why would I take you into a science lab and make you pay attention to this miracle? Yes, you, in the back row, the young man texting his girlfriend."

"Maybe brain cells need each other, as well as heart cells?"

"Yes, true, brain cells beat in a coherent harmony. Take your thinking deeper."

"Maybe all cells, in the whole body, need each other?"

"True again. Keep going."

"I know."

"Okay, the young lady in the front row with the intelligent eyes. What is it that you know?"

"We have 100 trillion cells in our bodies—I just looked it up on Wikipedia—and these cells evidently need each other to beat in harmony, otherwise, I guess, the body would get unglued. Perhaps human masses of beating cells affect other human masses of beating cells. Maybe we affect each other's frequencies? Maybe life needs life? If the community heartbeat goes on in a good way, then there will be harmony and health, but if the community heart beat is irregular or dysfunctional, we get disharmony and illness."

"Wow! Give me a hug, graduate student. That was brilliant, thank you. Yes, all cells need other cells, and they must beat together in harmony. We are part of a global heartbeat that is our reality. This is no small observation; remember what you witnessed here today. Now look back at the Petri dish. Reflect again at the drama playing out. Heart cells are special. Like other cells, they communicate through electromagnetic fields, frequencies that are not perceptible using eyes or ears, but they have a property that sets them apart; there is more to the story of the synchronously beating heart cells. Who can tell me more?"

"I have a question?"

"Yes, young man, what is your question?"

"If we are affected by each other's frequencies, if there is scientific evidence that proves this, then how is the mind just be in the head? I mean, shouldn't we reconsider the definition of mind?"

"Quick Igor, get me my smelling salts. I feel faint. Is the mind just in the head! A brilliant question, even though the ancients knew the answer about five thousand years ago. The brain is in the head, but the mind permeates the entire body and leaks out into the environment. Actually, it's more complicated than that, but you are on the correct path. By the way, where do you suppose there is a "mind organ," outside the brain?"

Silence. Wonderment.

"It is your heart. Scientists can hook up electrodes to your scalp anytime they want, maybe over happy hour and a couple beers. There is a huge body of evidence concluding that the brain has frequencies and these frequencies correspond to levels of consciousness. Now listen, this is important. When you hook up electrodes to the heart you get frequencies that put brain waves to shame. Heart frequencies are so powerful—5,000 times more powerful than brain waves—you can record from instruments three feet away from a beating heart, with no contact electrodes, enough electromagnetic energy to power a light bulb day in and day out. With strong enough instruments we could pick up heart frequencies a mile away.

"Here is the kicker: the heart has neurons, cells normally found in brains. The heart has so many neurons—half of all heart cells are neurons—we can speak of the heart as a brain organ. A heart brain! It's a nice pump, but it is also a marvelous brain.

"When scientists studied the electromagnetic energy that emanates from the heart, they saw that this cardiac energy had a characteristic shape, a figure-eight that ballooned outward/forward and also behind, with you as the center; this shape is called a torus. Toric energy fields are observed in atoms, the earth, the sun, the solar system, and the galaxy—there is something universal about this energy configuration.

"Amazingly, we can put internal energy into the torus, feed it so that it expands outward, or we can shrink the field so it barely encloses our bodies. Also, the toric field is holographic; information about the whole can be read from any location in the field. There are mysteries here awaiting the acute attention of a brilliant graduate student—just saying.

"When scientists traced nerve cells from the heart to the brain, they discovered that there was an uninterrupted expressway of nerve fibers going directly from the heart to the limbic region of the brain, and from the limbic region to the prefrontal cortex. Now listen to me. If you don't learn one thing, you will remember this:

"Love—an emotion associated with the limbic region—drives the heart frequency, and expands the size of the energy torus. Varieties of love drive different heart frequencies: kindness, nurturing, friendship, compassion, interest, touching, caring—all these stabilize the heart and project different flavors of toric energy fields. In turn, the heart-brain stabilizes the head-mind, the body-mind, and the mind that permeates the spaces around you.

"So, for you potential teachers, remember this: the heart's electromagnetic healing field can extend—scientifically, measurably—to fill an entire classroom. It can surround a single student, bathe them in loving warmth, or it can include an entire group of young humans within its loving environment. A good teacher can spread nurturing energy over students, bathing a classroom in compassion.

"A bad teacher, one with a Grinch heart, can suck energy out of a room and shrivel the minds of innocent little kids.

"When asked to explain his religion, the 14th Dalai Lama said:
"My religion is very simple. My religion is kindness."
"That is our religion as well.

"Let me repeat that. *Our religion is kindness.*

"Take a break; think it over."

Western education is a graded phenomenon. The system graduates students in a lock-step fashion. Kindergarten leads to first grade, on and on, graduate school, post-doc, internships, back for another PhD. Graded: A is wonderful; B means "good job;" C for the "average kids;" D means "You could do better;" E means, well, shame on you, you hapless nimrod. We march our children in lock step and give them graduated, simplistic feedback.

It is all egocentric, designed by old white guys wearing Penguin suits (suit and tie) with one eye on a stop watch and another on the rules. But life, from an allocentric perspective is a dance. It is music. It is meant to be experienced. "Modern" education in America has become a serious business with serious end points and serious goals—we call them accomplishments—getting the college degrees to get the jobs that require more marching from quota to quota, from sales pitch to sales pitch.

Meanwhile, a whole life goes by and we miss it. We forget to enjoy our moments and then, suddenly it seems, we have to depart—forever.

Kindness in words creates confidence.
Kindness in thinking creates profoundness.
Kindness in giving creates love.
~ Lao-Tzu

Something in our Western educational culture does not have kindness as the number one priority.

Physical Movement and Education

There is a clear understanding in special education that controlled and patterned movement is essential for physical and emotional development. Kids who can't move, or who are pushed and guided everywhere, are in constant danger of neuro-motor deterioration. "Move it or lose it" is the key phase, and it applies to mental capacity as well as physical capacity.

Observation and research show that movement is central to learning and normal development. But why would that be? Dual-process theory gives the answer. The need to move with purpose (in evolution) gave rise to two attention systems, two minds, and two kinds of consciousness. If there is no purposeful movement, then there is no normal development of attention, mind, and consciousness.

Regular education seems immune to the awareness of dual minds and for the need to purposefully move. In America, we cram large numbers of young kids in a small space with a single teacher. We do this for hours on end, keeping the kids confined to seats most of the time. Purposeful movement is reduced to fine-motor activity, moving eyeballs over a page, keeping the head pointed at the teacher, or blackboard, or screen—while keeping the body still. From the perspective of dual-process theory, *this is insanity.* The egocentric approach retards learning; it makes for dull minds and sluggish bodies. Inactivity goes against the entire evolutionary reason we have a mind. This non-movement strategy is the heartbeat of egocentric education. Ironically, egoic education is not logical, not supported by research, not supported by science, not kind, and not supported by evolution. Egoic education is ridiculous, yet it continues to worsen—to be the "law of the land"—rather than get better.

In practice, of course, in classrooms across the globe, the egocentric approach is rejected by teachers and students. There is no getting around our two minds. We cannot just toss out half our essence. Teachers routinely use their allocentric minds quite expertly. These teachers are kind, and they encourage sharing, cooperation, and respect for self and others. They *do* know that kids need to move. The egocentric, heavy-handed madness of modern education handcuffs, denigrates, and discourages innovation and love, but teachers push back every day and all the time. There is no physiological way to stop the oscillation between allocentricity and egocentricity. *Dual-process theory removes the justifications that the egocentric mind uses to militarize educa-tion.* Egocentricity is half a process—an important and *equal partner*—but it is *not* the whole show. There is an equal sister to the egocentric mind. Knowl-edge is only half the equation.

The other half of the educational equation is experience and self-discov-ery. Curiosity is encouraged and rewarded in an allocentric setting (like in Waldorf schools). Allocentric consciousness releases a host of positive ener-gies. The *joy of learning* is celebrated. Sharing and cooperation are high vir-tues. Innovation and creative energy are allowed to blossom at a developmen-tally appropriate pace. Cooperation and teamwork are part of daily routines. Good teachers know this in their heart centers and in their intuitive centers.

However, the demands of the egocentric mindset smother the best efforts of teachers. The curriculum, the incessant testing, the ridiculous obsession with accomplishment, is not supported by research, by evidence, by logic, by common sense, or by any glimmer of kindness. Dual-process theory is a tool to take into the Board of Education meeting—there are too many half-brained fools insisting that education be entirely egocentric. They have no case.

How is Egocentric Education Working Out?

What kind of population has the public education system of the United States produced? How is the egocentric approach to education working out for America?

In the United States, in 2016, we elected a president who was a bully. He insulted rather than debated. He called debating opponents idiots whenever they appealed to the challenges that a leader might face in government—he seemed not to grasp or care about the issues. Psychiatrists labeled him a malignant narcissist. He mocked the handicapped. He denigrated women. He insulted non-white populations. He had a paper-thin ego that counter-attacked anyone who dared question what he wanted to do or say. He had no experience in government. He probably paid few taxes, despite being a billionaire. He had little understanding of science, and he put in place a cabi-net that opposed centuries of progressive policies. He gathered around him a hateful, bigoted following that screamed and ranted at his speaking engage-ments—it looked like a return to the Mussolini era. He had close ties with Russia and their president—a former head of the KGB who interfered with the American election process. Trump got his information from Fox News and from a steady stream of hateful memes. Millions of Americans saw nothing wrong with any of this—Mr. Trump seemed to be the role model they had been searching for. He was a man in their own image—at the same very low level of consciousness.

If you tried to debate or confront the population of hateful people who followed the president-elect from rally to rally, there were three things off the table:

- You couldn't appeal to logic or facts. Trump followers rejected all appeals to reason.
- You couldn't appeal to history or experience. Wisdom was not one of this population's long suits.
- You couldn't appeal to compassion. Human suffering was not an issue of concern; it was irrelevant to this ego-focused population.

Indeed, it was the repetition of malicious memes, propaganda spewed by fake news networks and disinformation campaigns—some of it actually orchestrated by the KGB—that undermined the election process. How did this bloodless coup come about? The 45th president was elected by millions of American citizens—half the population—who did not find any of his tantrums objectionable. People who voted for Donald Trump wanted change—even if that change was massively destructive, illogical, unwise, and cruel.

One plausible reason for this disaster is an educational system in the United States crafted by the egocentric mind. The product of this system showed up at the polls to elect an ego-bound—some say mentally ill—president. This president was elected by an emotionally and academically illiterate populous. This population has many excuses, many defenses, but the bigotry, the allocentric blindness, and the low level of consciousness of the American population has been outed.

We have failed to teach our children that kindness, empathy, and cooperation *are equal to, if not greater than* individualism (systems-thinking) and competition. We have half-educated our population—the election of 2016 exposed the educational system bigly (to use a favorite adjective of the 45th president).

Our society insists that teachers be accountable to the egocentric mind. At the same time, we *do not* insist that teachers be accountable to the allocentric mind. If a teacher passes a child on to the next grade level—a child who cannot read, write, understand civics, or do math—the egocentric mind is outraged (but routinely passes the kid along anyway). However, if the teacher passes a child who is a bully, who has no empathy, no ability to cooperate, no tolerance (is a bigot, following hateful cultural role models), and has no sense of morality and spirituality, that is just fine—there are no checks and balances.

If teachers happen to bring allocentric characteristics to the classroom, then that is okay with the system, but it is not considered necessary. No one is holding the teacher accountable for educating civilized, sophisticated children with high levels of consciousness. It is not the teacher's role in an egocentric system to discuss values, consciousness, or soul (self). The ego goes nuts when this idea of "teaching values" is put forward. However, with the evidence coming from dual-process theory, the ego is naked and exposed— found guilty. We have created a democracy based around the ego, which is now busy eroding democracy. Irony drips from the ceiling.

Two Meanings of Life and the Role of Education

I said at the beginning that having two minds, two kinds of consciousness, must logically lead to two kinds of meaning for life. If the purpose of life derives from *navigation* with all its connotations, then the purpose of education is to assist with a dual—and equal—journey through the learning process.

The first principle, after *teacher be kind*, is that there is *nothing* in education that ought to be static. There is a flow, a process, a creative river to joyfully participate within. Egocentric education is *a striving to know*. This is the education we are familiar with since most of us went through the graded stages on our way to the life we are unfolding. However, egocentric education seems to be the only focus given serious credibility—backed up by power and funding. Allocentric meaning, on the other hand, is given lip service, little supportive funding, and is the first thing to go when dollars get scarce. We are willing to sacrifice kindness, creativity, and intuitive flow when the financial going gets rough. The egocentric mind cannot understand kindness; that is not the job of the egocentric mind. It saves itself first and tosses creativity, activity, innovation, and the joy of living out the nearest window when the millage fails.

405

If the purpose of life is dual, then the purpose of education is also dual:

- To help students become explorers. To teach them how to follow the mandate of the egocentric mind.
- To help students *experience* life to the fullest extent. To experience love, empathy, and tolerance. To feel the flow of existence, to be mindful, to be awake. To follow the mandate of the allocentric mind.

A healthy democracy is one in which the citizens have a balance between their two minds, and in which levels of consciousness are explored and embraced. Therefore, students need to learn to *explore and figure out, because that is the evolutionary job of the ego.* They need to have strong egos and a strong drive to transform the world around them into an environment that is financially and cognitively satisfying. However, equally important, we need to support *the job of the (allocentric) self: to understand and seek out wisdom, to taste life, to experience* moments—to *participate* rather than observe and judge.

The Work of the Educators

Educators *really do know* the primary importance of kindness, nurturing, and movement. They might not have an understanding of dual–process theory, and they may not use scientific terminology, but they certainly comprehend the essence of the theory. However, outside of academic circles the story is different. Political, cultural, and religious perspectives have historically forced egocentric dominance onto learning communities. The result has been the denigration of the allocentric mind. With the rise in awareness and acceptance of dual-process theory, educators now have a powerful new tool in their fight to bring balance to the academic world.

So what is it that educators down through the centuries have been trying to communicate to mankind? Here is a partial list:

- The hidden allocentric mind is the source of creation, innovation, spirituality, soul, compassion, wisdom—all the best of humankind's attributes reside hidden within. To leave these attributes out of education—creativity, innocence, wonderment—is criminal, stupid, and unsupported by evidence.
- The goal of allocentric education—some might call it eco-education—is to encourage positive, functional growth through relationship building. It is not about removing dysfunction through punishment.

Educators are the women and men charged with passing both knowledge *and* wisdom from one generation to the next. Educators are frontline human beings who face-off against ignorance on a daily basis. That is their job, to model compassion, to exult reason, to praise and model the quest for experience and the wisdom that arises from experience. If you are a teacher, you have an obligation to explore and explain the duality that is inherent in the human condition. Stand courageously against ignorance. Stand courageously for the attributes of the allocentric mind: love, peace, joy, wisdom, mindfulness, equanimity, and thankfulness. Half of your essence is being denied by the modern educational system and that can no longer continue—the stakes are too high in this high tech world of smart bombs and suitcase sized nuclear bombs. The modern world must have a population that is much more highly evolved than is presently the case.

Therefore, as an educator, *you have an obligation*:

Do as much good as you can in the time you have left

Get up.
This is a work day.

Music and Duality

Close the doors, light the lights.
We're stayin' home tonight,
far away from the bustle and the bright city lights.
Let them all fade away.
Just leave us alone.
And we'll live in a world of our own.
~ song lyric from *A World of Our Own*, by The Seekers, written by
Sonny James

The Seekers

For a few months in 2018 (during the final editing of this book), I became absorbed in the music of the Australian folk group, *The Seekers*. Their blended voices lifted up my spirit as I sailed happily through my days—I played their songs over and over again as I went about my daily tasks. The song lyric quoted above, from *A World of Our Own*, is sung at the right tempo for my nervous system; there is something healing in the music for my particular soul. I used the songs of *The Seekers* as a background beat as I composed this chapter.

There are certain songs that lift us right out of a chair, as if our heart was pulled upward toward the ceiling. We resonate with something primal in the universe. In the moments when music elevates our souls, we feel that our experience on this tiny blue planet is special and worth the many struggles that every human must endure. The right music resonates within us and harmonizes with our soul. It is of no small interest that the kind of music that lifts up our soul is unique—every person experiences music in a different way. There is some kind of fundamental body/mind resonance that occurs when the *right* harmonics are played within vibrational range of a human body.

The song lyric above ends with this line: "I know you will find there will be peace of mind when we live in a world of our own." This sentence suggests that music is both personal and shared. Music creates a resonance that is uniquely our own, and yet we find camaraderie and relationship when music is shared. Music is social, emotional, and healing. The right song leaves us with tears of joy and appreciation. We are stunned by the magic. Our soul dances in a space that is pure joy. The right music brings us peace of mind.

Therefore, we are grateful for the strange wondrous musical gift that comes from someplace outside of us. And *music is a gift* from the writers, composers, and performers to an audience, to you. Joy arises whenever we disappear into the music we love.

❧⎯⎯⎯⎯⎯❧

As I wrote this final chapter, I struggled with a question: How do I express the power of music without using music? A breakthrough *came for me* when I started looping the greatest hits of *The Seekers*. Judith Durham's perfectly-pitched angelic voice, backed by the exquisite harmony of Keith Potger, Bruce Woodley, and Athol Guy, swept me off my feet and set me on a march into a mysterious somewhere—I didn't care where. I am familiar with this feeling of energy being channeled through me—it happens when I write. Music opens the channel wider and makes the messages that flow through my pen clearer and more enjoyable to receive.

408

In an interview for British TV, Judith Durham was asked if she was nervous before performances. She told the interviewer that she was overcome with fear as she waited to be introduced. However, once on stage she became a channel for the music. She felt that she was the right instrument through which the musical magic could speak. All Judith had to do was let the music flow through her. This is a lesson for all of us. We can be vessels for something powerful, compassionate, and uplifting if we relax and allow positive energy to flow into and through us. Follow the music.

I hold in my hand a baton, a magic wand that artists pass from generation to generation. I know that this book is about done—I did my best to write down what was channeled by minds much more sophisticated and compassionate than my own. I now hand the wand to you. I ask that *you do one better than was done before*: compose your art and then pass the baton forward. You come from the dreamtime, my friend; you are the keeper of the flame:

> *I come from the dream-time*
> *From the dusty red-soil plains*
> *I am the ancient heart*
> *The keeper of the flame*
~A song lyric from the song *I AM AUSTRALIAN*, written by Bruce Woodley of The Seekers

Artists and musicians are the keepers of the flame. *You*, as you create, are the keeper of the flame. Let the fire rage in your underground mind. Do not go gentle into that cold dark night. Take the fire with you wherever you go.

Could there be a better name for what is happening to us: *The Seekers*. We are seekers after the music that brings harmony and love to this universe. It starts here, on this tiny blue sphere in the middle of this cold, heartless void. We sing into the void and the void shrinks back in fear. Look at the power we have! We can be outrageously defiant against cruel and impossible odds. We are a song composed by a universe, and that universe sings through each of us: One of the most famous of *The Seekers'* melodies says it best; we need each other: "I know I will never find another you."

> *There's a new world somewhere*
> *They call the promised land.*

And I'll be there someday
If you could hold my hand.
I still need you there beside me
No matter what I do
For I know I'll never find another you.
There is always someone
For each of us, they say.
And you'll be my someone
Forever and a day.
I could search the whole world over
Until my life is through
But I know I'll never find another you.
~ song lyric from *I KNOW I'LL NEVER FIND ANOTHER YOU*, by The Seekers,
written by Tom Springfield.

This song has resonated with three generations now because it is how the self feels about its companions. We are so deeply grateful that others stand with us— even though each of us is a unique miracle, we are not alone. There will never be another you.

Actually, there will never be another duality like you, because you are both an ego *and* a soul vibrating in harmony. My story in this book is that there are two of you sitting there. But you don't see the *hidden you*, the real you, the girl with the magic voice, the boy with the magic words. You are like *Georgy Girl*:

Hey there, Georgy girl
There's another Georgy deep inside
Bring out all the love you hide
And, oh, what a change there'd be
The world would see
A new Georgy girl
Hey there, Georgy girl
Dreamin' of the someone you could be
Life is a reality
You can't always run away
Don't be so scared of changing
And rearranging yourself

It's time for jumping down from the shelf
A little bit
~ song lyric from *Hey There Georgy Girl*, by The Seekers, written by
Jim Dale, music by Tom Springfield.

There is another Georgy girl deep inside you, a feminine energy that defines you, an allocentric mind hidden below the egoic mind. Don't be so scared of changing and rearranging yourself. You can let your poetic soul out of prison. You can feel free. It's time for jumping down from the shelf (a little bit—don't leap off a cliff in your exuberance).

You don't need to be a famous singer standing in front of 200,000 people in Melbourne, Australia. You don't need to be Rumi, or Pete Seeger, or the Dalai Lama. You are enough. Let me say that again because you didn't hear me: *You are enough.* The miracle, the music, the seed has already been planted in you because you were born on earth as a human being. Indeed, it is statistically impossible for you to be here in this fleeting moment; there is mathematically *no way* you can be here. But then, here you are, the keeper of the flame, holding the magic wand.

I'm the hot wind from the desert
I'm the black soil of the plain
I'm the mountains and the valleys
I'm the drought and flooding rains
I am the rock
I am the sky
The rivers when they run
The spirit of this great land
I am Australian
We are one,
But we are many
And from all the lands on earth we come
We'll share a dream
And sing with one voice
I am,
you are,
we are Australian

411

~A song lyric from the song *I AM AUSTRALIAN*, written by Bruce Woodley of The Seekers

Bruce Woodley of *The Seekers*, the man who wrote this unofficial national anthem of Australia, stood on stage with tears rolling down his face as an audience of thousands of Australians in Melbourne rose to their feet and give this wonderful man a collective hug. This is a song for Australians, a gift of love from an Australian folk group, *The Seekers*, from Bruce Woodley, but it was intended for all of humanity—*we are one, but we are many, and from all the lands on earth we come.*

Put your favorite soul-inspiring music on in the background as you sail through this final section. Like I have done throughout the book, I will oscillate between the sing-song harmonies and dialogues of the allocentric mind and the puzzle-solving logic of the egocentric mind. Sit within the protective shell of your music and allow my words to be whatever they are for you.

Swan Song

"Aren't you supposed to be writing about music and duality, Dutch? What brings you back to the Third-Eye Café so soon?"

> "I feel a triumphant sadness, Surge. I sat at my desk for days writing dry prose, never satisfied. Music is about emotions, something primal, but I couldn't capture it in words. So, I have come to you for help, Surge—like I always do when I get stuck. If nothing else, we can talk it over one last time and leave it at that—dialogue suffices, or will have to."

"However, there is more that saddens you, isn't there, Dutch?"

> "Yes. This is our last dialogue, Surge. I am back for our final communication. You have been with me for 500 pages, two books, over a span of three years. This is the end of the line. I have to let you go. I just wanted to say thank you for your guidance. I happily send this book into the noosphere to join its cousins in the folds of time— because you helped craft it. Thank you, Surge. My books are memoirs now, a few hundred pages about the contents of my mind as it grew to maturity on planet earth—I leave the books as gifts to my kids and grandkids. And, of course, I leave these musings to the

reader—to you, a most unusual and fascinating person—in the hope that your mind will do one better than was done before. I am counting on you to make a contribution. I am pleased that we were able to meet, dear reader—in this strange but delightful way."

"Your swan song has begun, Dutch—the end of human consciousness awaits you. But don't rush off—linger over your coffee and appreciate our friendship. And you are welcome—the pleasure was also mine. I will miss you, too. Although, I have been your muse for many years now, so don't be in a hurry to say goodbye unless you are done being creative. But let's speak of something happier. Let's speak of music. How can I help? Why was it so hard to explain something as wonderful as music—surely you know."

"Yes, I know. I can lay the dry knowledge out on the tabletop. But music goes directly to the human heart and then it radiates throughout the whole body. Music *actually is* the message of the spheres; the vibrations that create music are as ancient as the universe. I can talk about background and foreground, egocentric and allocentric, the mother oneness divided into the primal duality—the yin and yang—but words are not the music, Surge. Words filter and judge. The ego is all about the words, but the ego is deaf to music. The ego can hear but only the self can listen. Something absolutely essential is lost when words try to explain something as primal as music."

"And why is that, Dutch?"

"I suppose it is because music is about relationships and participating. Music is shared harmony, rhythm, and melody *in the moment*; it is not about essays and judgments. We don't stand apart from music, we become music. We are energy bodies; music is pure food for us."

"So, if I understand you correctly, you don't want to end the book by talking and explaining. Is that right? You want the dialogue to sing instead."

"Yes. I am choking on abstract words."

"None of the words please you?"

"A few do, but they are random."

"For example?"

"Other minds have found eloquence, outrage, or exuberance when the subject turns to music. Here's a quote I like because it is an emotional voice disgusted with what our culture has done to the sacredness of music:

Where music once nourished a healthy appetite, whether in the concert hall or the village square, now a perpetual banquet of song serves only to soothe a blunted palate. We live in an age of widespread musical obesity.

~ *MUSIC, THE BRAIN, AND ECSTASY*, Robert Jourdain, 1997.

"I love that phrase *musical obesity*. Not only are we physically turning into a nation of Teletubbies, so also has our morality become overweight and unpleasant. Our music reflects our collective level of consciousness: we are a culture bloated with unsophisticated music. We have lost connection with the mystery and message of music. I am thankful that Robert Jourdain was able to capture this dilemma in his quote."

"Be careful with your criticism. There is something in a rant that destroys music, Dutch. However, I think you will find that it is wrong to avoid the egoic dissection of music. Yes, play the soul-filling harmonies in the background, but also manifest and share useful patterns. Both minds are needed, Dutch; you said so yourself. Remember that a song has a beat, but it also has words—a song that penetrates to your core does so because it is a blend of ego and self. Music is a rare gift because it is harmony, a balance between chaotic extremes. It is okay to go forward from here using dry words and cold logic; it's okay to make the ego happy as well as the soul."

"Thank you, Surge."

"Let the Swan Song continue, Dutch."

Happy Hour/Friday Night/Getting out of Dodge

It's Friday evening after a long week of deadlines, edginess, and disharmony. Your body is not happy with the state of your mind—your face is stuck in some kind of pre-death mask and every muscle in your neck and shoulders is knotted. You need a glass of wine (or not) and some background music to drown out the thoughts in your mind that insist on looping. You head over to the local pub without rinsing the dishes or opening the mail. Your body needs Happy Hour; right now.

This is a lucky day for you. The music in the pub is your kind of music. The beat is perfect for resetting your jangly nervous system; the lighting is perfect too, and the people in the room have established an emotional reverberation that feels healing and happy.

It takes a while but slowly your body becomes entrained to the environment; you begin to resonate, reverberate, flow, and then suddenly, for no reason, you feel happy. Smiles come. Your facial muscles relax. Your friends look healthy and attractive. You feel younger, and hope is mildly restored. Why would happy hour music have this effect on your whole body?

My answer is the same as it was throughout this journey: the non-verbal, whole-body allocentric mind is driven by flow—music is flow, without egoic attachments. The allocentric mind does not think or solve problems—it longs for relationships and oneness. Music is the primary language of the allocentric mind, and as such, it is the antidote to egocentricity. Music, especially lyric-free music, inhibits the overheated egocentric mind. Music is cool energy; we chill out in the presence of our much needed primal rhythms.

But there is more to this happy hour evening. The band you love has packed up their guitars and drums. A new group is set to play next. A jazz artist is about to perform. If you are like me, you find jazz somewhat unsettling—maybe your level of consciousness has not yet reached a level sufficient to embrace jazz—so you think maybe you will leave. But a friend leans over and whispers "Stick around; I think you might enjoy John Coltrane."

Brace yourself, my friend. John Coltrane (1926-1967) is a musical genius—he knows how to merge into the music of the cosmos. In his thirties, Coltrane became passionate about the connections between music, philosophy, physics, and spirituality—he knew that his own musical genius wanted more than dim-lighted bars and drug-induced highs. Jazz became his medium for exploring the fundamental language of the universe—he had a tool, an instrument, for exploring beyond the mundane. During a transformational spiritual experience, Coltrane asked for help from the musical spirits that inhabit the harmonic spheres (my hyperbole). He did not see this harmonic force as coming from one religion, but from the shared core of all religions. In other words, Coltrane found his muse and together they made music.

During his transformational experience, Coltrane asked for his gift as a musician to contain sufficient power to make others happy through music. Furthermore, Coltrane wished to use his musical abilities to help move people upward through levels of consciousness. Coltrane was very much aware of Einstein's intuitive work and he wanted to understand how intuition could be used to arrive at greater levels of awareness and love. When composing and playing music, Coltrane tried to use the same intuitive powers that helped Einstein create new knowledge.

Coltrane felt that the universe was trying to communicate through music. Primal sounds were the notes that formed harmonies, rhythms, and melodies. Coltrane used his musical genius to express these transformative frequencies. He was searching for a resonance that would melt the sun. That same fiery frequency could transform consciousness from the everyday sleepwalking state to something magical and ethereal. Coltrane's followers called his compositions *spiritual music*; Coltrane's audience could feel that he was reaching for something that was transformative, and, indeed, they often felt transformed after a performance.

From John Coltrane and *The Seekers*, from the freedom of Friday night, we turn next to Monday morning—we must go back to work doing the bidding of cognitive evolution. The ego is eager to play with patterns.

The Evolution and History of Music

There is a theory, one that fits like a glove with my understanding of the evolution of the brain, which postulates that music came about because human beings became bilateral creatures. If we hadn't stood upright and moved our bilateral bodies in a forward direction, music would never have evolved to be the sophisticated expression it has become. Walking upright changed the structure of the throat. Language and singing could only have developed when the larynx was positioned deep in the throat—compared to other primates. The hands were also free to evolve greater sophistication. The use of our upper limbs to craft and use tools (including musical instruments) only came about after upright bipedal navigation evolved.

To walk in a forward direction, to run after prey or escape from a predator, or to dance, required a harmonic oscillation between the two sides of the body. A steady, reliable, alternating rhythm had to evolve. Therefore, our ability to create, using harmonic rhythms, came about because of the need to navigate, which required bilaterality and oscillation between dualities (movement versus no-movement). When the hands and arms alternated in a bilateral pattern, human beings discovered drum beats—a way to express the fundamental oscillations at the heart of the universe. Sophisticated human hands and arms eventually evolved to play the piano, the flute, and stringed instruments like the guitar.

The brain is the instrument that is played by the egocentric mind—the ego performs using head-bound space. However, for the allocentric mind, the whole body is the instrument. We feel the urge to move, to dance, to tap our feet, to sway from side to side, to animate and oscillate with our whole bodies when we hear music. We do this because of proprioception—the muscular sense that animates the body. Nietzsche said we listen to music with our muscles—he intuitively knew over a hundred years ago that muscles held the key to understanding consciousness.

Rhythms set our whole body resonating. Our faces and posture mirror the emotional narrative of the melody. The ability to speak and the ability to play music—to sing or use an instrument—requires the learning of sequential motor patterns that are then retained as sensorimotor (proprioceptive) memories. Both oral communication and music are forms of purposeful movements. When we speak, the phonemes expressed are egocentric, while the gaps (the silence) between the words are allocentric—silence is the background canvas. When we express music, the chords (notes) are also separated by allocentric gaps. Therefore, the allocentric background also provides a grammar or punctuation for both oral and musical expression—the gaps and rests expose the silence of the background canvas.

Just as the evolution of language is a guide to the evolution of consciousness, so too is the evolution of music a guide to the evolution of consciousness. Like language, the history of music is the history of the evolution of mankind's cognitive and spiritual development. Music is a language just as oral communication is a language. As such, music cannot be considered a single entity; it needs to be decomposed into different component operations or levels of processing. At the deepest evolutionary level, music and oral communication arise from proprioception and both operate using a mutually exclusive alternation between an allocentric frame of reference and an egocentric frame of reference. Oral language uses words composed of phonemes. Music uses chords composed of notes. Language is a form of music or music is a form of language—they both have a flow, both communicate, both alter frequencies, and both change the biochemistry of the mind.

Music is the way the allocentric mind communicates, just as spoken language is used by the egocentric mind to communicate. Because music is allocentric, it has the attributes of the allocentric mind. Therefore, allocentric music is about relationships, participating, healing, wholeness, and flow. Music dissolves the ego, keeps us centered in the moment and releases us from past regrets and future angst. Music is non-verbal, although it can manifest patterns—so it can give rise to egocentric forms, like tunes, songs, melodies, chords, and scales. Just as verbal language has gotten ever more complex throughout history, so also has allocentric communication through music become ever more complex.

Music as Sacred

Human beings have found music to be sacred since the dawn of sophisticated language—this was apparent 5,000 years ago in China and India. The ancients knew that music directly alters emotions, contributes to overall health and wellbeing, and fills the energy reserves of the human body. Music was prescribed as energy-medicine for thousands of years. In addition, music has long been considered an avenue for elevating spirituality. Here is a quote by David Tame from his book *The Secret Power of Music* (1984) that supports these observations; the farther back in time we journey, the more sacred music becomes:

> Almost three thousand years before the birth of Christ, at a time when the music of European man may have amounted to the beating of bones on hollow logs, the people of China were already in possession of the most complex and fascinating philosophy of music of which we know little today . . . the tradition of Chinese classical music is so ancient that its origins are described today only in legend.
>
> In the case of China the rule holds true that the further we go back in history, the more sacred and vital a significance we find to the phenomenon of sound itself. In the viewpoint of the ancient Chinese, the notes of all music contained an essence of transcendent power. A piece of music was an energy-formula . . . The particular

mystical influences of a piece of music depended upon such factors as its rhythm, it melodic patterns, and the combination of instruments used.

~ THE SECRET POWER OF MUSIC, David Tame, 1984.

Therefore, music is more related to healing and spiritual evolution than it is to modern incarnations like music-as-entertainment, or music-as-background for media commercials and murder mysteries. The Chinese associated cosmic sound with exalted consciousness. Human beings can attune themselves with the cosmic source. Spirituality was a question of vibration. Chinese music was directed at the raising and purification of consciousness. Historically, music was a powerful medicine that needed be wisely administered.

Author Gary Lachman wrote an article for *Quest Magazine* in 2002 called "Concerto for Magic and Mysticism: Esotericism and Western Music." As I said earlier, Lachman is a student of human consciousness and has written extensively about the evolution of mind. In this article, he discusses the connection between esotericism, mysticism, and the evolution of Western music:

> . . . the link between music and esotericism goes back at least as far as the sixth century BCE, when the Greek sage Pythagoras discovered the laws of harmony and developed a mystical brotherhood around them. But Mozart can surely stand as an archetype for an association that runs throughout Western music. In fact, as the musicologist Wilfrid Mellers suggests, the central musical forms of the Classical and Romantic traditions, the sonata and symphony, are musical initiation journeys, sonic spiritual pilgrimages in which the hero undergoes trials and challenges to arrive at a new level of integration. The symphony orchestra itself is an example of Masonic brotherhood, fellows "banding" together for a common cause.
>
> Mozart took to Masonry with a passion, introducing Masonic elements into many of his works, and writing music specifically for Masonic affairs, like his moving Masonic Funeral Music. His final three symphonies form a triptych of Masonic initiation symbolism (Mellers). But his greatest Masonic work is his opera The Magic Flute (1791), which centers on the archetypal struggle between Darkness and Light. Masonic, Egyptian, and hermetic elements crowd the

work—too many to elucidate here—but by the time Mozart wrote it, Freemasonry was on the defensive. The Illuminati had been suppressed in Bavaria, and Mozart had to muffle his esoteric themes in the innocuous fabric of a fairy tale. Nevertheless, much of the message got through, and one wonders if it is just by chance that Prague, the city that loved Mozart most, was the age-old home of alchemists, Rosicrucians, astrologers, and esotericists?

~"Concerto for Magic and Mysticism: Esotericism and Western Music," QUEST MAGAZINE, Gary Lachman, 2002.

What Gary Lachman is saying in this article is that composers throughout history have worked to elevate the level of consciousness of the audience. Many composers have deliberately used the universal codes of music, the primal patterns, frequencies, and rhythms, to awaken the populous. Musicians are still doing this work—even if they are not aware of how their art impacts culture. Lachman says that Mozart's *Magic Flute* "centers on the archetypal struggle between Darkness and Light." In other words, Mozart saw the battle between the allocentric and the egocentric, and he knew this oscillating polarity was the primal understanding that defined humanity.

In a 1906 lecture in Berlin, Rudolf Steiner discussed what he called "The Inner Nature of Music and the Experience of Tone." Steiner's focus in the article was on the role of music as a medium for communication with the spiritual world:

> The creative musician translates what he has experienced in the spiritual world into harmonies, melodies, and rhythms of music that is physically manifest. Music, therefore, is a messenger from the spiritual world, speaking to us through tones as long as we are unable to partake in super-sensible events directly.
> ~ "The Inner Nature of Music and the Experience of Tone," a lecture given in Berlin, Rudolf Steiner. 1906.

Steiner is saying that the avenues for developing higher levels of consciousness are not easily reached—the common man is not prone to discipline and esoteric studies. However, music is available in every culture and is the primary way that the common man is reached and elevated. In other words, we can grow our spirituality through music, whether that is our intention or not. It is as if the universe sends waves of instructions that gradually cause the evolution of cognition even in beings not working on spiritual growth. I suggest that we are drawn to certain rhythms, certain harmonies, and certain melodies because these resonate with the level of consciousness that we inhabit. Rhythms that draw our attention and that resonate with our bodies set us up for transformation.

The Work of Musicians

If you want money more than anything,
you will be bought and sold.
If you have a greed for food,
you will become a loaf of bread
This is a subtle truth.
Whatever you love, you are.
~ "A Subtle Truth," *A Year with Rumi*, 2006.

Following the flow of the above Rumi quote, if you are a musician, you become music. If you play music while filled with love, the music will contain love and spread that love to everyone who listens. If you play music with your egocentric mind, the music will be precise, mechanically perfect, pleasing to the mathematicians in the room—pleasing to the lovers of patterns and shapes. However, if you play music using your allocentric mind, the music will create relationships, the plants in the room will blossom, the animals in the room will sigh with contentment, and the human beings in the room will feel your loving touch. Using your intuitive talents, you can help others escape the dry habits of the egocentric mind; you can help the audience to *become* music rather than judge music. Indeed, if you are a musician, *you have a responsibility* to heal and relieve suffering.

Fundamentally, music is a kind of *flow*; a flow that human beings need— just as we need food and water. Flow is allocentric, non-verbal, not about a search for objective meaning. Music, like all of the nutrients that the allocentric mind needs for survival, is immediate, in the moment; music inhibits the egocentric mind. For the listener, there is no analysis, no problems to solve, no monkey mind to cause angst, doubt, dread, and despair. Music, as composed and shared by the musically talented among us connects human beings to the outermost edge of nature's creative flow where love resides. A musician must know about human cognitive duality to create satisfying and complete musical moments. Once a musician reaches a sophisticated level of consciousness, *an obligation, a responsibility arises* to help others spiritually evolve.

Musicians, poets, philosophers, priests, psychologists, writers, artists, all are justice warriors. The egocentric mind fights its battles using logic, determination, reason, and intention; the allocentric mind flows with compassion, intuition, imagination, and love. Both minds can and should unite against injustice. A great song requires a background beat to support the melody. To create a great song, the artist must combine the best of both minds. The allocentric frame of reference is the soulful background rhythm; the poetic words of the song and the patterned-melody are egocentric manifestations.

A great responsibility arises for people who have reached a high level of spiritual evolution. These people know—because they are spiritually evolved—about human duality; they know about levels of consciousness; they know about muses, channeling, and spirit guides. They know that help is available through the asking. So they ask. They stand and they deliver.

Responsibilities arise because of your unique gift. Therefore, you will share because you must share. Musicians, artists, intuitives, empaths—those who create and relate through the primal frequencies—will step up now. The world needs to evolve to become happier, healthier, more peaceful, more loving, and wiser.

Therefore, as a musician, an artist, *you have an obligation*:
Do as much good as you can in the time you have left
Get up.
This is a work day.

The Confusion Caused By Being Your Own Twin

Be grateful for whoever comes
because each has been sent
as a guide from beyond.
~ Rumi

I have made the case now in two books: We are dual creatures. We are crafted by quantum and universal laws to have two kinds of cognition—we are cognitively and fundamentally split in half. In this dawning Age of Aquarius, we have become intuitively aware of this split. The hard evidence has mounted to such an extent that most variables are falling under two headings, which I have defined as allocentric and egocentric. We are in a transition phase, a moment of pause while we let this new awareness sink in.

You are reading this sentence because you have evolved cognitively to a level of sophistication that draws you to this examination of mind—especially your own mind, which is complex and self-reflective. You are not alone. There are others who also want kindness, peace, and curiosity to blossom. Like you, there are others who are also confused by being their own twin, and yet they are determined to see this inherent complexity as a gift.

Despite so much turmoil in the world, I am heartened by the balancing forces arising in modern societies: there is a growing level of cooperation, kindness, enthusiasm, and intention, which is combining with discipline and spiritual growth. The evolution of this life force is as strong and loving as ever—perhaps exponentially more evolved than at any other moment in the history of the earth. Hope is stubbornly balancing despair.

However, our planet is still suffering. There is massive hunger, massive ignorance, and starkly low levels of consciousness in most of the cultures on earth. The powers that rule seem to be corporations without a conscience, many of them dealing in the manufacture and distribution of weapons. Wars constantly appear and then rage unchecked. Stockpiles of weapons wait for their turn to commit murder. Excessively patriarchal attitudes prevail even in the most democratic and socially advanced societies. Science, our great ally in the fight against ignorance, does not have a creed of kindness, at least not one that is well-known and dependable. Robotic mega-corporations and narcissistic billionaires decide what is good and what is not good for the rest of us—they decide what needs to be elevated and what needs to be fiercely put down.

The forces of entropy wage a constant cosmic war against the forces of syntropy. The Devil fights against God every moment of eternity. Darkness fights the Light. This is a battle greater than human beings, and yet we can decide which side to support. Are we good guys or are we bad guys? Do we want a planet of love, tolerance, and peace, or are we okay with hate, intolerance, and war? Why is it even necessary to ask such a question? It is because, at a low level of consciousness, intolerance and hatefulness bubble over into unrest and war. Are we on a spiritual journey; a path to higher consciousness, or are we stuck in the mud, our cognitive wheels spinning and our factories overflowing with weapons?

Here is a formula that I find helpful: Wisdom equals Mindfulness + Intention + Discipline. To be wise you need first to be awake (mindful). You need to look around at the world and take stock of the situation. You need to be aware of relationships. You need to be able to perceive where there are problems, where there is suffering. The way we fight against suffering is through our personal and collective intentions. Find a problem—there are lots to choose from—and then make an intention to help with the solution. Finally, bring the discipline—day in and day out—to do the work that transforms suffering into happiness. Do your part. We need you. And you need us. Do as much good as you can in the time you have left. Remember always the words of Rumi:

> *Get up.*
> *This is a work day.*

I will conclude the book with a short summary of the waystations we visited on our journey to ports of duality. Everywhere we look, we can find duality. From the very design of our body, with its mirror image halves, to the two perceptual systems that had to evolve to enable purposeful movements—allocentric processing and egocentric processing—and in every discipline of study, we find a mutually exclusive oscillation, a polarity. Every expert finds this duality and every profession has generated different terminology for the same concept.

When we look at the quantum world we find *Heisenberg's uncertainty principle*, which is a confession that a duality exists at the most fundamental of physical levels: Movement alternates with no-movement, *on* oscillates with *off*, dark with light, life with death, syntropy with entropy, being with nothingness. You cannot freeze an energy wave at the very same time you measure the flow of the same wave. Therefore, logic itself dictates a fundamental duality.

In esoteric language we say, as did ancient people long before our time: *as above, so below.* In other words, if we find duality at the most basic of quantum levels, we would expect to find duality at all successive macro levels. And we do. We find polarized entities held together and yet repulsed throughout chemistry, biology, and physics.

In our very physiology, we find the universal duality expressed: inhibition (no-movement) is balanced by activation (movement). This is true throughout the entire nervous system and within every cell and cell-assembly of the body. I used the human eye, a quantum receptor, to show that *Heisenberg's uncertainty principle* is employed in the retina of the eye. The operation of the retinal rod cells (concerned with movement) and the retinal cone cells (concerned with freezing movement) mirror Heisenberg's mathematics. I also pointed out another (integrally related) feature of the rod-cone balance: The rod system generates gestalts (the background), while the cone system generates the features (patterns, forms) that arise within the gestalt (scene). Using *as above so below*, we would expect to find this background-foreground dichotomy wherever we look—and we do.

If we scale up to everyday humanity, we find our great artists and thinkers cataloging, analyzing, explaining, and creating duality. Below are some examples I mentioned in the book. You can generate a much longer list if you probe deeper into the history of any discipline.

The philosopher Descartes is commonly understood to be the father of what came to be called Cartesian Dualism, the theory that the mind is a different substance than physical reality. For some, this is a truth, for others it is a falsehood. For esoteric thinkers and those who see the correspondences (*as above so below*), Descartes simply come upon the universal paradox where two opposites can be true.

Owen Barfield, a philologist, philosopher, and writer, contrasted *spirit* with *matter*. He said that *spirit* was the background out of which the feature we call *matter* emerged. Barfield used the poet Coleridge and the Renaissance man Goethe to support his observations. Samuel Coleridge crafted a Law of Polarity (Essential Dualism) and concluded that human beings had a primary and secondary imagination—there were two kinds of processing systems inherent within human cognition. Goethe used the phrase *expand and expire* to explain the duality. Goethe might have gotten this perspective from the mystic Jacob Boehme who said that there was a force of contraction that was sucking everything inward. This was balanced by a force of expansion that was driving everything outward. Goethe also said we needed to transform *knowledge by description* into *knowledge by acquaintance*. The American writer and mystic Ralph Waldo Emerson, a contemporary of Barfield and Coleridge, spoke of "Me and Not Me," to explain the duality. In my terminology, Barfield, Coleridge, Goethe, Boehme, and Emerson were discussing the allocentric mind and the egocentric mind.

The philosophers of the Western world and the philosophers of the Eastern world all found duality to be at the heart of their theories. G. I Gurdjieff, who spent thirty years studying the practices and philosophies of Eastern esoteric schools, stated that the first premise of these schools of thought was duality. Gurdjieff said that every religion consisted of two parts—an exoteric and an esoteric. The exoteric teaches *what is to be done*. The esoteric teaches *how to be*. Gurdjieff spoke of "I am" (the egocentric mind) and "I Am" (the allocentric mind). Gurdjieff's follower and biographer Jean De Salzmann said that in each event in life there is a double movement of *involution* and *evolution*. This terminology captures both the mutually exclusive duality of human cognition as well as the oscillation between foreground and background.

Shakespeare told us what the primal question was: "To be, or not to be; that is the question." Gurdjieff said: "Remember yourself." *To be* is to remember your self. *Not to be* is what happens when you forget your self. The self is the allocentric mind which is a whole-body mind. To be is to feel your embodied animation. Are we awake and *being* or are we asleep and invisible to ourselves.

Islamic philosopher Averroes (1126-1198) wrote of an inherent mental struggle to reconcile faith with reason. Averroes's ideas are called the *theory of the double truth*. Sprinkled throughout this book are the poems of Rumi, who tried in various ways to expose humanities fundamental duality—the split between matter and soul.

Buddhists speak of the *absolute* and the *relative*. They mean that there is an unchanging background (the absolute) that is the canvas for an ever-changing (relative) world. The *absolute* is free of concepts, free of dualistic ideas of ego and other. However, the *relative* is dual. The ego must have a foreground and a background to exist.

The ultimate insight of the Hindu ascetics was that *Atman* equaled *Brahman*. In other words, the essence of you is the very same as the essence of the universe (*as above so below*), and that essence is dual. The mutually exclusive polarities *are equal*. We have forgotten this wisdom in the Western world.

Atman is the individual soul. *Brahman* is the cosmic soul, the background that holds all the individual souls.

The early church fathers, according to Cynthia Bourgeault in her lovely book *The Wisdom Jesus* (2008) postulated two ways of knowing: *epinoia*, knowing through intuition, direct revelation, through experience; and *dianoia*, knowing through the use of logic, doctrine, and dogma.

The German philosopher Immanuel Kant differentiated *phenomenon* and *noumenon*. Kant (1724-1804) was a contemporary of Goethe (1749-1832). In his philosophical writings, Kant said that we swim in an illusory sea created by our senses, a fabricated world of phenomenon. The *thing-in-itself* (the noumenon) cannot be accurately perceived. Therefore, we can never know what is objectively real. *Phenomenon* and *noumenon* are just new words for the same duality: allocentric and egocentric.

The philosopher Schopenhauer defined the same duality as a struggle between "the world out there" (space, time, cause, effect, events, and objects), versus our mental state, "a world in here." The German philosophers saw the fundamental duality arising because "mind" is such a foreign strangeness compared to everything else in the cosmos. Each Western philosopher seems to recirculate the same duality using a new set of dual words. Friedrich Nietzsche used the concepts behind the Greek myths of Apollo and Dionysus to demonstrate the universal duality. The Phenomenologists, like Husserl and Heidegger, wrote of being and becoming, being and nothingness, being and doing—it is the same duality dressed in different clothing.

The German philosopher Leibniz (1646-1716) compared what he called "pure or bare perception" with "conscious appreciation." This is yet another version of allocentric (pure) perception contrasted with egocentric (conscious) perception.

Eventually, historians postulated a cultural evolution in which the *Greek material worldview* was contrasted with the *spiritual worldview of Christianity*, a battle between reason and faith. This is just another incarnation of the fundamental duality: reason is the operation of the egoic mind, faith the operation of the allocentric soul. The Wisdom Tradition is based on the reality of a dual truth, the polarity between *being* and *doing*. The modern secular world is one of *becoming*. The distinction was first posited by the Pre-Socratic thinker Parmenides centuries before Plato, and has occupied philosophers ever since.

From the onset of modern psychology, our fundamental duality has been known and explored. The pioneering American psychologist William James said that we could follow the *path of incremental learning*, taking *one thing-at-a-time*, accumulating knowledge, or we could gain insights *all-at-once*, suddenly seeing the big picture. In his book *Principles of Psychology*, James theorized "coexisting and split modes of consciousness, an "upper self" and an "under self." According to James, these two modes of consciousness were mutually unaware of each other, and yet they had complementary effects on each other.

The poets knew about duality instinctively. Collin Wilson studied the evolution of poetry and contrasted that history with the evolution of consciousness. Wilson concluded that there are "two sorts of poetry" that correspond to two types of cognitive processing. In Theodore Roethke's poems we find a primal dichotomy between analysis and intuition, flesh and spirit, *I and otherwise*. Poets know that we have mutually exclusive minds.

Science also found duality and has learned to work with the conundrum. Quantum physicist David Bohm postulated the existence of an implicate order and an explicate order. He could see that at the quantum level it was necessary to theorize the existence of two fundamentals—a duality. Roboticist Rodney Brooks could see that a robot needed two operating systems, a work-place mind and a task mind. Rudolf Steiner, who called himself a spiritual scientist, compared Jesus (an egocentric manifestation), with Christ (an allocentric manifestation). We have in Steiner's philosophy another terminology for our two minds. Just as we have egocentric belief, ego, and the concept of *Jesus*, we also have faith, soul, and the concept of *Christ*. Steiner also saw the connection between background and foreground. He said that *evolution* is manifestation, the outward breath of the cosmos, while involution ("backward evolution," Steiner called it) is the in-breath of the cosmos, which pulled its perfections back into the chaos of the background.

Blaise Pascal (1623-1662), a mathematician and philosopher, wrote in his unfinished notes that there was a contrast between the *analytical mind* of the mathematician and the *intuitive mind* of the artist. This contrast between intuition and the analytical has become a scientific meme, a common usage in our modern world. This division is often supported by the half-complete notion of a left-brain and a right brain.

The evidence for duality, for levels of consciousness, and for "perception based on foreground and background," is pervasive in our literature. I hope that this book has made the case sufficiently for you to at least entertain the possibility that dual-process theory should be taken seriously. I made my arguments for human cognitive duality in my previous book *Consciousness, A New Slant on an Old Conundrum*. I had to do a quick restatement as I began this work.

I wrote this book to emphasize that there are responsibilities and consequences that arise once we adopt dual-process theory. It is *not okay* to continue crafting our culture, our educational system, our political system, our religious traditions, and our science on a purely egocentric base. This is wrong; the purely egocentric perspective is not supported by logic, history, common sense, wisdom, or compassion.

I will say goodbye with some lines from a song by *The Seekers* "The Carnival is Over:"

> *High above, the dawn is waiting*
> *And my tears are falling rain*
> *For the carnival is over*
> *We may never meet again*
> *Now the harbour light is calling*
> *This will be our last goodbye*
> *Though the carnival is over*
> *I will love you till I die*
> *(and after that)*
> ~A song lyric from "The Carnival is Over by The Seekers.

BIBLIOGRAPHY

He who has been,
from then on cannot
not have been
~Vladimir Jankelevitch

A

Alexander, Stephon
- *The Jazz of Physics; The Secret Link Between Music and the Structure of The Universe*, 2016.

Ambrosio, Francis
- *Philosophy, Religion, and the Meaning of Life*, a Great Works Lecture, 2009.

Armstrong, John
- *Love, Life, Goethe*, 2007.

B

Baldwin, Doug
- *Bugs Blindness, and the Pursuit of Happiness*, 2016.
- *Consciousness, A New Slant on an Old Conundrum*, 2017.

Barfield, Owen

For a complete review of all Owen Barfield's publications, see the "Bibliography of the Published Writings of Owen Barfield, 1917 – 2015," by Jane W. Hipolito. It is available on line. I reviewed the following Barfield publications:
- *History in English Words*. London: Methuen and Co., Ltd., 1926.
- *Poetic Diction: A Study in Meaning*. London: Faber and Gwyer, Ltd.,

1928.
- *Romanticism Comes of Age*. London: Anthroposophical Publishing Co., 1944.
- *This Ever-Diverse Pair*, 1950.
- *Saving the Appearances: A Study in Idolatry*. London: Faber and Faber, 1957. New York: Harcourt, Brace and World, 1965.
- *What Coleridge Thought*, 1971.
- *The Rediscovery of Meaning and Other Essays*, Owen Barfield, 1977.
- "Coleridge in the Twenty-First Century." Review of Poetry Realized in Nature: Samuel Taylor, Coleridge and Early Nineteenth-Century Science, by Trevor H. Levere. Towards 2.2 (Spring, 1982): 38-41.
- *A Barfield Reader: Selections from the Writings of Owen Barfield*. Edited by G. B. Tennyson, 1999.
- *History, Guilt, and Habit*. Middletown, CT: Wesleyan University Press, 1979. Wesleyan Paperback, 1981. San Rafael, CA: Sophia Perennis, The Barfield Press, 2006.
- *The "Great War" of Owen Barfield and C. S. Lewis: Philosophical Writings 1927-1936*. With C. S. Lewis. Edited by Norbert Feinendegen and Arend Smilde. Journal of Inklings Studies, Inklings Studies Supplements No. 1, 2015.

Barks, Coleman
- *A Year With Rumi, Daily Readings*, (with John Moyne, Nevit Ergan, A. J. Arberry, and Reynold Nicolson), 2006.
- *Rumi The Big Red Book, The Great Masterpiece Celebrating Mystical Love and Friendship*, 2011.

Beck, Don Edward and Christopher Cowan
- *Spiral Dynamics: Mastering Values, Leadership and Change*, Don Edward Beck and Christopher Cowan, 2005.

Brentano Franz
- *Psychology from an Empirical Standpoint*, 1874.

Brooks, Rodney
- *Cambrian Intelligence: The Early History of AI*, 1999.
- *Flesh and Machines: How Robots Will Change Us*, 2003.
- *Fields of Color: The Theory That Escaped Einstein*, 2010.

Brown, Neal
- *Theodore Roethke: The Journey From I to Otherwise*, 1982.

C

Carey, Nessa
- *The Epigenetics Revolution, How Modern Biology is Rewriting Our Understanding of genetics, Disease, and Inheritance*, 2013.

Chesterton, G. K.
- *What's Wrong with the World*, 1910.

Chopra, Deepak
- *How to Know God: The Soul's Journey into the Mystery of Mysteries*, 2001.
- *Life After Death: The Burden of Proof Life After Death*, 2008.
- *The Third Jesus: The Christ We Cannot Ignore*, 2009.
- *Spiritual Solutions: Answers to Life's Greatest Challenges*, 2012.
- *Super Genes: Unlock the Astonishing Power of Your DNA for Optimum Health and Well-Being* (with Rudolph E. Tanzi Ph.D.), 2015.
- *The Future of God: A Practical Approach to Spirituality for Our Times*, 2015.
- *Quantum Healing (Revised and Updated): Exploring the Frontiers of Mind/Body Medicine*, 2015.

Collins, Billy
- *Horoscopes for the Dead*, 2011.
- *Aimless Love*, 2013.
- *The Rain in Portugal*, 2016.

Cook-Greuter, Susanne
- *Transcendence and Mature Thought in Adulthood*, Melvin E. Miller and Susanne R. Cook-Greuter, 1994.
- *Creativity, Spirituality, and Transcendence: Paths to Integrity and Wisdom in the Mature Self*, Melvin E. Miller and Susanne R. Cook-Greuter, 1999.
- *Action Inquiry: The Secret of Timely and Transforming Leadership:* William R. Torbert, Susanne Cook-Greuter, Dalmar Fisher, Erica Foldy, Alain Gauthier, 2004.
- *Post-autonomous Ego Development: A Study of Its Nature and Measurement*, Susanne R. Cook-Greuter, 2010.

D

David-Neel, Alexandra and lama Yongden
- *The Secret Oral Teachings in Tibetan Buddhist Sects*, 1967.

Dawkins, Richard
- *The Blind Watchmaker: Why the Evidence of Evolution Reveals a Universe without Design*, 1996.
- *River Out of Eden: A Darwinian View of Life*, 1996.
- *Unweaving the Rainbow: Science, Delusion and the Appetite for Wonder*, 2000.
- *A Devil's Chaplain: Reflections on Hope, Lies, Science, and Love*, 2004.
- *The Ancestor's Tale (A Pilgrimage to the Dawn of Life)*, 2005.
- *The Selfish Gene: 30th Anniversary Edition*, 2006.
- *The God Delusion*, 2008.
- *The Oxford Book of Modern Science Writing*, 2009.
- *The Greatest Show on Earth: The Evidence for Evolution*, 2010.
- *The Magic of Reality: How We Know What's Really True*, 2012.
- *Appetite for Wonder, An: The Making of a Scientist*, 2014.
- *Brief Candle in the Dark: My Life in Science*, 2015.
- *The Ancestor's Tale: A Pilgrimage to the Dawn of Evolution*, 2016.
- *The Extended Phenotype: The Long Reach of the Gene*, 2016.
- *Climbing Mount Improbable*, 2016.

De Hartmann, Thomas and Olga
- *Our Life with Gurdjieff*, 1964.

De Salzmann, Jeanne
- *The Reality of Being; The Fourth Way of Gurdjieff*, 2011.

Deutsch, Diana
- *The Psychology of Music*, 2nd ed., 1999.

E

Edelman, Gerald
- *Neural Darwinism*, 1978
- *The Mindful Brain: Cortical Organization and the Group-selective Theory*, with Vernon Lee, 1982
- *How We Know*, 1985
- *Topobiology*, 1988
- *The Remembered Present: A Biological Theory of Consciousness*, 1989

- *Bright Air, Brilliant Fire: On the Matter of the Mind*, 1991
- *A Universe of Consciousness: How Matter Becomes Imagination*, 2000
- *Wider than the Sky*, 2004
- *Second Nature*, 2006

Evans, Jonathan St B.T.
- *In Two Minds: Dual Processes and Beyond*, with Keith Frankish, 2009
- *Thinking Twice: Two minds in one brain*, 2010
- *Thinking and Reasoning: A Very Short Introduction*, 2017
- "The duality of mind: An historical perspective," with Keith Frankish, available on line.

F

Flanagan, Owen
- *Science of the Mind: 2nd Edition*, 1991.
- *Varieties of Moral Personality: Ethics and Psychological Realism*, 1991.
- *Consciousness Reconsidered*, 1992.
- *Self-Expressions: Mind, Morals, and the Meaning of Life*, 1996.
- *Dreaming Souls: Sleep, Dreams and the Evolution of the Conscious Mind*, 2000.
- *The Problem Of The Soul: Two Visions Of Mind And How To Reconcile Them*, 2002.
- *The Really Hard Problem: Meaning in a Material World*, 2007.
- *The Bodhisattva's Brain: Buddhism Naturalized*, 2013.

Frankish, Keith
- *Mind and Supermind*, 2007
- The duality of mind: An historical perspective, with Jonathan Evans, available on line.

G

Galton, Sir Frances
- *Inquiries into Human Faculty and Its Development*, 1883.
- *Memories of my Life*, 1908.

Gebser, Jean
- *The Ever-present Origin*, 1991.

Gilligan, Carol
- *In a Different Voice*, 1982.
- *Mapping the moral domain: a contribution of women's thinking to psychological theory and education*, 1989.
- *Making connections: the relational worlds of adolescent girls at Emma Willard School*, 1990.
- *Meeting at the crossroads: women's psychology and girls' development*; with Brown, Lyn, 1992.
- *Between voice and silence: women and girls, race and relationships*, with McLean Taylor, Jill; Sullivan, Amy M., 1997.
- *The birth of pleasure*, 2002.
- *Kyra: a novel*, 2008.
- *The deepening darkness: patriarchy, resistance, & democracy's future*, with Richards, David A.J., 2009.
- *Joining the resistance*, 2011.

Gladstone, William
- *Studies on Homer and the Homeric Age*, 1858.

Great Courses—various professors
- Great Minds of the Western Intellectual Tradition in three volumes, 2000.

Grof, Stanislav
- *The Consciousness Revolution, A Transatlantic Dialogue, Two days with Ervin Laszlo, Stanislav Grof, and Peter Russell*, 2003.
- *What is Reality? The New Map of Cosmos, Consciousness, and Existence*, 2016.

H

Hardy, Grant
- *Great Minds of the Eastern Intellectual Tradition*, The Great Courses, 2011.

Hermanns, William
- **Einstein and Poet; In Search of the Cosmic Man**, 1983.

Hocks, Richard A.
- "The "Other" Postmodern Theorist: Owen Barfield's, Concept of the Evolution of Consciousness," available on line at http://polanyisociety.

org/TAD%20WEB%20ARCHIVE/TAD18-1/TAD18-1-fnl-pg27-38-pdf.pdf.

Hulme, Kathryn
- *Undiscovered Country, A Spiritual Adventure*, 1966.

J

James, William
- *Principles of Psychology*, 1890
- *The Dilemma of Determinism, The Will to Believe*, 1896.
- *Journal of Philosophy, Psychology and Scientific Methods*, "Does Conscious Exist?" vol. 1, No. 18, September 1, 1904.

Jaspers, Karl
- *The Origin and Goal of History*, 1953.

Jaynes, Julian
- *The Origin of Consciousness in the Breakdown of the Bicameral Mind*, 1976.

Jourdain, Robert
- *Music, The Brain, and Ecstasy*, 1997.

K

Kegan, Robert
- *The Evolving Self*, 1982.
- *In Over Our Heads*, 1994.

Klein, Daniel B.
- *Plato and a Platypus Walk Into a Bar*, 2007.

Kurzweil, Raymond
- *Visions of the Future*, 1985.
- *The Age if Intelligent Machines*, 1990.
- *The Age of Spiritual Machines*, 1999.
- *The Singularity is Near*, 2005.
- *How to Create a Mind*, 2012.

L

Lachman, Gary

- "Meditation and Spiritual Perception," Classics from the Journal for Anthroposophy, 2011.
- *A Secret History of Consciousness*, with Colin Wilson, 2003.
- *Turn Off Your Mind: The Mystic Sixties and the Dark Side of the Age of Aquarius*, 2003.
- *A Dark Muse: A History of the Occult*, 2004.
- *Rudolf Steiner: An Introduction to His Life and Work*, 2007.
- *Classics from the Journal for Anthroposophy, Meditation and Spiritual Perception*, "Rudolf Steiner, Jean Gebser and the Evolution of Consciousness," Gary Lachman, 2011.
- *The Quest for Hermes Trismegistus: From Ancient Egypt to the Modern World*, 2011.
- *Swedenborg: An Introduction to His Life and Ideas*, 2012.
- *Madame Blavatsky: The Mother of Modern Spirituality*, 2012.
- *Jung the Mystic: The Esoteric Dimensions of Carl Jung's Life and Teachings*, 2012.
- *Politics and the Occult: The Left, the Right, and the Radically Unseen*, 2012.
- *The Caretakers of the Cosmos: Living Responsibly in an Unfinished World*, 2013.
- *Aleister Crowley: Magick, Rock and Roll, and the Wickedest Man in the World*, 2014.
- *In Search of P. D. Ouspensky: The Genius in the Shadow of Gurdjieff*, 2014.
- *Revolutionaries of the Soul: Reflections on Magicians, Philosophers, and Occultists*, 2014.
- *The Secret Teachers of the Western World*, 2015.
- *Beyond the Robot: The Life and Work of Colin Wilson*, 2016.

Laszlo, Ervin
- *The Systems View of the World: A Holistic Vision for Our Time*, 2002.
- *The Consciousness Revolution, A Transatlantic Dialogue, Two days with Ervin Laszlo, Stanislav Grof, and Peter Russell*, 2003.
- *Science and the Akashic Field: An Integral Theory of Everything*, 2007.
- *Quantum Shift in the Global Brain: How the New Scientific Reality Can Change Us and Our World*, 2008.
- *The New Science and Spirituality Reader*, 2012.
- *Dawn of the Akashic Age: New Consciousness, Quantum Resonance, and the Future of the World*, 2013.

- *The Self-Actualizing Cosmos: The Akasha Revolution in Science and Human Consciousness*, 2014.
- *The Immortal Mind; Science and the Continuity of Consciousness Beyond the Brain*, 2014.
- *What is Consciousness? Three Sages Look Behind the Veil*, 2016.
- *What is Reality? The New Map of Cosmos, Consciousness, and Existence*, 2016.
- *The Intelligence of the Cosmos: Why Are We Here? New Answers from the Frontiers of Science*, 2017.

Llinas, Rodolfo R.
- *i of the vortex, From Neuron to Self*, 2001.

Loevinger, Jane
- Measuring Ego Development, Jane Loevinger and Le Xuan Hy, 2nd edition 2015.
- Technical Foundations for Measuring Ego Development, Jane Loevinger, 1998.

M

Maitri, Sandra
- *The Spiritual Dimension of the Enneagram; Nine Faces of the Soul*, 2001.

Mansfield, Victor
- *Synchronicity, Science, and Soul-Making*, 1995.

McIntosh, Steve
- *Integral Consciousness and the Future of Evolution: How the Integral Worldview is Transforming Politics, Culture, and Spirituality*, 2007.

Meier, C. A. editor
- *Atom and Archetype, The Pauli/Jung Letters 1932-1958*, 2001.

Mitchell, Edgar
- "Nature's Mind: The Quantum Hologram," Edgar Mitchell, Sc.D. Available on line at http://www.experiencer.org/natures-mind-the-quantum-hologram-by-edgar-mitchell-sc-d/

N

Nott, C. S.

- *Teachings of Gurdjieff: A Pupil's Journal*, 1961.
- *Journey Through This World: Meetings with Gurdjieff, Orage and Ouspensky*, 1969.

O

O'Fallon, Terri
- "The Evolution of the Human Soul: Developmental Practices in Spiritual Guidance." May 3, 2010. Available online at https://lorian.org/ThesisLibrary/TerriOFallon-thesis.pdf

Osho (Bhagwan Shree Rajneesh)
- *The Book of Secrets*, 1974.
- *Meditation: The Art of Ecstasy*, 1979.
- *The Hidden Harmony*, 1976.
- *The Path of Love*, 1978
- *Beyond Enlightenment*, 1986
- *Death, The Greatest Fiction*, 1988.
- *In Search of the Miraculous*, 1992.
- *Awareness, The Key to Living in Balance*, 2001.

Ouspensky, P. D.
- *The Fourth Way: An Arrangement by Subject of Verbatim Extracts from the Records of Ouspensky's Meetings in London and New York*, 1971.
- *The Psychology of Man's Possible Evolution*, 1973.
- *A New Model of the Universe*, 1997.
- *In Search of the Miraculous*, 2001.
- *Tertium Organum*, 2010.

P

Patterson, William Patrick
- *Eating the "I": An Account of the Fourth Way: the Way of Transformation in Ordinary*, 1992.
- *Struggle of the Magicians: Exploring the Teacher Student Relationship*, 1996.
- *Gurdjieff in Egypt: The Origin of Esoteric Knowledge*, (DVD), 1997.
- *Taking With the Left Hand: Enneagram Craze, People of the Bookmark, & The Mouravieff "Phenomenon*, 1998.

- *Ladies of the Rope: Gurdjieff's Special Left Bank Women's Group*, 1999.
- *Georgi Ivanovitch Gurdjieff: The Man, the Teaching, His Mission*, 2014.

Peirce, Penney and Laural Merlington
- *Frequency: the Power of Personal Vibration*, 2012.

Piaget, Jean
- *Play Dreams & Imitation in Childhood*, Jean Piaget, 1962.
- *Six Psychological Studies*, Jean Piaget and Anita Tenzer, 1968.
- *The Psychology Of The Child*, Jean Piaget and Barbel Inhelder, 1969.
- *The Construction of Reality in the Child*, Jean Piaget. 1971 .
- *The Essential Piaget*, Jean Piaget and Howard E. Gruber, 1977.
- *Behavior and Evolution*, Jean Piaget, 1978.
- *Language and Learning: The Debate Between Jean Piaget and Noam Chomsky*, Noam Chomsky and Jean Piaget, 1980.
- *The Moral Judgment of the Child*, Jean Piaget, 1997.
- *The Psychology of Intelligence* (Routledge Classics), Jean Piaget, 2001.
- *The Language and Thought of the Child*, Jean Piaget, 2001.

Plotkin, Bill
- *Nature and the Human Soul, Cultivating Wholeness and Community in a Fragmented World*, 2008.

R

Ricoeur Jean Paul
- *Freud and Philosophy*, 1965.
- *The Conflict of Interpretations*, 1969.
- *Time and Narrative*, 1983.
- *Oneself as Another*, 1992
- *Memory, History, Forgetting*, 2004.

Rhodes, Susan
- *The Positive Enneagram: A New Approach to the Nine Personality Types*, 2009.
- *Archetypes of the Enneagram: Exploring the Life Themes of the 27 Enneagram Subtypes from the Perspective of Soul*, 2010.
- *The Integral Enneagram*, 2013.

Rilke, Rainer Maria
- *Rilke's Book of Hours: Love Poems to God*, 1905.

- *Letters to a Young Poet*, 1929.

Robertson, Steve
- "Wisdom and Inspiration," an essay by Steve Robertson. Available online at: https://steverobertsonspeaker.wordpress.com/music-affects-consciousness/

Roethke, Theodore
- *The Waking*, 1953.
- *The Far Field*, 1964.
- *Straw for the Fire*, 1972
- *On Poetry and Craft*, 2001
- *Selected Poems*, 2005

Ruffini, Giulio
- An algorithmic information theory of consciousness Neuroscience of Consciousness; Volume 2017, Issue 1, 1 January 2017, referenced from: https://academic.oup.com/nc/pages/About.

Russell, Peter
- *The Consciousness Revolution, A Transatlantic Dialogue, Two days with Ervin Laszlo, Stanislav Grof, and Peter Russell*, 2003.

S

Sacks, Oliver
- *Musicophilia: Tales of Music and the Brain*, 2008.

Schneider, Michael S.
- *A Beginners Guide to Constructing the Universe, The Matghematical Archetypes of Nature, Art, and Science*, 1952.

Schopenhauer, Arthur
- *The World as Will and Representation*, 1818.
- *On Human Nature*,1897.

Sharp, H. J.
- *Sacred Geometry and the Enneagram*, 2003.
- *My End is My Beginning*, 2007.

Sheldrake, Rupert
- *A New Science of Life*, 1981.
- *The Presence of the Past*, 1988.

Steiner, Rudolf

- *The Essential Rudolf Steiner*, available online or as an ebook, 2012.

T

Tame, David
- *The Secret Power of Music: The Transformation of Self and Society Through Musical Energy*, 1884.

Three Initiates
- *The Kybalion: A Study of the Hermetic Philosophy of Ancient Egypt and Greece*, 2016.

Tolle, Eckhart
- *The Power of Now*, 1997.
- *A New Earth*, 2005.

Tononi, Giulio
- *PHI: A Voyage from the Brain to the Soul*, 2012.

V

Versluis Arthurs
- *Awakening the Contemplative Spirit*, 2004.
- *Restoring Paradise: Western Esotericism, Literature, and Consciousness*, 2004.
- *The New Inquisitions: Heretic-hunting, the Origins of Modern Totalitarianism*, 2006.
- *Magic and Mysticism: An Introduction to Western Esotericism*, 2007.
- *The Secret History of Western Sexual Mysticism*, 2008.
- *The Mystical State: Politics, Gnosis, and Emergent Cultures*, 2011.
- *American Gurus: From Transcendentalism to New Age Religion*, 2014.
- *Perineal Philosophy*, 2015.

W

Walker, Kenneth
- *Venture with Ideas: Meetings with Gurdjieff and Ouspensky*, 1952.

Wilber, Ken
- *The Spectrum of Consciousness*, 1977.

Bibliography

- *No Boundary: Eastern and Western Approaches to Personal Growth*, 1979/2001.
- *The Atman Project: A Transpersonal View of Human Development*, 1980.
- *Up from Eden: A Transpersonal View of Human Evolution*, 1981.
- *The Holographic Paradigm and Other Paradoxes: Exploring the Leading Edge of Science*, 1982.
- *A Sociable God: A Brief Introduction to a Transcendental Sociology*, subtitled *Toward a New Understanding of Religion*, 1983/2005.
- *Eye to Eye: The Quest for the New Paradigm*, 1984.
- *Quantum Questions: Mystical Writings of the World's Great Physicists*, 1984/2001.
- *Transformations of Consciousness: Conventional and Contemplative Perspectives on Development* (co-authors: Jack Engler, Daniel Brown), 1986.
- *Spiritual Choices: The Problem of Recognizing Authentic Paths to Inner Transformation* (co-authors: Dick Anthony, Bruce Ecker), 1987.
- *Grace and Grit: Spirituality and Healing in the Life of Treya Killam Wilber*, 1991.
- *Sex, Ecology, Spirituality: The Spirit of Evolution*, 1995/2001.
- *A Brief History of Everything*, 1996/2001.
- *The Eye of Spirit: An Integral Vision for a World Gone Slightly Mad*, 1997/2001.
- *The Essential Ken Wilber: An Introductory Reader*, 1998.
- *The Marriage of Sense and Soul: Integrating Science and Religion*, 1998/1999.
- *One Taste: The Journals of Ken Wilber*, 1999/2000.
- *Integral Psychology: Consciousness, Spirit, Psychology, Therapy*, 2000.
- *A Theory of Everything: An Integral Vision for Business, Politics, Science and Spirituality*, 2000.
- *The Simple Feeling of Being: Visionary, Spiritual, and Poetic Writings*, 2004.
- *The Integral Operating System*, 2005.
- *Integral Spirituality: A Startling New Role for Religion in the Modern and Postmodern World*, 2006.
- *The Integral Vision: A Very Short Introduction to the Revolutionary Integral Approach to Life, God, the Universe, and Everything*, 2007.
- *Integral Life Practice: A 21st-Century Blueprint for Physical Health, Emotional Balance, Mental Clarity, and Spiritual Awakening*, 2008.

- *The Pocket Ken Wilber*, 2008.
- *Integral Meditation: Mindfulness as a Way to Grow Up, Wake Up, and Show Up in Your Life*, 2016.
- *Trump and a Post-Truth World*, Ken Wilber, an online book, 2017.

Wilson, Colin

- *The Outsider*, 1956.
- *The Essential Colin Wilson*, 1986 .
- *The Strange Life of P.D. Ouspensky*, 1993.
- *Poetry & Mysticism*, 2001.
- *Super Consciousness: The Quest for the Peak Experience*, 2009.
- *The Philosopher's Stone* (with Colin Stanley), 2013.

Index

www.ingramcontent.com/pod-product-compliance
Lightning Source LLC
Chambersburg PA
CBHW051437170526
45166CB00001B/17